Ethical, Regulatory, and Intellectual Property Impacts on AI Development

Iris-Panagiota Efthymiou
https://orcid.org/0000-0001-9656-8378
University of Greenwich, UK

Konstantinos Kouroupis
https://orcid.org/0000-0003-1193-8224
Frederick University, Cyprus

IGI Global
Scientific Publishing
Publishing Tomorrow's Research Today

Vice President of Editorial Melissa Wagner
Director of Acquisitions Mikaela Felty
Director of Book Development Jocelynn Hessler
Production Manager Mike Brehm
Cover Design Jose Rosado

Published in the United States of America by
IGI Global Scientific Publishing
701 East Chocolate Avenue
Hershey, PA, 17033, USA
Tel: 717-533-8845 | Fax: 717-533-7115
Website: https://www.igi-global.com E-mail: cust@igi-global.com

Library of Congress Cataloging-in-Publication Data

LCCN: (CIP Data Pending)
ISBN13: 9798260000007
Isbn13Softcover: 9798260000014
EISBN13: 9798260000021
British Cataloguing in Publication Data
A Cataloguing in Publication record for this book is available from the British Library.

Table of Contents

Detailed Table of Contents

Rapid technological advancement and the widespread digitalization of contemporary society have profoundly reshaped democratic governance. Artificial Intelligence (AI) plays a central role in this transformation, influencing institutional functioning, civic participation, and the protection of fundamental rights. Within the emerging framework of e-democracy and e-governance, AI-driven tools enhance political communication and citizen engagement, while simultaneously posing risks to democratic participation and freedom of expression through disinformation and automated content generation. This chapter examines the relationship between AI, democracy, and freedom of expression, analyzing relevant legal and ethical frameworks at the European and international levels. It adopts a critical perspective on AI-enabled disinformation and proposes regulatory and policy responses aimed at preserving democratic values while fostering trustworthy and inclusive e-democratic systems.

Over the period 2015–2025, scholarship on ethical and equitable artificial intelligence converged around three interdependent goals: (1) identifying and classifying algorithmic harms that flow from biased data, model design, and socio-technical deployment; (2) developing technical and organisational mitigation strategies (pre-processing, in-processing, post-processing, and socio-technical interventions) to reduce disparate impacts; and (3) operationalising independent and regulatory audit mechanisms (algorithmic audits, assurance frameworks, and public reporting) that enable accountability, transparency, and remedial governance. Early work (2015–2018) emphasized detection and formal fairness metrics and revealed fundamental trade-offs between fairness criteria and predictive utility; mid-period research (2019–2022) broadened attention to domain-specific harms, human factors, and measurement practices; and recent contributions (2023–2025) focus on standardized auditing protocols, governance playbooks, and scalable bias-aware pipelines integrated into organisational lifecycles.

Generative AI systems are fast changing the research, industry and society. These systems are used to generate new content, decisions and knowledge with minimum human involvement. This possibility brings to the fore serious ethical issues. This chapter explores some of the most important ethical aspects of the establishment and application of generative AI. Its interests lie in areas like prejudice, openness, responsibility, confidentiality and abuse. The chapter brings out the uncertainty surrounding the model behaviour, outcome, and long term social impact. It talks about the inadequacy of the current ethical frameworks in generative systems. Another concept that is discussed in the chapter is responsible innovation as a guiding technique. Responsible innovation focuses on foresight, human control and inclusive governance. The risks may be mitigated by incorporating ethical thinking in design and deployment. The chapter also intends to guide the researchers, developers, and policymakers to make informed and ethical decisions in the face of uncertainty.

This chapter per the authors, examines the role of behavioural law in shaping AI governance, integrating ethical principles with regulatory frameworks. Drawing on behavioural economics, it analyses how cognitive biases and human-AI interactions create ethical challenges in AI deployment. The chapter aims to develop a hybrid model that bridges fairness, transparency, and accountability with enforceable regulations to mitigate risks like algorithmic bias and privacy breaches. Through case studies on data privacy, autonomous systems, and decision-making algorithms, it demonstrates how behavioural tools, nudges, defaults, and transparency mandates enhance regulatory efficacy while promoting innovation. It advocates for empirical research to refine these approaches, proposing adaptive frameworks that incorporate stakeholder input and global standards to foster trust and equity in AI ecosystems. By blending theory and practice, this chapter per the authors, provides policymakers, developers, and scholars with actionable strategies for sustainable AI governance in a technology-driven world.

The emergence of foundation models in the recent past has completely changed the environment of artificial-intelligence governance. Unlike the strictly application-specific systems, foundation models are general-purpose infrastructures implemented in various sectors and jurisdictions, which creates systemic and changing risks. The available legal systems, based on sector regulation, data-protection law, and general liability principles, are poorly adapted to the magnitude, obscurity, and accountability dispersal of these technologies. This chapter has performed an extensive research of the legal and ethical dilemmas posed by foundation models through both doctrinal and comparative studies. It promotes an organization, life cycle regulatory framework which balances the enforceable legal requirements and documented ethical requirements. Through the concepts of proportionality, accountability by design, transparency, human controls and strong system of governance mechanisms, the chapter suggests a consistent and practical answer to governance of foundation models.

In an era where algorithms increasingly influence justice, governance, and human relationships, this paper reimagines law as more than a regulator of technology, it becomes its moral compass. "Law in the Loop: Reimagining Justice and Regulation in the Age of Artificial Intelligence" argues for a human-centric AI governance framework that embeds ethics, accountability, and dignity within code itself. Through a synthesis of jurisprudence and technological ethics, the study explores how legal personhood, fairness, and privacy must evolve to confront the rise of autonomous systems. It advocates for a model of co-regulation where law, ethics, and innovation coalesce to preserve humanity amidst automation. By drawing from thinkers like Hart and Finnis and global frameworks such as UNESCO's AI Ethics and the EU AI Act, the paper envisions a future where technology enhances, rather than eclipses, human judgment transforming AI from a tool of control into an ally of justice, empathy, and collective flourishing.

Iris-Panagiota Efthymiou, University of Piraeus, Greece

Artificial intelligence is now central to economic activity, public governance, and everyday decision-making, yet its rapid growth exposes tensions between innovation, ethics, and the law. This chapter examines how legal systems can protect human rights, fairness, accountability, and transparency while supporting technological progress. It analyses current regulatory models, gaps in liability, bias and discrimination risks, data governance, and the responsibilities of developers, institutions, and states. It also considers ethical principles and their limits when translated into binding rules. The chapter argues for clear standards, proportionate regulation, and shared accountability frameworks that align AI development with public interest, democratic values, and social trust.

Alanamol Shinto, Christ University, India

Intellectual property rights, or IPRs, are crucial in the digital age for promoting innovation, protecting the rights of creators, and driving economic growth. The paper examines the challenges that intellectual property rights (IPR) face, including digital piracy, online copyright infringement, and cross-border enforcement issues, as globalization and digital transformation continue to transform industries. AI has created serious problems and posed challenges in the areas of intellectual property rights, more particularly in copyright law. Analyzes the models used in the USA, EU, China, UK, India, and other countries, identifying their strengths and weaknesses. The present paper explores how AI plays a vital role in creating artistic works such as art, music, poems, and novels, among other forms. Moreover, the research examines the legal structures of various countries and provides a summary of different authorship models for AI-generated content.

Khushi Sood, Christ University, Delhi NCR, India
Khushboo Malik, Christ University, Delhi NCR, India

Artificial intelligence development in this digital era has questioned the fundamental concepts of intellectual property law, whose foundation was based on and centred on human creativity, inventiveness, and authorship. AI being able to generate outputs such as artistic and literary works, symbols and patterns, and inventions raises a question on the traditional notions of intellectual property. This chapter critically

Preface

INTRODUCTION

Artificial Intelligence (AI) has evolved from a specialized field of computational research into a transformative force shaping democratic institutions, economic systems, legal frameworks, and creative industries. As AI systems increasingly influence decision-making processes, public discourse, market behavior, and even cultural production, the urgency of examining their ethical, regulatory, and intellectual property implications has never been greater. This edited volume, *Ethical, Regulatory, and Intellectual Property Impacts on AI Development*, brings together interdisciplinary perspectives that critically explore how law, governance, and ethics must evolve alongside technological innovation.

The development of AI presents a paradox. On one hand, it offers unprecedented opportunities to enhance efficiency, expand access to knowledge, improve public services, and foster innovation. On the other, it introduces complex risks: algorithmic bias, opacity in automated decision-making, erosion of democratic discourse through disinformation, privacy intrusions, and fundamental challenges to established legal doctrines such as authorship, inventorship, and liability. These tensions demand not only technical solutions but also normative clarity and regulatory foresight. AI governance must therefore move beyond reactive regulation toward proactive, principled, and adaptive frameworks.

This volume responds to that need by situating AI within broader societal and legal contexts. Rather than treating ethics, regulation, and intellectual property as isolated domains, the chapters collectively examine their interdependence. Ethical design principles must be translated into enforceable standards. Regulatory instruments must account for behavioral realities and technological uncertainty. Intellectual property regimes must adapt to hybrid models of human–AI collaboration. Through doctrinal analysis, comparative legal study, empirical insight, and normative reflection, this book seeks to contribute to a more coherent and sustainable architecture for AI governance.

CHAPTER OVERVIEW

The volume is structured to move from democratic and ethical foundations to regulatory innovation and, ultimately, to intellectual property challenges in the age of AI.

Chapter 1, authored by Konstantinos Kouroupis, examines AI as a catalyst for democracy in the digital era. The chapter explores the dual role of AI in enhancing civic engagement through e-governance tools while simultaneously posing risks through automated disinformation and manipulation. By analyzing European and international legal frameworks, it proposes policy responses aimed at preserving freedom of expression and democratic integrity within AI-enabled systems.

Chapter 2, by Rachna Rana and Pankaj Bhambri, provides a comprehensive examination of ethical and equitable AI, focusing on bias mitigation and algorithmic auditing. Tracing developments from 2015 to 2025, the chapter highlights the evolution from fairness metrics to standardized auditing protocols and governance playbooks. It emphasizes the importance of integrating bias-aware processes into organizational lifecycles to operationalize accountability and transparency.

Chapter 3, written by Azamat Ali, turns to the ethical complexities of generative AI. Addressing uncertainty in model behavior and long-term societal impact, the chapter underscores the limitations of existing ethical frameworks when applied to generative systems. By advancing responsible innovation as a guiding approach, it advocates for foresight, inclusivity, and human oversight in AI design and deployment.

Chapter 4, by Iris-Panagiota Efthymiou, bridges ethics and regulation through the lens of behavioral law. Drawing on behavioral economics, the chapter demonstrates how cognitive biases and human–AI interactions complicate governance. It proposes a hybrid regulatory model that incorporates nudges, transparency mandates, and adaptive mechanisms to enhance both compliance and innovation.

Chapter 5, authored by Kumar Ranjan, addresses the governance of foundation models. Recognizing their systemic and cross-sectoral impact, the chapter critiques traditional sector-based regulation and advances a lifecycle-oriented regulatory framework grounded in proportionality, accountability by design, and robust oversight mechanisms.

Chapter 6, by Nehal Saxena, reimagines law as a moral compass in the age of AI. Through jurisprudential analysis and engagement with global frameworks such as UNESCO's AI Ethics and the EU AI Act, the chapter proposes a human-centric governance model in which law, ethics, and innovation operate in a co-regulatory partnership to safeguard dignity and justice.

Chapter 7, authored by Iris-Panagiota Efthymiou, examines the broader balance between innovation and responsibility. It analyzes liability gaps, data governance, bias risks, and the limits of ethical principles when translated into binding legal standards. The chapter argues for proportionate regulation and shared accountability to align AI development with democratic values and public trust.

The final three chapters focus on intellectual property challenges posed by AI-generated works and inventions.

Chapter 8, by Alanamol Shinto, surveys global intellectual property regimes and examines how AI-generated artistic and creative outputs challenge traditional copyright doctrines. Through comparative analysis of the USA, EU, UK, China, India, and other jurisdictions, the chapter identifies strengths, weaknesses, and emerging authorship models.

Chapter 9, by Khushi Sood and Khushboo Malik, further explores the redefinition of ownership, authorship, and inventorship in the AI era. By analyzing regulatory approaches in the EU, United States, India, and landmark cases such as DABUS, the chapter proposes balanced reform frameworks aimed at harmonizing accountability with innovation incentives.

Chapter 10, authored by Junaidi Junaidi, addresses the urgency of regulating AI-generated works through the lens of development law theory. Focusing on fair use and copyright limitations within the Indonesian context, the chapter emphasizes the need for comprehensive regulation aligned with national philosophical values while accommodating technological advancement.

CONCLUSION

Taken together, the contributions in this volume underscore a central insight: AI governance cannot be fragmented. Ethical reflection without regulatory enforceability remains aspirational; regulation without ethical grounding risks rigidity or injustice; intellectual property reform without consideration of technological realities may stifle innovation or undermine creators' rights. Sustainable AI development requires an integrated framework that connects these domains in a coherent and adaptive manner.

The global landscape of AI regulation is rapidly evolving, with initiatives such as the EU AI Act, UNESCO's ethical guidelines, and emerging national frameworks signaling a new era of normative engagement with technology. Yet regulation alone cannot resolve all tensions. Behavioral insights, institutional accountability, technical safeguards, and international cooperation must complement formal legal rules. Moreover, as foundation models and generative systems continue to expand in capability and influence, governance models must remain flexible enough to address unforeseen risks while preserving space for responsible innovation.

This volume aims to serve scholars, policymakers, practitioners, and students seeking to understand and shape the future of AI governance. By engaging deeply with questions of democracy, fairness, transparency, accountability, authorship, and ownership, the contributors collectively affirm a guiding principle: AI must remain anchored in human values. Technological progress should enhance democratic participation, protect rights, foster creativity, and promote social flourishing. The challenge before us is not merely to regulate AI, but to ensure that its development reflects our highest legal and ethical commitments.

Chapter 1
AI as a Catalyst for Democracy in the Current Digital Era

Konstantinos Kouroupis
https://orcid.org/0000-0003-1193-8224
Frederick University, Cyprus

ABSTRACT

Rapid technological advancement and the widespread digitalization of contemporary society have profoundly reshaped democratic governance. Artificial Intelligence (AI) plays a central role in this transformation, influencing institutional functioning, civic participation, and the protection of fundamental rights. Within the emerging framework of e-democracy and e-governance, AI-driven tools enhance political communication and citizen engagement, while simultaneously posing risks to democratic participation and freedom of expression through disinformation and automated content generation. This chapter examines the relationship between AI, democracy, and freedom of expression, analyzing relevant legal and ethical frameworks at the European and international levels. It adopts a critical perspective on AI-enabled disinformation and proposes regulatory and policy responses aimed at preserving democratic values while fostering trustworthy and inclusive e-democratic systems.

1. INTRODUCTION: THE DIGITAL TRANSFORMATION OF DEMOCRATIC GOVERNANCE

The democratic setting in the contemporary world is undergoing a revolutionary transformation, driven by the unstoppable pace of technological advancement and the malicious proliferation of digitalization across society at large. It should be underlined

DOI: 10.4018/979-8-2600-0000-7.ch001

that digitalization marks the adoption of the EU's Digital Agenda and the nature of modern societies (Kouroupis, 2023). Recent global crises, such as the COVID-19 pandemic, have further highlighted the critical importance of protecting human rights during periods of rapid social and technological change, underscoring the need for robust governance frameworks that can adapt to evolving challenges (Karagkouni et al., 2023). This digital revolution has radically altered the traditional model by which the citizenry engages with democratic processes, creating new opportunities for participation at the same moment that it unleashes new threats to the integrity and effectiveness of democratic institutions. As the European Parliament has explicitly declared through an official press release (European Parliament, 2021), among the most fundamental components of the digital transformation are the integration of digital technologies by companies as well as the massive use of digital platforms.

Undoubtedly, Artificial Intelligence (AI) constitutes the milestone of the current digital transformation bringing not only powerful innovation and productivity but also serious concerns and risks regarding the relationship between human and machines. AI penetrates almost every fundamental area of public life: justice, economy, and technology (Russell & Norvig, 2020). According to the computer scientist Nils John Nilsson, it corresponds to a technology that "functions appropriately and with foresight in its environment" (Nilsson, 2009). The High-Level Expert Group on Artificial Intelligence (AI HLEG) of the European Commission (EC) defined AI as "systems that display intelligent behaviour by analysing their environment and taking actions - with some degree of autonomy - to achieve specific goals" (Sheikh et al., 2023). The key term is "intelligence," which is attributed from human to machines.

The integration of AI into democratic processes is more than a matter of technological advancement; it is an interactive multi-level process where information is transmitted and managed to the mechanisms through which citizens are involved in political discourse and decision-making. Thus, the impact on fundamental rights, such as privacy, security, dignity, and freedom of expression, is inevitable (Bajgar & Horenovsky, 2023).

In terms of politics and social sciences, democratization through digital means corresponds to what is called "e-democracy" a unified approach to governance that employs digital technology to enhance democratic engagement, increase government responsiveness and strengthen connections between citizens and elected officials (Fountain, 2001). Furthermore, it encompasses the implementation of democratic government processes by electronic means in the business environment and public procurement bringing new challenges at procedural and organizational level (Ronchi, A. M. (2019). *e-Democracy*. Cham: Springer International Publishing). Fundamental elements of the term are e-government and e-participation since political actors make an extensive use of Information and Communication Technologies (Clift, 2004; OECD, 2003; United Nations, 2003).

However, the relationship between AI and democracy is not entirely beneficial. While AI technologies offer unprecedented potential to promote democratic processes and engage citizens via the facilitation of access to information, the innovative and attractive means of communicating political messages and campaigns, they also raise very significant challenges that require to be rightly addressed. These challenges refer to algorithmic bias and manipulation potential as well as to issues of privacy, surveillance and the accumulation of technological power in the hands of political leaders or large corporations.

The first section of the chapter aims to provide a close examination of how AI functions as a democratic stimulus within the contemporary digital world with particular emphasis on the regulatory schemes and policy models which reflect this relation. By considering both the potential and the risk of AI in democratic society, the ultimate goal is to demonstrate how societies can harness the capability of these technologies to change while maintaining the fundamental principles and institutions upon which democratic government is based. The second section adopts a more critical approach on the matter. By investigating the potential threats to democracy and fundamental rights due to the massive invasion of artificial intelligence tools, it is attempted the suggestion of several measures, following a multi-dimensional and multi-disciplinary perspective. Therefore, the responses suggested are laying on both normative and ethical level, requesting the undertaking of actions by all involved stakeholders, namely the state, institutions and the private sector.

SECTION I. INTERPRETING E-DEMOCRACY IN CURRENT DIGITAL ERA

2.1 Conceptual Foundations of E-Democracy

The principle of e-democracy involves much more than just transferring conventional democratic mechanisms onto the internet. It is often related to other terms such as electronic administration, electronic government or governance, and digital democracy (Hague & Loader, 1999; Hacker & van Dijk, 2000a). The use of Information and Communication Technologies (ICT) penetrates the perception of the phenomenon. According to Hacker and van Dijk (2000b), "digital democracy is the use of information and communication technology (ICT) and computer-mediated communication (CMC) in all kinds of media (e.g., the internet, interactive broadcasting, and digital telephony) for purposes of enhancing political democracy or the participation of citizens in democratic communication".

The most decisive attempt to define the term e-democracy has been implemented by the European Council through the adoption of the *Recommendation of the*

Committee of Ministers to Member States on Electronic Democracy (E-Democracy). In this text, the aforementioned term is strongly connected with e-governance. As such it corresponds to:

the use of information and communication technologies (ICTs) to enhance public services, democratic involvement, and public policy-making, and organizational change and skills development.

This definition highlights the wider processes of digital transformation within governance, noting that true e-democracy requires not just technological adoption but actual changes to how government institutions operate and citizens engage with the democratic process. According to Efthymiou e-government is a key pillar for the development of e-democracy, providing the infrastructure for effective citizen participation (Efthymiou, 2025).

Another issue that should be underlined within the above interpretation derives from the Appendix to Recommendation, where it is explicitly declared that:

E-democracy is one of several strategies for supporting democracy, democratic institutions, and democratic processes and spreading democratic values. It is additional, complementary to, and interlinked with traditional processes of democracy.

Therefore, e-democracy incorporates all elements and issues of democracy as an institution. As such, the rule of law and fundamental rights and freedoms are (or shall be) respected. The use of electronic means modernizes or updates the nature of democracy.

E-democracy can be observed in practice in a variety of main dimensions. In its more classic form, it involves the computerization of public services in order to make them more available, efficient, and convenient to citizens. As such, we are leading to e-administration, which refers to the provision of online administrative services and public (Muir & Oppenheim, 2002) via the use of technology and the internet (Tambouris et al, 2001). In any case, electronic administration appears as the radical change in the management of public information for the benefit of public interest.

At the institutional level, e-democracy receives a more proper political concept as e-voting. According to the European Committee on Democracy and Governance (ECDG) of the Council of Europe, to date, the Council of Europe is the only international organization to have set standards in the area of e-voting, which is defined as the use of electronic means to cast and/or count a vote. The specific reference text is Recommendation CM/Rec(2017)5 of the Committee of Ministers to member states on standards for e-voting, which is supplemented by guidelines on its implementation and accompanied by an explanatory memorandum (Council of Europe, 2017). In the European Union, Estonia is the only EU country with full-scale e-voting in political elections at both local and national levels. Pilots began in 2005 for municipal elections and expanded to parliamentary elections in 2007. Progressively, Estonia advanced the deployment of Internet voting and in 2023, for the first time, more than 50% of voters in parliamentary elections cast their ballots online, Alvarez et al. (2009), BBC News (2005) and ERR News (2023)

2.2 The Role of AI in Promoting E-Democracy

AI technologies are increasingly central to advancing e-democracy initiatives across multiple domains. In citizen services, AI-based chatbots and virtual assistants provide 24/7 government support in multiple languages, helping navigate complex procedures and offering personalized recommendations. These technologies enhance citizen engagement and help prevent autocracy (Cupać et al., 2024; Novelli & Sandri, 2024; Jungherr, 2023). AI is also transforming democratic participation and policy-making by enabling mass deliberation and consultation, significantly impacting fundamental rights like access to justice and freedom of expression (Bajgar & Horenovsky, 2023). Notable examples include Denmark's Synthetic Party, led by AI to develop its political program and raise awareness about AI's role in governance and accountability (Futurism, 2022). In the U.S., ABC News live fact-checked statements during the 2024 presidential debates between Donald Trump and Kamala Harris, demonstrating technology's role in combating misinformation (Sarnoff & Burton, 2024). The presence of major tech CEOs like Elon Musk, Mark Zuckerberg, and Jeff Bezos at Trump's inauguration underscores technology's strategic importance in political agendas (Mock, 2025).

The European Union's member states are adopting comprehensive AI policies for public life, with Greece playing a significant role. Greece established the High-Level Advisory Committee on Artificial Intelligence in November 2023, coordinated by the Special Secretariat of Foresight (Capital.gr, 2023). Greece's national AI strategy, *A Blueprint for Greece's AI Transformation*, presents a four-pillar plan promoting innovation and entrepreneurship, strengthening education and research, designing responsible regulatory frameworks, and deploying AI in the public sector for transparency and efficiency. The strategy centers on the "Pharos" project and establishing a national AI center of excellence, supported by Law 4961/2022 encouraging AI adoption. Developed through collaboration with government leaders, researchers, entrepreneurs, and international experts, the strategy includes six flagship projects and emphasizes data governance, education, research, and strategic investment (Special Secretariat for Foresight, 2024). Additionally, AI enables unprecedented transparency and accountability by monitoring government performance, identifying ethical violations, and providing citizens real-time information on public spending and results.

2.3. AI and Democratic Participation Enhancement

Artificial intelligence (AI) technologies are more extensively discussed as the means of increasing democratic participation by removing barriers to accessing information, civic participation, and being connected with the public institutions. Research suggests that with the help of AI-powered tools such as chatbots and intelligent civic platforms, citizens can better understand the public policies, the election process, and the mechanisms involved in running the government to become more engaged (Sharma, 2024). They allow multilingual communication which is particularly important in communication with marginalized or linguistically diverse groups, personalization of political news according to the interests of users, and real-time communication between citizens and authorities (Zidouemba, 2025). Similarly, AI-driven applications can transform the once-in-a-while process of democracy that is centered on voting into a more participatory and engaging process.

AI also fosters democratic participation by improving the decision-making process and discussions by analyzing data. Machine learning and natural language processing can help policymakers meet the needs of society since they can gather vast volumes of feedback provided by citizens, analyze the mood of the population, and identify tendencies in civic dialogs (Kuziemski and Misuraca, 2020; Sharma, 2024). The participatory systems are allowing citizens to directly articulate preferences, negotiate the policy alternatives, and influence the allocation of resources by means of participatory democracy mechanisms such as AI-enhanced online forums, liquid democracy design, and participatory budgeting sites (Jungherr, 2023). Another instance of how AI is used to structure deliberative spaces, as exemplified by the IGI Global chapter, is the emphasis placed on points of agreement and disagreement, the moderation of conversations, and bettering citizens' reasoning in a more democratic process.

Nevertheless, AI needs to be ethically designed, governed responsibly, and well-regulated to realize its democratic potential (Vousinas, et al., 2022). Although accountability and transparency may be enhanced with the help of AI tools such as open-data analytics and automatic fact-checker, when it is not controlled, there are issues with the aspects of algorithmic bias, surveillance, and political influence (Sharma, 2024; Zidouemba, 2025). Thus, AI systems should be open, responsible, and participatory so as to actually promote democratic participation and ensuring that the fundamental rights (such as equality, privacy, and freedom of thought) are respected. By doing so, AI is framed as an enabling infrastructure that, when handled appropriately, can lead to more citizen engagement, a greater level of trust in the democratic processes and allow more inclusive and robust democratic systems to be built instead of a replacement of democratic institutions (IGI Global, 2024; Jungherr, 2023).

2.4. The Regulatory Framework: European and International Perspectives

2.4.1. The European Strategy of AI Regulation

The European Union has emerged as a global leader in developing comprehensive regulatory guidelines for AI technologies, particularly focusing on their impact on democratic governance and human rights. The EU approach emphasizes human-centered AI development, strong privacy protection, and explicit consideration of AI's democratic implications (European Commission). While the General Data Protection Regulation (GDPR) does not specifically address AI, it has established important precedents for AI systems handling personal data. The GDPR's requirements for transparency, consent, and data portability have significant implications for AI systems in democratic governance, ensuring citizens maintain control over their personal data while accessing AI-supported government services. These principles apply when implementing the EU AI Act, with the major challenge being the safeguarding of fundamental rights (privacy, data protection, and non-discrimination) which represent the three main intersections between AI and fundamental rights (Ufert, 2020).

Building on these foundations, the EU has developed specific AI guidelines, including the Ethics Guidelines for Trustworthy AI, which establish seven key requirements: human agency and governance, technical robustness and safety, privacy and data governance, transparency, diversity and fairness, societal and environmental well-being, and accountability (European Commission, 2019). These requirements are particularly crucial for AI systems operating in democratic societies where fairness, transparency, and accountability are paramount. The guidelines stipulate that trustworthy AI must comply with the law, fulfill ethical principles, and be technically robust. The EU's AI Act represents the most ambitious regulatory attempt to date, adopting a risk-based approach with stricter standards for AI systems that pose higher risks to human rights and democratic processes. Democratic processes themselves, including voting, political campaigning, and government decision-making demand specific regulations. As stated in the Act, AI systems intended to influence election outcomes or voting behavior are classified as high-risk, except for tools used for administrative and logistical campaign organization (Recital 62, ANNEX III).

2.4.2 International Standards and Cooperation

The global reach of AI technologies and their democratic implications have prompted increasing international collaboration in developing standards and best practices, with different international institutions addressing the challenge of democratic AI governance from varying perspectives. The Council of Europe emphasizes establishing a safe, human-centric framework for governing AI. The Council of Europe Framework Convention on Artificial Intelligence and Human Rights, Democracy and the Rule of Law represents the first global legally binding instrument designed to ensure AI upholds common standards in human rights, democracy, and the rule of law while minimizing risks to these principles (Council of Europe, ibid). An ad-hoc Committee on Artificial Intelligence (CAI) has been established under the Committee of Ministers to develop harmonized AI regulations based on the rule of law, human privacy and dignity, and fundamental rights and freedoms. The committee has published guidelines emphasizing the importance of maintaining human rights protections in AI-enabled democratic governance, stressing the need for transparent, accountable, and non-discriminatory AI systems, particularly those affecting political participation or government decision-making. The Framework Convention closely aligns with the EU AI Act, as both adopt a human-centric approach (Presno Linera & Anne Meuwese, 2025).

The Organisation for Economic Co-operation and Development (OECD) has developed AI principles adopted by over 40 countries worldwide, aiming to make AI systems secure, safe, and reliable while stimulating innovation and economic growth. The OECD principles specifically address the need for AI governance frameworks that protect democratic values and institutions (OECD, 2023). UNESCO has similarly developed ethics for AI that take an expansive perspective on the relationship between AI and human flourishing, including democratic governance. Through its Recommendation on the Ethics of Artificial Intelligence published in November 2021, UNESCO provides specific guidelines for powerful and trustworthy AI that respects human rights and dignity while promoting fundamental principles such as transparency and fairness (UNESCO, 2021).

3. LEGAL AND ETHICAL REQUIREMENTS FOR TRUSTWORTHY AI IN DEMOCRATIC SYSTEMS

3.1 Transparency

Transparency is probably the most fundamental requirement for AI systems employed in democratic contexts. People should be entitled to learn about how AI systems that affect their lives operate, what they take as information, and how they make decisions. This transparency need serves a number of functions: it enables democratic accountability, allows for more enlightened citizen participation, and can help identify and counteract concealed biases or errors in AI systems. Transparency and fairness play in an interactive form in order to safeguard privacy, rule of law and human dignity (Radanliev, 2025).

The requirement for transparency in democratic AI systems is at various levels. At minimum, citizens should be informed when they are interacting with AI systems rather than human officials. This notification mechanism safeguards citizens against uncertainty regarding the character of their interaction and allows them to make an informed decision about what to do next (Helbing et al., 2018).

Additionally, transparency obligations extend to accounts of how AI systems operate and arrive at decisions. For government decision-making AI systems, for example, deciding whether a person is eligible for social welfare, the amount of tax one owes, or if citizens should be issued a permit, citizens must be able to access clear explanations of how the AI system comes to the decision. This entails details on what is considered by the AI system, how the considerations are balanced, and what sources of information are used. An indicative example of the deployment of AI tools in public administration, as it has been previously demonstrated, refers to the use of a platform in the Netherlands which aims to tackle fraud. In fact, the national authorities used the SyRI (System Risk Indication) algorithm to detect welfare fraud, which quietly flagged thousands of citizens as high-risk by interconnecting and processing information from various government services across social security, tax collection, and labor law enforcement. However, in 2020, a Dutch court banned the system due to lack of transparency and ordered the government to stop using it in such areas (Rachovitsa & Johann, 2022; Toh, 2020).

3.2 Fairness and Non-Discrimination

Maintaining fairness and preventing discrimination are critical requirements for AI systems applied to democratic settings. AI systems can reinforce or even deepen existing social prejudices, causing discriminatory treatment to some sections of citizens. In democratic systems, where equality of treatment and opportunity is the key, prevention of AI-caused discrimination is a requirement.

The call for democratic AI systems to be equitable is far-reaching. First, AI systems should never discriminate based on citizens' protected characteristics such as race, gender, religion, or political opinion. This entails precision in the training data, the algorithmic structure, and also vigilant monitoring of system output to catch and rectify any possible bias (Zhang, 2024).

Furthermore, the AI technologies utilized in democratic processes should also provide access and treatment to every citizen. This means that government services based on AI should be accessible to disabled citizens, offered in more than one language if necessary, and made to work well for citizens who might have varying degrees of digital competence or degrees of access to technology (Chemnad and Othman, 2024; Kumar et al., 2023).

Provisions for fairness also apply to avoiding AI systems operating in a way that is unintentionally biased towards certain political opinions or ideologies over others. This is especially crucial for AI systems operating in information sharing, content moderation, or the election process, where seemingly biased behavior can be behind public skepticism of democratic processes (Kus, 2025; Kumar et al., 2023). Additionally to the aforementioned remark, it should be underlined that in the framework of operating gig economy platforms, diversity and inclusion issues must be integrated into the regulation of digital tools to avoid the reproduction of inequalities (Efthymiou, 2025).

3.3 Human Oversight and Accountability

Accountability is a necessary requirement for AI systems in democratic environments, requiring open recourse channels whenever systems make mistakes or unfair judgments. Such a framework for accountability operates on multiple levels: as citizens, they ought to have access to appeal means and human review of automated decisions, entitlements to appeal important decisions to human administrators who are capable of making judgments based on contextual factors beyond the capabilities of AI systems. At an institutional level, the government and agencies that use AI systems are accountable for them, requiring transparent procurement processes, monitoring systems in place by design, and regular reporting on system performance and outcomes. Human control is also required to ensure AI systems augment and not replace human judgment in important democratic processes, with ultimate decision-making authority resting with accountable human officials (Camilleri, 2023; Robles & Mallinson, 2023).

Enforcement of responsibility for AI in democratic nations necessitates critical analysis and adaptation of existing legal and regulatory regimes to address new challenges posed by automated systems. Existing law and practice must be greatly updated to establish clear chains of responsibility for automated activities, impose appropriate human oversight provisions, and stipulate significant performance requirements for AI system reliability and effectiveness. Although AI systems offer valuable potential in data manipulation, pattern recognition, and decision-making support, democratic accountability necessitates that the final authority of decision remain in human officials who can be held accountable by institutionalized democratic means. This way, technological progress serves democratic ideals rather than undermining the fundamental accountability mechanisms that underpin democratic government and public trust.

Conclusions: Toward a Framework for Democratic AI Governance

The implementation of AI technologies within democratic practices is simultaneously a compelling opportunity and a daunting challenge for contemporary societies. As it has been thoroughly illustrated through this section, AI can function as an effective catalyst of democracy, increasing the participation of citizens, streamlining government services, and improving the ties between citizens and their representatives. But realizing that potential requires careful attention to AI system design, deployment, and regulation so that they promote, rather than undermine, democratic values and institutions.

The policy and regulatory frameworks addressed in this chapter are essential pillars of democratic AI governance but also the reality of the challenge. The requirement that certain AI systems be transparent, fair, accountable, and respect for privacy will rest on ongoing technical research, policy development, and participation by affected parties. It will also hinge on the awareness that the regulation of AI in democracies is not technically but politically a matter that requires wide social reflection and democratic engagement.

The path forward is through ongoing engagement between technologists, policymakers, civil society organizations, and citizens to design AI systems that actually do work in the interests of democratic demands. This collaboration must be guided by a profound commitment to democratic values and principles and by recognition of the fact that the modernization of democracy via the deployment of AI tools shall be based both on ethical and legal principles. The latter correspond to the respect of the following requirements which equally receive an ethical and normative perception: those of transparency, accountability, and privacy (Ramya et al., 2024).

SECTION II: AI, MISINFORMATION, AND THE PROTECTION OF DEMOCRATIC SPEECH: CHALLENGES AND SOLUTIONS

Introduction: Politics and Artificial Intelligence: Democracy in Crossroads

The introduction of artificial intelligence into the political arena has shifted from future prophecy to undeniable reality, pervasive across all domains of contemporary democratic governance. AI is now an active participant in contemporary politics rather than merely a future technology. While regulatory frameworks for AI's political use continue to emerge at national and international levels, these efforts lack homogeneity and require strong institutional and state-level commitment for implementation. The critical question this technological revolution poses concerns democracy and social interest rather than scientific progress: Are we witnessing an enhancement of democratic processes through novel devices, or are we heading toward societal de-democratization?

Democracy is grounded in shared facts, but misinformation corrupts this foundation by creating disjointed realities (Lewandowsky et al., 2020). Presenting citizens with falsehoods incapacitates their decision-making ability, demolishing the quality of democratic deliberation (Lazer et al., 2018). Research shows misinformation spreads faster and wider than accurate information, particularly on social media platforms (Vosoughi et al., 2018), fueling political polarization and making democratic compromise more difficult (Guess et al., 2019). This damages trust in institutions including media, science, and electoral processes (Ognyanova et al., 2020), making citizens more vulnerable to populist or authoritarian promises (Waisbord, 2018). Exposure to fake news reduces media trust while increasing trust in partisan organizations (Guess et al., 2020). False election information discourages participation and delegitimizes democratic processes (Amin, 2024), with populist far-right leaders particularly adept at exploiting media to manipulate public opinion (Mazzoleni, 2008). Algorithmic amplification exacerbates this problem (Tufekci, 2017), while fragmented media systems create echo chambers reinforcing existing biases (Sunstein, 2017) and weakening collective action (Norris & Inglehart, 2019). While citizen engagement and fact-checking can counter some misinformation effects (Nyhan & Reifler, 2010), and Russian propaganda techniques exploit societal divisions to destabilize democratic institutions, fact-checking alone is insufficient, active regulatory mechanisms, knowledgeable citizenship, and media plurality, as demonstrated by Germany, are necessary to combat these threats (Bayer et al., 2020). Unrestrained misinformation perverts truth, confidence, and responsibility, accelerating societal de-democratization.

2.1. The Impact of AI on Democratic Engagement

The most visible manifestation of AI becoming a part of politics occurs in the arena of political campaigning, where AI has changed traditional ways of voter engagement. Modern political campaigns rely more and more on AI-based solutions to create compelling multimedia, such as sophisticated videos featuring virtual candidates or AI-generated spokespersons, interactive online environments, and personalized messaging platforms that adjust their reply according to the voter interests and concerns.

These innovations offer unprecedented political candidates' avenues for interfacing with constituents in ways heretofore unimaginable. Campaign strategies based on AI enable rapid rollout of tailored messages to multiple platforms at once, reducing significantly the time and expense required for successful political engagement. The cost savings are especially meaningful to smaller political movements or candidates with fewer resources, potentially leveling the playing field for access to advanced campaign technologies previously within the reach only of financially well-supported political entities.

In addition, AI systems possess enhanced real-time processing and analysis capabilities of citizen queries, problems, and grievances, enabling political leaders to respond faster and more accurately to new needs and priorities of their citizens. This increased responsiveness may have the potential to strengthen the link between elected officials and citizens, creating more flexible and participatory models of democratic engagement. This might be the responsibility of all relevant competent stakeholders who shall apply the most appropriate AI tools for public benefit (Efthymiou, 2026).

Other than traditional campaigning, artificial intelligence also provides new methods of engaging citizens in the democratic process. AI-enabling citizens' assemblies are one of the highest points for this technology, providing structured environments where political elites and citizens can collaborate on hard policy trade-offs together (Curtis, 2023). These virtual forums can manage vast amounts of data, identify places of commonality among vastly diverging viewpoints, and facilitate well-informed decision-making on society's hard issues.

These technologies might even address some of the deep-rooted problems of democratic governance, such as overload of information, polarization, and the lack of ability to coordinate multifaceted policy debates across divergent interests. Through providing objective platforms for dialogue and scrutiny, AI tools might usher in more harmony and improved understanding of policy implications among different sections of society.

The predictive power of AI systems also offers valuable insights into the likely policy effects, allowing citizens and policymakers to have a better understanding of the likely consequences of different political choices. Increased foresight can translate into better-informed voter choices and more effective government, as policy could be piloted and refined according to empirical forecasts of its real-world effects.

2.2. The Dark Side of AI in the Political Arena

Beneath the supportive applications lies a sinister truth undermining democratic foundations of self-determination. The tremendous computing power involved in large-scale data processing that makes AI systems possible to understand citizen taste and predict political behavior also creates new levels of control and manipulation. When massive databases containing people's data, behavior patterns, and psychological profiles are fed into processors and political strategy is generated, the line between responsive governance and manipulative control easily becomes erased.

Sophisticated AI algorithms can analyze the online footprint of individual citizens with unprecedented precision, extracting conclusions about their political leanings, hot buttons, personal weaknesses, and vulnerability to changing their minds on issues. The information becomes a valuable commodity for political campaigns seeking to craft their message to maximize persuasive influence. The targeting ability of such systems makes political advertising possible through micro-targeting, where different messages are delivered to different groups based on their expected vulnerability to certain arguments or emotional appeals.

The move away from mass public discourse to tailored persuasion campaigns fundamentally alters the nature of democratic discourse. Instead of reacting to open forums where ideas are discussed freely, political leaders now are able to craft highly targeted messages to resonate with particular people or small demographic segments without subjecting those messages to broader scrutiny or debate.

The chilling potential for AI-generated political content appears evident and inevitable through a research implemented by Stanford University. Researchers conducted an experiment to see if artificial intelligence had the potential to create persuasive arguments that would change human minds on divisive political issues. Using advanced machine learning techniques, they wrote persuasive pieces on various social issues of high interest, such as the question of gun owenership, that stimulate public opinion.

The test design involved thousands of volunteers randomly assigned to groups to read either human- or AI-provided arguments, unaware of the source of the content. Volunteers indicated their views on the topics in question before and after exposure to the persuasive material. The results were startling and truly disturbing: in each of the tested comparisons, messages generated by the AI were "consistently persuasive to human readers", often trumping their human-generated counterparts in their ability to shift attitudes (Myers, 2023).

This finding suggests that artificial intelligence systems can possess more effective skills to craft compelling political content compared to human communicators. The implications are profound since it suggests that artificial intelligence can be utilized in a way that influences public opinion on a mass scale, and the effect may be more than with traditional means of political persuasion.

The demonstrated ability of AI-generated persuasive content raises fundamental questions about the integrity and authenticity of political discourse in democratic states. Since artificial intelligence is capable of generating more persuasive messages than human beings, the foundational pillars of democratic discourse are questioned. Democracy operates under the assumption that citizens have the capacity to engage with genuine ideas and arguments advanced by fellow citizens and representatives and make decisions on the basis of real human values and ideas.

The integration of AI-generated text into political discourse undermines this premise by including arguments and stances that are not necessarily a product of human belief but algorithmic optimization for persuasiveness. This shift signals away from politics as a sphere of human interaction and actual representation towards politics as a realm of technological manipulation and fake influence.

The consequences extend beyond the sphere of individual persuasion to affect the wider democratic culture. Political speech defined by AI-tuned messaging loses the spontaneous, immediate, and authentic features of human political exchange. Citizens are left to respond to synthetic stimuli rather than face genuine human opinion, and a democratic engagement that is necessarily shallow and manipulative ensues.

2.3. AI and Misinformation

The issue of disinformation and misinformation in political environments has been breathtakingly amplified by the capabilities of artificial intelligence systems. The internet has already been a rich ground for the spread of fabricated information, propaganda, and manipulated content, but AI tools have increased tenfold the scale and sophistication of these processes.

AI can review vast volumes of social media content, news articles, comments, and other web material looking for emotional triggers, psychological vulnerabilities, and convincing rhetorical strategies across different audience groups. This analytical potential enables highly targeted campaigns of disinformation that exploit distinctive fears, biases, and assumptions that are present in certain communities or demographic cohorts.

Moreover, AI sites can participate actively in web political debate, generating comments, messages, and responses that seem to come from actual human users but are actually designed to sway public opinion along predetermined agendas. Such artificial players can deluge web forums with copy designed to promote some views, suppress others, or create false impressions about the believability or popularity of different political positions.

Arguably the most ominous development in AI-disinformation is the emergence of deepfake technology, which enables the creation of highly realistic but artificial audio and video. According to European Parliamentary Research Service analysis, politicians are the most vulnerable to deepfake targeting, especially those who lack the means necessary to monitor and protect their online reputation adequately (Adam & Hockuard, 2023).

Deepfake technology can be utilized to produce convincing videos of political leaders saying or doing something they never did or said. These materials can be disseminated quickly on social media and other digital media, and before anyone can establish or question their authenticity, they may have been viewed by millions of people. The effect of such information is potentially disastrous, especially during particularly important times like election campaigns when people's opinions are most malleable and impactful. Additionally, Efthymiou and Egleton claim that Deepfakes not only threaten political discourse but also markets and economies, multiplying the consequences of disinformation (Efthymiou & Egleton, 2025).

The European Parliament's research service warned that generative AI content of this kind "can contribute to large-scale deception and undermine public confidence in democratic institutions" (OECD, 2023). The erosion of trust in political communication being authentic undermines one of the conditions for democratic government to work: citizens' ability to educate themselves and make well-informed decisions on the basis of accurate information about their representatives and what is at stake.

The dissemination of AI-based disinformation creates a broader crisis of trust that transcends individual examples of counterfeit content to affect the broader information environment upon which democratic decision-making depends. When citizens increasingly realize that there are sophisticated manipulation efforts in play, they will likely no longer have faith in the authenticity of any political message, producing generalized distrust that makes well-informed democratic participation extremely difficult.

Earlier Eurobarometer surveys have also documented declining trust in the sources of internet information reflecting growing public awareness of manipulation and disinformation attacks (European Commission, 2022). Loss of confidence sets a vicious cycle of reduced influence for genuine political communication while manipulative media can even gain confidence from their audience by profiting from citizens' shame and uncertainty regarding information validity.

The crisis of confidence also aims at democratic representation as such. When citizens cannot distinguish actual political positions from AI-created political content designed to influence their opinions, the connection between political representation and the will of the people becomes tenuous. The representatives may then respond to artificially generated seeming public opinion rather than the actual preferences of citizens, thereby developing policies that do not reflect the actual desires of the people. Therefore, manipulation of public opinion becomes easier and quality of democracy erodes.

The scenario under which a law might be enacted on the strength of artificially created public opinion is no longer speculative but is already an existing and immediate danger to democratic governance (Kreps & Kriner, 2023; Schneier & Sanders, 2023).

Furthermore, increasing levels of corporate and even law enforcement access to significant amounts of citizens' personal data from social networks and other online sources present additional threats (Lageson, 2020). The data would be potentially used to identify and target political dissidents, monitor citizens, and stifle opposition voices, transforming AI technology from democratic participation tools into instruments of political repression.

One of the major concerns engendered by the integration of AI in political systems is how artificial intelligence systems can be made accountable for what they do and what they decide. Machine learning algorithms can make decisions that profoundly affect individual citizens as well as alter public perception, and yet the process of how such systems make decisions is often unclear and even incomprehensible to their own creators (Busuioc, 2021; Jackson, 2021).

This concealment creates a fundamental accountability deficit under democratic rule. In traditional political arrangements, citizens can hold representatives accountable for decisions and actions through various devices such as elections, public media, and judicial recourse. In the situation where political decisions are informed or made by AI systems whose decision-making rationale is not necessarily knowable or explainable, such traditional mechanisms of accountability don't apply.

The accountability for the behavior and effect of AI systems within political contexts remains very controversial issue (Kgoale & Odeku, 2023; Soyer & Tettenborn, 2022). Should accountability be assigned to the programmers who created the systems, the firms applying them, the politicians consenting to their application, or some combination thereof? The formidability of modern AI systems makes it difficult to correlate specific outcomes with determinate human decisions, and so generates a diffusion of responsibility that demeans democratic responsibility.

The accountability deficit created by AI systems is a direct assault on fundamental democratic principles and human rights (Raso et al., 2018; Marwala, 2023). If the source of political forces that influence the lives of citizens cannot be traced or if it cannot be protested against under the processes to which crucial decisions are reduced, their basic right of political participation and representation is lost.

Besides, political uses of AI systems are often involved in the processing and harvesting of masses of personal information in the absence of real consent or regulation. This practice is a danger to privacy rights and presents risks of discrimination and manipulation that can disproportionately affect marginalized communities. The opacity of these systems prevents those upon whom their influence is acted from redress or protection from the harmful employment of AI technology.

The integration of AI in political decision-making also poses a threat to the very nature of democratic representation (Arsenault & Kreps, 2022). Political decisions are increasingly being influenced by algorithmic processes rather than human judgments and ethics.

2.4 Regulatory and Policy Responses

Governments worldwide have taken various policy and regulatory actions to address the threat of AI-driven disinformation, differing in extent, modalities, and effect of their actions. The divergence reflects varying traditions of the law, varying constitutional orders, and varying cultural values about speech and platform regulation. It has been already mentioned that almost all EU member states have adopted an AI strategy which aims to draw a roadmap towards the establishment of a safe, powerful, and trustful digital environment.

Content regulation approaches revolve around requiring platforms to remove or restrict access to known disinformation. To that direction, special emphasis should be given on the European Union's Digital Services Act, which requires large platforms to have in place systems for detecting and removing illegal material, including some forms of disinformation (Articles 4-7 of the DSA as well as the notice-action mechanisms which are set up in Articles 16 and 22).

Disclosure and transparency requirements are another kind of regulatory approach that aims to compel platforms and creators to reveal more about their activity and funding rather than directly regulating content. These kinds of approaches have the option of requiring platforms to publish regular reports of their content moderation policies, store political advertisements in databases, or implement identity verification standards on certain account types (Wolff, 2025).

Platform liability frameworks aim to counter disinformation by modifying the legal immunity that platforms enjoy for user-posted content. Conditional immunity regimes have been proposed in certain jurisdictions where platforms remain immune from liability provided they are in compliance with particular conditions regarding content moderation or disclosure. These frameworks aim to instill platform self-regulation but avoid explicit government content control.

Platform liability models are often applied in the case of political ad libraries set up by the aforementioned platforms. However, in January 2025, in a video posted alongside a blog post by the company on Tuesday, chief executive Mark Zuckerberg said third-party moderators were "too politically biased" and announced the abolition of independent fact-checkers, replacing them with X-style "community notes" where commenting on the accuracy of posts is left to users (BBC News, 2025a). Meta's sweeping changes to the management of its social networks, including Facebook, Instagram, and Threads, have put the company on a collision course with European Union and United Kingdom lawmakers, according to experts (The Guardian, 2025). Legislators in Brussels and London have criticized Zuckerberg's decision to eliminate fact-checkers in the U.S., characterizing the move as "particularly concerning" (Politico, 2025). Meta's new policy now allows users to refer to transgender individuals as "it" and make claims about "mental illness or abnormality" based on gender or sexual orientation (NBC News, 2025).

Meta announced it will rely on a "community notes" system, similar to the one adopted by Elon Musk on platform X (formerly Twitter), sparking fears about the spread of misinformation regarding elections, health issues, pandemics, and armed conflicts. Nobel Prize-winning journalist Maria Ressa warned of "extremely dangerous times" for journalism and democracy, arguing that "Meta is allowing lies, fear, and hatred to contaminate every user of the platform" (The Guardian, 2025).

Meta is expected to face serious regulatory pressure in the EU and UK if it extends its new policies beyond the U.S. The European Commission responded to Zuckerberg, rejecting claims that the Digital Services Act (DSA) imposes "censorship". MEP Valérie Hayer emphasized: "*Europe will not accept manipulation and misinformation as the norm of society. The elimination of fact-checkers in the U.S. represents a strategic and moral error by Meta*" (Politico, 2025). Experts further warn that Meta's changes may expose teenagers to harmful content, such as violence or pornography, as the company will rely on users to flag violations. The NSPCC, a children's protection charity, has requested clarification on how Meta intends to safeguard children from these risks (BBC News, 2025). Overall, experts argue that the clash between Meta and European regulators over content management policies is inevitable, as the company's new approach may directly conflict with European legislation.

2.4.1. Civil Society and Industry Initiatives.

It becomes obvious that building an AI ethical and legal environment demands the mobilization of all relevant competent key-players, such as civil society organizations, universities, and industry groups which shall organize an intense campaign in order to cultivate digital consciousness among citizens. At this point, specific practical measures against disinformation have been published in a recent report conducted by the Global Digital Human Rights Network (GDHRNet), the name of the COST Action 19143, which is placed under the auspices of the European Commission and systematically explored the theoretical and practical challenges posed by the online context to the protection of human rights (COST Action, 2020).

As such, practical steps against disinformation include:

1. **Monitoring and fact-checking**: carried out by internet communications companies, academia, media, civil society, and independent fact-checking organizations.
2. **Investigative responses**, which establish the accuracy of online content and provide insights into disinformation campaigns, including its origins, key actors, degree of spread, and affected communities.

3. **Technical and algorithmic responses**, implemented by the social media platforms, video-sharing and search engines themselves, but also through third-party tools (e.g., browser plug-ins) or experimental methods from academic research, using algorithms and/or **Artificial Intelligence (AI)** to detect and limit the spread of disinformation, or provide context or additional information on individual items and posts.
4. Finally, the establishment of **expert units/teams**, which should be located outside the state or should not belong to the public sector for purposes of impartiality, neutrality, and transparency, would decisively contribute in the battle against disinformation. Those expert units/teams would develop strong cooperation with organs of state to handle crises and identify emerging threats.

Additional measures, which focus specifically on the targets of the disinformation, include:

1. **Ethical and normative responses** carried out at international, regional, and local levels involving public condemnation of acts of disinformation or recommendations and resolutions aimed at thwarting these acts and sensitizing the public to the issues.
2. **Educational responses** which promote citizens' media and information literacy, critical thinking, and verification in the context of online information consumption, as well as journalist training.

Conclusions: Balancing AI Innovation with Democratic Preservation.

The analysis reveals that addressing AI-enabled disinformation requires a multifaceted approach that transcends simple technological or regulatory solutions. The fundamental conclusion is that democratic societies must develop adaptive governance frameworks that can evolve with technological advancement while preserving core democratic values (Helbing et al., 2019). This necessitates moving beyond binary approaches of either unrestricted technological development or heavy-handed censorship toward more nuanced, graduated response mechanisms.

The research demonstrates that effective anti-disinformation strategies must incorporate proportional responses based on threat severity rather than applying uniform standards to all potentially false information (Gorwa et al., 2020). This graduated approach includes content tagging, reduced algorithmic distribution, contextual information provision, and selective content removal, with each measure calibrated to the specific nature and intent of the disinformation. Such systems require sophisticated assessment mechanisms that can distinguish between deliberate manipulation, good-faith errors, and legitimate political disagreement (Benkler et al., 2018).

Transparency and accountability emerge as critical safeguards against the weaponization of anti-disinformation efforts for political suppression. The analysis emphasizes that platforms and intermediaries must disclose comprehensive information about their content policies, enforcement practices, and decision-making processes to enable public scrutiny (Suzor et al., 2019). This transparency requirement extends to providing accessible appeals procedures and external oversight mechanisms to prevent mission creep in content moderation practices.

The protection of vulnerable populations represents a particularly complex challenge requiring targeted interventions. Electoral processes, minority communities, and youth populations face heightened risks from disinformation campaigns that exploit cultural, linguistic, or economic vulnerabilities (Howard & Hussain, 2013). Effective protection strategies must balance enhanced safeguards with respect for political speech rights, requiring close collaboration with community organizations and culturally appropriate communication approaches.

International cooperation emerges as essential given the transnational nature of disinformation campaigns, yet must navigate significant challenges related to sovereignty, legal pluralism, and varying cultural approaches to speech regulation (Bradshaw & Howard, 2019). The analysis suggests that successful international frameworks require mutual legal assistance agreements, information-sharing mechanisms, and diplomatic coordination while preventing authoritarian exploitation of cooperation mechanisms.

The concept of "political consciousness development" represents perhaps the most significant long-term conclusion, extending beyond traditional media literacy to encompass understanding of information system vulnerabilities, manipulation techniques, and democratic participation strategies (Bennett & Livingston, 2018). This broader educational approach requires sustained investment in civic engagement, critical thinking development, and institutional capacity building.

Critically, the analysis acknowledges that technological solutions alone are insufficient without accompanying social and institutional transformation. The reconceptualization of freedom of expression in algorithmic environments requires considering both negative rights against censorship and positive rights to diverse information access and meaningful participation in democratic discourse (Tufekci, 2017). This expanded understanding necessitates active measures to ensure AI systems enhance rather than undermine democratic expression preconditions.

The ultimate conclusion emphasizes that maintaining democratic values while harnessing AI benefits represents an ongoing political endeavor rather than a technical problem with final solutions. Success requires sustained commitment from all stakeholders—governments, technology companies, civil society, academia, and citizens—to continuously adapt institutional frameworks as technologies and social arrangements evolve. This adaptive approach must maintain faith in democratic governance while acknowledging institutional limitations and working toward continuous improvement in democratic effectiveness and responsiveness.

REFERENCES

Action, C. O. S. T. (2020). *Global digital human rights network* (CA19143). European Cooperation in Science and Technology. https://www.cost.eu/actions/CA19143/

Adam, M., & Hockuard, C. (2023). *Artificial intelligence, democracy and elections.* European Parliamentary Research Service. https://www.europarl.europa.eu/RegData/etudes/BRIE/2023/751478/EPRS_BRI(2023)751478_EN.pdf

Alvarez, R. M., Hall, T. E., & Trechsel, A. H. (2009). Internet Voting in Comparative Perspective: The Case of Estonia. *PS, Political Science & Politics*, *42*(3), 497–505. DOI: 10.1017/S1049096509090787

Amin, P. (2024). Disinformation and the Impact on Democracy. In Hewage, C., Yasakethu, L., & Jayakody, D. N. K. (Eds.), *Data Protection.* Springer., DOI: 10.1007/978-3-031-76473-8_12

Arsenault, A. C., & Kreps, S. (2022). *AI and International Politics.* The Oxford Handbook of AI Governance.

Bajgar, O., & Horenovsky, J. (2023). Negative human rights as a basis for long-term AI safety and regulation. *Journal of Artificial Intelligence Research*, *76*, 1043–1075. Advance online publication. DOI: 10.1613/jair.1.14020

BBC News. (2025a). Facebook and Instagram get rid of fact checkers. https://www.bbc.com/news/articles/cly74mpy8klo

Benkler, Y., Faris, R., & Roberts, H. (2018). *Network propaganda: Manipulation, disinformation, and radicalization in American politics.* Oxford University Press. DOI: 10.1093/oso/9780190923624.001.0001

Bennett, W. L., & Livingston, S. (2018). The disinformation order: Disruptive communication and the decline of democratic institutions. *European Journal of Communication*, *33*(2), 122–139. DOI: 10.1177/0267323118760317

Bradshaw, S., & Howard, P. N. (2019). *The global disinformation disorder: 2019 global inventory of organised social media manipulation.* Computational Propaganda Research Project. https://comprop.oii.ox.ac.uk

Busuioc, M. (2021). Accountable artificial intelligence: Holding algorithms to account. *Public Administration Review*, *81*(5), 825–836. DOI: 10.1111/puar.13293 PMID: 34690372

Camilleri, M. A. (2023). Artificial intelligence governance: Ethical considerations and implications for social responsibility. *Expert Systems: International Journal of Knowledge Engineering and Neural Networks*, *41*(7), e13406. Advance online publication. DOI: 10.1111/exsy.13406

Capital.gr. (2023, October 19). High Level Advisory Committee for Artificial Intelligence (AI) established. https://www.capital.gr/english/3744823/high-level-advisory-committee-for-artificial-intelligence-ai-established/

Chemnad, K., & Othman, A. (2024). Digital accessibility in the era of artificial intelligence—Bibliometric analysis and systematic review. *Frontiers in Artificial Intelligence*, *7*, 7. DOI: 10.3389/frai.2024.1349668 PMID: 38435800

Clift, S. L. (2004) E-Government and Democracy: Representation and Citizen Engagement in the Information Age. Research provided for the UN UNPAN/DESA for the 2003 World Public Sector Report. Available at: https://www.publicus.net/articles/cliftegovdemocracy.pdf

Council of Europe. (2017). Recommendation CM/Rec(2017)5 of the Committee of Ministers to member states on standards for e-voting. https://search.coe.int/cm?i=0900001680726f6f

Council of Europe. Committee on Artificial Intelligence (CAI). https://www.coe.int/en/web/artificial-intelligence/cai

Council of Europe. Framework Convention on Artificial Intelligence and Human Rights, Democracy and the Rule of Law. https://www.coe.int/en/web/artificial-intelligence

Cupać, J., Schopmans, H., & Tuncer-Ebetürk, İ. (2024). Democratization in the age of artificial intelligence: Introduction to the special issue. *Democratization*, *31*(5), 899–921. DOI: 10.1080/13510347.2024.2338852

Curtis, P. (2023, July 28). Artificial intelligence is powering politics—but it could also reboot democracy. *The Guardian*. https://www.theguardian.com/commentisfree/2023/jul/28/artificial-intelligence-powering-politics-reboot-democracy

Efthymiou, I. (2025). Diversity, Inclusion, and Equity in Gig Economies. In Doon, R. (Ed.), *Sustainability and Adaptability of Gig Economies in Global Business* (pp. 231–260). IGI Global Scientific Publishing., DOI: 10.4018/979-8-3693-9385-7.ch009

Efthymiou, I. (2026). Bridging the Divide: Addressing the Leadership Gap. In Montaudon-Tomas, C., & Pinto-López, I. (Eds.), *Bridging the Leadership Gap: Policies and Workplace Equality* (pp. 103–138). IGI Global Scientific Publishing., DOI: 10.4018/979-8-3373-6345-5.ch004

Efthymiou, I., & Egleton, T. E. (2025). The Impact of Deep Fakes in Markets and Economies. In Gupta, G., Bohara, S., Kovid, R., & Pandla, K. (Eds.), *Deepfakes and Their Impact on Business* (pp. 19–50). IGI Global Scientific Publishing., DOI: 10.4018/979-8-3693-6890-9.ch002

Efthymiou, I. P. (2025). The role of e-government and e-governance in modern societies. In Wongmahesak, K., Ahmad, J., & Mustanir, A. (Eds.), *Public governance practices in the age of AI* (pp. 45–68). IGI Global Scientific Publishing., DOI: 10.4018/979-8-3693-9286-7.ch003

European Commission. (2019). Ethics guidelines for trustworthy AI. https://ec.europa.eu/futurium/en/ai-alliance-consultation.1.html

European Commission. (2022). Eurobarometer on media and news. https://europa.eu/eurobarometer/surveys/detail/2832

European Parliament. (2021, April 22). Digital transformation: Importance, benefits and EU policy. https://www.europarl.europa.eu/news/en/headlines/society/20210414STO02010/digital-transformation-importance-benefits-and-eu-policy

Fountain, J. E. (2001). *Building the virtual state: Information technology and institutional change*. Brookings Institution Press.

Futurism (2022), New European Political Party Is Led by an Artificial Intelligence, https://futurism.com/political-party-led-by-artificial-intelligence

Gorwa, R., Binns, R., & Katzenbach, C. (2020). Algorithmic content moderation: Technical and political challenges in the automation of platform governance. *Big Data & Society*, *7*(1). Advance online publication. DOI: 10.1177/2053951719897945

Guess, A., Nagler, J., & Tucker, J. (2019). Less than you think: Prevalence and predictors of fake news dissemination on Facebook. *Science Advances*, *5*(1), eaau4586. Advance online publication. DOI: 10.1126/sciadv.aau4586 PMID: 30662946

Guess, A. M., Nyhan, B., & Reifler, J. (2020). Exposure to untrustworthy websites in the 2016 US election. *Nature Human Behaviour*, *4*(5), 472–480. DOI: 10.1038/s41562-020-0833-x PMID: 32123342

Hacker, K., & van Dijk, J. (2000a). *Digital democracy: Issues of theory and practice*. Sage publications. DOI: 10.4135/9781446218891

Hacker, K., & van Dijk, J. (2000b). What is digital democracy? In *Digital democracy: Issues of theory and practice* (pp. 1–22). Sage. DOI: 10.4135/9781446218891.n1

Hague, B. N., & Loader, B. D. (1999). *Digital democracy: Discourse and decision making in the information age*. Routledge.

Helbing, D., Frey, B. S., Gigerenzer, G., Hafen, E., Hagner, M., Hofstetter, Y., Van den Hoven, J., Zicari, R. V., & Zwitter, A. (2019). Will democracy survive big data and artificial intelligence? In *Towards digital enlightenment: Essays on the dark and light sides of the digital revolution* (pp. 73–98). Springer., DOI: 10.1007/978-3-319-90869-4_7

Howard, P. N., & Hussain, M. M. (2013). *Democracy's fourth wave?: Digital media and the Arab Spring*. Oxford University Press. DOI: 10.1093/acprof:oso/9780199936953.001.0001

Jackson, M. C. (2021). Artificial intelligence & algorithmic bias: The issues with technology reflecting history & humans. *Journal of Business & Technology Law, 16*, 299–330.

Jungherr, A. (2023). Artificial Intelligence and Democracy: A Conceptual Framework. *Social Media + Society, 9*(3), 20563051231186353. Advance online publication. DOI: 10.1177/20563051231186353

Karagkouni, V., Kouroupis, S., Parlalis, I., Revolidis, M., Stylianidou, A., & Sykiotis, [First initial]. (2023). *The impact of the Covid-19 pandemic on human rights*. Logos Verlag Berlin GmbH.

Kgoale, T. J., & Odeku, K. O. (2023). An analysis of the international and European Union legal instruments for holding artificial intelligence accountable. *Juridical Tribune, 13*(3).

Kouroupis, K. (2023). *Digital transformation and the EU's Digital Agenda*. Logos Verlag.

Kreps, S., & Kriner, D. (2023). How AI threatens democracy. *Journal of Democracy, 34*(4), 122–131. https://www.journalofdemocracy.org/articles/how-ai-threatens-democracy/. DOI: 10.1353/jod.2023.a907693

Kumar, V., Barik, S., Aggarwal, S., Kumar, D., & Raj, V. (2023). The use of artificial intelligence for persons with disability: A bright and promising future ahead. *Disability and Rehabilitation. Assistive Technology*. Advance online publication. DOI: 10.1080/17483107.2023.2288241 PMID: 38079540

Kus, Z. (2025). Digital citizenship and the future of AI engagement, ethics, and privacy. In *Digital citizenship and the future of AI engagement, ethics, and privacy*. IGI Global.

Lageson, S. E. (2020). *Digital punishment: Privacy, stigma, and the harms of data-driven criminal justice*. Oxford University Press. DOI: 10.1093/oso/9780190872007.001.0001

Lazer, D. M. J., Baum, M. A., Benkler, Y., Berinsky, A. J., Greenhill, K. M., Menczer, F., Metzger, M. J., Nyhan, B., Pennycook, G., Rothschild, D., Schudson, M., Sloman, S. A., Sunstein, C. R., Thorson, E. A., Watts, D. J., & Zittrain, J. L. (2018). The science of fake news. *Science, 359*(6380), 1094–1096. DOI: 10.1126/science.aao2998 PMID: 29590025

Lewandowsky, S., Ecker, U. K. H., Seifert, C. M., Schwarz, N., & Cook, J. (2020). Misinformation and its correction: Continued influence and successful debiasing. *Psychological Science in the Public Interest, 13*(3), 106–131. DOI: 10.1177/1529100612451018 PMID: 26173286

Marwala, T. (2023). Artificial intelligence in politics. In *Artificial intelligence, game theory and mechanism design in politics* (pp. 1–20). Palgrave Macmillan., DOI: 10.1007/978-981-99-5103-1_4

Mazzoleni, G. (2008). Populism and the Media. In Albertazzi, D., & McDonnell, D. (Eds.), *Twenty-First Century Populism*. Palgrave Macmillan., DOI: 10.1057/9780230592100_4

Mock, C. (2025). Tech CEOs, Trump and society's new AI age. *Australian Journalism Review, 47*(1), 7–15. DOI: 10.1386/ajr_00176_7

Muir, A., & Oppenheim, C. (2002). National information policy developments worldwide I: Electronic government. *Journal of Information Science, 28*(3), 173–186. DOI: 10.1177/016555150202800301

Myers, A. (2023, February 27). AI's powers of political persuasion. *Stanford HAI*. https://hai.stanford.edu/news/ais-powers-political-persuasion

News, N. B. C. (2025), Matt Lavietes, Meta's new hate speech guidelines permit users to say LGBTQ people are mentally ill, https://www.nbcnews.com/tech/social-media/meta-new-hate-speech-rules-allow-users-call-lgbtq-people-mentally-ill-rcna186700

Nilsson, N. J. (2009). *The quest for artificial intelligence*. Cambridge University Press. DOI: 10.1017/CBO9780511819346

Norris, P., & Inglehart, R. (2019). *Cultural backlash: Trump, Brexit, and authoritarian populism*. Cambridge University Press. DOI: 10.1017/9781108595841

Novelli, C., & Sandri, G. (2024). Digital Democracy in the Age of Artificial Intelligence. SSRN Electronic Journal. DOI: 10.2139/ssrn.4901264

Nyhan, B., & Reifler, J. (2010). When corrections fail: The persistence of political misperceptions. *Political Behavior, 32*(2), 303–330. DOI: 10.1007/s11109-010-9112-2

OECD. (2003). *Promise and Problems of e-Democracy. Challenges of Citizen Engagement*. OECD.

OECD. (2023). *AI language models: Technological, socio-economic and policy considerations*. OECD Digital Economy Papers. https://www.oecd-ilibrary.org/science-and-technology/ai-language-models_13d38f92-en

OECD. (2023). *The state of implementation of the OECD AI Principles four years on*. OECD Publishing. https://www.oecd-ilibrary.org/science-and-technology/the-state-of-implementation-of-the-oecd-ai-principles_13d38f92-en

Ognyanova, K., Lazer, D., Robertson, R. E., & Wilson, C. (2020). Misinformation in action: Fake news exposure is linked to lower trust in media, higher trust in government when your side is in power. *Harvard Kennedy School Misinformation Review, 1*(4). Advance online publication. DOI: 10.37016/mr-2020-024

Perez-Morote, R., Rosa, C. P., & Cortes, E. A. (2022). Exploring the relation between the digital divide and government's effort to develop e-participation. In *Research anthology on citizen engagement and activism for social change* (pp. 997–1018). IGI Global., DOI: 10.4018/978-1-6684-3706-3.ch054

Politico. (2025). EU lawmakers blast Meta's elimination of fact-checkers in U.S. https://www.politico.eu

Rachovitsa, A., & Johann, N. (2022). The human rights implications of the use of AI in the digital welfare state: Lessons learned from the Dutch SyRI case. *Human Rights Law Review*, *22*(2), ngac010. Advance online publication. DOI: 10.1093/hrlr/ngac010

Radanliev, P. (2025). AI ethics: Integrating transparency, fairness, and privacy in AI development. *Applied Artificial Intelligence*, *39*(1), 2463722. Advance online publication. DOI: 10.1080/08839514.2025.2463722

Ramya, R., Priya, S., Thamizhikkavi, P., & Anand, M. (2024). The pillars of AI ethics: Transparency, accountability, and privacy. In *Responsible implementations of generative AI for multidisciplinary use*. IGI Global Publishing. DOI: 10.4018/979-8-3693-9173-0.ch004

Raso, F. A., Hilligoss, H., Krishnamurthy, V., Bavitz, C., & Kim, L. (2018). *Artificial intelligence & human rights: Opportunities & risks*. Berkman Klein Center Research Publication, 2018(6).

Robles, P., & Mallinson, D. J. (2023). Artificial intelligence technology, public trust, and effective governance. *The Review of Policy Research*. Advance online publication. DOI: 10.1111/ropr.12555

Ronchi, A. M. (2019). e-Democracy. Cham: Springer International Publishing.

Russell, S., & Norvig, P. (2020). *Artificial intelligence: A modern approach* (4th ed.). Pearson.

Sarnoff, L., & Burton, C. (2024, January 15). ABC News live fact-checks Trump-Harris debate. *ABC News*. https://www.abcnews.com

Schneier, B., & Sanders, N. E. (2023, July 28). Six ways that AI could change politics. *MIT Technology Review*. https://www.technologyreview.com/2023/07/28/1076756/six-ways-that-ai-could-change-politics/

Sharma, M. (2024). *The impact of artificial intelligence (AI) on the future of democracy and civic participation*. DOI: 10.2139/ssrn.4703439

Sheikh, H., Prins, C., & Schrijvers, E. (2023). Artificial intelligence: Definition and background. In Mission, A. I. (Ed.), *Research for policy* (pp. 11–34). Springer., DOI: 10.1007/978-3-031-21448-6_2

Soyer, B., & Tettenborn, A. (2022). Artificial intelligence and civil liability—Do we need a new regime? *International Journal of Law and Information Technology, 30*(4), 385–397. DOI: 10.1093/ijlit/eaad001

Special Secretariat for Foresight. (2024). *A blueprint for Greece's AI transformation*. Government of Greece. https://foresight.gov.gr/en/studies/A-Blueprint-for-Greece-s-AI-Transformation/

Sunstein, C. R. (2017). *Republic: Divided democracy in the age of social media*. Princeton University Press. DOI: 10.1515/9781400884711

Suzor, N. P., West, S. M., Quodling, A., & York, J. (2019). What do we mean when we talk about transparency? Toward meaningful transparency in commercial content moderation. *International Journal of Communication, 13*, 1526–1547.

Tambouris, Efthimios & Gorilas, S. & Boukis, G. & A, Archetypon. (2001). Investigation of Electronic Government.

The Guardian. (2025), Meta is ushering in a 'world without facts', says Nobel peace prize winner, https://www.theguardian.com/world/2025/jan/08/facebook-end-factchecking-nobel-peace-prize-winner-maria-ressa

The Guardian. (2025). Meta's changes to policing will lead to clash with EU and UK, say experts. https://www.theguardian.com/technology/2025/jan/08/metas-changes-to-social-media-policing-will-lead-to-clash-with-eu-and-uk-say-experts

Toh, A. (2020, February 6). Dutch ruling a victory for rights of the poor: Court orders government to stop use of welfare surveillance program. *Human Rights Watch.* https://www.hrw.org/news/2020/02/06/dutch-ruling-victory-rights-poor

Tufekci, Z. (2017). *Twitter and tear gas: The power and fragility of networked protest.* Yale University Press.

Ufert, F. (2020). AI Regulation Through the Lens of Fundamental Rights: How Well Does the GDPR Address the Challenges Posed by AI? *European Papers, 5*(2), 1087–1097.

UNESCO. (2021). *Recommendation on the ethics of artificial intelligence.* UNESCO Publishing. https://unesdoc.unesco.org/ark:/48223/pf0000377993

Vosoughi, S., Roy, D., & Aral, S. (2018). The spread of true and false news online. *Science, 359*(6380), 1146–1151. DOI: 10.1126/science.aap9559 PMID: 29590045

Vousinas, G. L., Simitsi, I., Livieri, G., Gkouva, G. C., & Efthymiou, I. P. (2022). Mapping the road of the ethical dilemmas behind Artificial Intelligence. *Journal of Politics and Ethics in New Technologies and AI, 1*(1), e31238. DOI: 10.12681/jpentai.31238

Waisbord, S. R. (2018). Truth is what happens to news: On journalism, fake news, and post-truth. *Journalism Studies, 19*(13), 1866–1878. DOI: 10.1080/1461670X.2018.1492881

Wolff, J. (2025). Policy approaches to defining and enforcing responsibilities for online platforms. In *Defeating disinformation.* Cambridge University Press. DOI: 10.1017/9781009438636.011

Zhang, W. (2024). AI fairness in practice: Paradigm, challenges, and prospects. *AI Magazine, 45*(3), 386–395. DOI: 10.1002/aaai.12189

Zidouemba, M. T. (2025). Governance and artificial intelligence: The use of artificial intelligence in democracy and its impacts on the rights to participation. *Discover Artificial Intelligence, 5*(1), 12. DOI: 10.1007/s44163-025-00229-5

Chapter 2
Ensuring Ethical and Equitable AI:
From Bias Mitigation to Algorithmic Auditing

Rachna Rana
Ludhiana Group of Colleges, Ludhiana, India

Pankaj Bhambri
https://orcid.org/0000-0003-4437-4103
Guru Nanak Dev Engineering College, Ludhiana, India

ABSTRACT

Over the period 2015–2025, scholarship on ethical and equitable artificial intelligence converged around three interdependent goals: (1) identifying and classifying algorithmic harms that flow from biased data, model design, and socio-technical deployment; (2) developing technical and organisational mitigation strategies (pre-processing, in-processing, post-processing, and socio-technical interventions) to reduce disparate impacts; and (3) operationalising independent and regulatory audit mechanisms (algorithmic audits, assurance frameworks, and public reporting) that enable accountability, transparency, and remedial governance. Early work (2015–2018) emphasized detection and formal fairness metrics and revealed fundamental trade-offs between fairness criteria and predictive utility; mid-period research (2019–2022) broadened attention to domain-specific harms, human factors, and measurement practices; and recent contributions (2023–2025) focus on standardized auditing protocols, governance playbooks, and scalable bias-aware pipelines integrated into organisational lifecycles.

DOI: 10.4018/979-8-2600-0000-7.ch002

1. INTRODUCTION

1.1.1. Algorithmic Fairness

Algorithmic fairness refers to the principles, metrics, and computational strategies designed to ensure that automated systems produce equitable outcomes across diverse social groups. Early research highlighted that predictive models often mirror or amplify social inequities embedded in historical data (Barocas & Selbst, 2016; Angwin et al., 2016). Scholars such as Hardt, Price, and Srebro (2016) formalized fairness as *equality of opportunity* in supervised learning, introducing mathematical metrics that assess outcome parity across sensitive attributes like race or gender (Bhambri & Kant, 2025b). Over time, fairness expanded beyond statistical parity to encompass contextual and procedural notions of justice (Kleinberg et al., 2017; Caton & Haas, 2020).

Between 2019 and 2025, literature emphasized fairness as a *socio-technical construct* that cannot be solved solely through algorithmic adjustments (Mehrabi et al., 2021; Suresh & Guttag, 2019). Empirical studies across finance, recruitment, and healthcare revealed that fairness metrics often conflict—improvements in one dimension may worsen another (Raji et al., 2020; Liu et al., 2022). Consequently, fairness frameworks now advocate transparency in metric selection, stakeholder participation in defining fairness goals, and continuous auditing to detect drift over time (NIST, 2022; Royal Society, 2024). Ultimately, algorithmic fairness stands as a cornerstone of ethical AI—its evolution from technical parity checks to governance-embedded accountability illustrates a shift toward responsible, inclusive computation (Lam et al., 2024).

Critical Synthesis and Evolving Debates: The evolution of algorithmic fairness from technical parity to socio-technical governance reveals unresolved tensions and debates. Early mathematical formalizations, while crucial for measurability, often abstract away from the structural and historical inequalities that produce biased data (Crawford & Paglen, 2019). A key ongoing debate centers on the conflict between competing fairness metrics—such as demographic parity and equal opportunity—which are often mathematically irreconcilable (Kleinberg et al., 2017). Furthermore, the field grapples with the critique that "fairness" is frequently defined by dominant actors in technology and policy, potentially marginalizing community-defined notions of justice (Benjamin, 2019). The shift toward contextual fairness and participatory governance represents an attempt to address these limitations, moving the field from a search for universal technical solutions to an embrace of situated, negotiated ethical practice.

1.1.2. Bias Mitigation Techniques

Bias mitigation techniques encompass a spectrum of methodological interventions applied throughout the machine-learning lifecycle to identify and reduce discriminatory patterns (Mohana Sundari et al., 2024). The literature groups these interventions into three main categories: *pre-processing*, *in-processing*, and *post-processing* (Mehrabi et al., 2021; Hort et al., 2022). Pre-processing approaches adjust or rebalance datasets through sampling, re-weighting, or synthetic data generation (Gonzalez-Sendino et al., 2024). In-processing methods embed fairness constraints directly into learning algorithms—e.g., adversarial debiasing or regularization penalties—to balance predictive accuracy and equity (Wang et al., 2021). Post-processing corrections calibrate model outputs, applying fairness metrics after training to ensure equitable decisions (Ferrara, 2023).

From 2015 to 2025, research revealed both progress and limitations. While algorithmic interventions can reduce measurable disparities, they may fail when biased representations persist in data collection or labeling (Buolamwini & Gebru, 2018; Gebru et al., 2021). Recent scholarship stresses *bias prevention* rather than *retroactive correction* by improving dataset documentation ("datasheets for datasets") and contextual awareness in design (Gebru et al., 2021; Li et al., 2023). Hybrid socio-technical models now combine technical debiasing with organizational reforms—ethics boards, impact assessments, and domain-specific fairness policies (Brookings Institution, 2019; AI Now Institute, 2023). Emerging standards like NIST SP 1270 (2022) and ISO IEC TR 24028 promote bias management as a continuous quality-assurance process rather than an isolated step. Accordingly, bias mitigation has evolved from mathematical optimization into a holistic governance framework that aligns data stewardship, model design, and ethical accountability (Kazim et al., 2021; Lam et al., 2024).

1.1.3. Algorithmic Auditing

Algorithmic auditing denotes systematic evaluation and verification of AI systems to ensure that their design, implementation, and deployment comply with ethical, legal, and performance standards (Thirumalaiyammal et al., 2024). The concept gained prominence after high-profile cases of biased decision support—such as predictive policing and recruitment systems—demonstrated the need for independent scrutiny (Angwin et al., 2016; Dastin, 2018). Early methodological foundations were laid by Sandvig et al. (2014), who proposed audit experiments to detect discrimination on digital platforms. Later, Raji and Buolamwini (2019) and

Kroll et al. (2017) advanced *actionable auditing* frameworks integrating technical testing with organizational accountability.

From 2019 onward, the field diversified into *internal*, *external*, and *regulatory* audits (Vecchione et al., 2021; Royal Society, 2024). Internal audits focus on developer-led assessments embedded in product pipelines (Raji et al., 2020). External audits involve third-party evaluators who test algorithmic systems under realistic conditions, while regulatory audits are performed under legal mandates such as the EU AI Act or national data-protection authorities (DRCF, 2021; Cech, 2021). Recent literature underscores that technical fairness checks alone are insufficient; audits must also examine governance structures, documentation, and the socio-economic context of deployment (Hacohen, 2023; Lam et al., 2024). The move toward standardized audit protocols—supported by NIST SP 1270 (2022) and policy frameworks in multiple jurisdictions—reflects a maturing consensus that algorithmic auditing is essential for operationalizing AI ethics (AI Now Institute, 2023). By 2025, scholars advocate *continuous* and *participatory* auditing, integrating stakeholders and impacted communities into the oversight process to ensure accountability throughout the AI lifecycle (Royal Society, 2024).

1.1.4. Accountable AI Governance

Accountable AI governance represents the institutional, procedural, and ethical frameworks that define responsibilities for the design, deployment, and oversight of artificial intelligence systems (Bhambri & Bajdor, 2025). Between 2015 and 2025, research gradually shifted from abstract debates about algorithmic fairness to the practical question of who is accountable when AI harms occur (Kroll et al., 2017; Selbst, 2019). Early work in this period emphasized that accountability must extend beyond model creators to include the organizations, regulators, and social systems that use AI for decision-making (Barocas & Selbst, 2016; Veale, Van Kleek, & Binns, 2018). Scholars defined *accountable algorithms* as systems whose operations are transparent, verifiable, and subject to oversight (Kroll et al., 2017).

From 2019 onward, multi-level governance frameworks emerged, incorporating ethics boards, impact assessments, and model documentation practices such as *model cards* and *datasheets for datasets* (Gebru et al., 2021). These instruments ensure traceability across the AI lifecycle by clarifying design intent, training data, and performance limitations. Regulatory bodies—such as the European Commission with its AI Act and the National Institute of Standards and Technology (NIST, 2022)—began embedding accountability principles into technical standards and risk management frameworks. The literature emphasizes that true accountability cannot rely on voluntary compliance alone; it requires enforceable governance structures, independent auditing, and public transparency (AI Now Institute, 2023; Royal Society, 2024).

Recent works advocate algorithmic assurance ecosystems, where governance integrates human oversight, technical audits, and continuous monitoring (Lam et al., 2024; Cech, 2021). The emphasis on procedural accountability underscores that responsibility must be distributed yet traceable: every stage—data collection, model development, deployment, and post-market monitoring—must have identifiable owners and documentation trails (Veale & Binns, 2017; Raji et al., 2020). This shift marks the evolution of AI governance from principle-based ethics to operational accountability frameworks. By 2025, *accountable AI governance* stands as a cornerstone of ethical and equitable AI, aligning technical reliability with social legitimacy and regulatory enforceability.

1.1.5. Transparency and Explainability

Transparency and explainability have become critical components of ethical AI because they enable users, regulators, and affected individuals to understand, question, and contest algorithmic decisions (Bajdor & Bhambri, 2024). Transparency refers to the disclosure of how AI systems function and make predictions, while explainability pertains to the human interpretability of those processes (Goodman & Flaxman, 2017; Wachter, Mittelstadt, & Russell, 2017). Research during 2015–2020 primarily focused on *interpretable machine learning*, proposing models like LIME and SHAP to reveal feature importance or decision boundaries. However, subsequent work revealed that technical transparency alone does not guarantee accountability; explanations must be *contextually meaningful* to non-experts and aligned with legal rights such as the GDPR's "right to explanation" (Wachter et al., 2017; Selbst & Barocas, 2018).

By 2021, transparency was redefined as a *relational practice*—a bridge between developers, users, and affected communities rather than a one-way disclosure (Kossow, 2021; Royal Society, 2024). Scholars argued that opacity often arises not only from model complexity but from institutional secrecy and commercial confidentiality (Pasquale, 2015). Consequently, explainability became a multidisciplinary domain involving law, design, ethics, and human–computer interaction (Selbst, 2019; Li et al., 2023). Technical solutions like counterfactual explanations (Wachter et al., 2017) and documentation tools (Gebru et al., 2021) were complemented by procedural mechanisms—impact assessments, transparency reports, and third-party audits—to ensure public accountability.

Between 2023 and 2025, the conversation matured toward *assurance-based transparency* (Lam et al., 2024), where explanations are validated through audit trails, governance protocols, and standardized reporting frameworks (NIST, 2022).

Scholars emphasize that explainability should not merely simplify complex models but should facilitate trust, allow contestability, and expose underlying assumptions or biases (Ferrara, 2023; AI Now Institute, 2023). Transparency thus evolves from a descriptive attribute of systems to an ethical commitment by institutions to act responsibly. Within equitable AI, explainability ensures that algorithmic decisions are not only technically sound but also *socially intelligible* and open to scrutiny.

1.1.6. Global and Epistemic Inclusivity in Socio-Technical Design

The socio-technical perspective necessitates moving beyond a Western-centric view of AI ethics. Research and policy frameworks from the Global South—such as India's NITI Aayog approach to responsible AI or the African Union's Continental AI Strategy—emphasize different priorities, including digital sovereignty, infrastructural equity, and post-colonial data justice (Mohamed et al., 2020). These perspectives challenge the universal applicability of fairness metrics developed in North American and European contexts, highlighting how local cultural norms, legal traditions, and economic conditions shape ethical requirements. Incorporating these viewpoints is not merely additive but essential for deconstructing the power dynamics embedded in global AI supply chains and for developing truly inclusive socio-technical systems.

1.1.7. Socio-Technical Systems Perspective

The socio-technical systems perspective recognizes that fairness and ethics in AI cannot be achieved solely through mathematical or algorithmic solutions; they depend on the social, cultural, and institutional environments in which technologies operate (Bhambri & Pawełoszek, 2025). This perspective gained traction after empirical findings showed that bias often stems from systemic inequalities reflected in data and institutional practices (Buolamwini & Gebru, 2018; Barocas & Selbst, 2016). Scholars such as Selbst et al. (2019) and Suresh and Guttag (2019) demonstrated that harms emerge throughout the AI pipeline—from problem formulation to data collection, model training, and deployment—making purely technical fixes insufficient.

Research between 2019 and 2025 emphasized *socio-technical alignment*, where fairness is treated as a collaborative process involving engineers, domain experts, policymakers, and affected communities (Veale et al., 2018; Vecchione et al., 2021). Interdisciplinary frameworks encourage participatory design and *algorithmic impact assessments* to evaluate social consequences before deployment (DRCF, 2021; Royal Society, 2024). The socio-technical approach also informed the development of algorithmic audits and governance structures, positioning ethics as an ongoing negotiation among stakeholders (Kazim et al., 2021; Lam et al., 2024).

Furthermore, the socio-technical paradigm reframes fairness from a *metric-based property* to a *contextual practice*—acknowledging that equity means different things across domains like healthcare, education, and criminal justice (Liu et al., 2022; Pham et al., 2025). Scholars advocate embedding social science expertise in AI development teams to anticipate downstream effects and ethical risks (AI Now Institute, 2023). This human-centered reframing allows AI to evolve within cultural, legal, and moral boundaries rather than outside them. By 2025, the socio-technical systems perspective stands as a unifying framework linking bias mitigation, algorithmic auditing, and governance—asserting that equitable AI must integrate human judgment, institutional accountability, and contextual understanding alongside technical innovation.

1.1.8. Data Ethics and Representation

Data ethics and representation address the moral principles and practical norms guiding how data are collected, curated, labeled, and used in artificial intelligence systems (Rana & Bhambri, 2024). Scholars have long noted that algorithmic bias often originates not in models themselves, but in *the social and historical conditions encoded in data* (Barocas & Selbst, 2016; Crawford & Paglen, 2019). Between 2015 and 2020, landmark studies such as Buolamwini and Gebru (2018) revealed that facial recognition systems exhibited racial and gender disparities due to skewed training data. This prompted a paradigm shift toward *data ethics*—ensuring that datasets reflect fairness, inclusion, and accountability from the outset (O'Neil, 2016; Gebru et al., 2021).

Contemporary research recognizes that representation is a sociopolitical challenge: datasets are never neutral, and the act of labeling embeds the perspectives and values of annotators (Suresh & Guttag, 2019). Ethical data practices therefore require transparency about data provenance, sampling methods, and potential exclusions (Lepri et al., 2018). Gebru et al. (2021) introduced Datasheets for Datasets, advocating standardized documentation that discloses dataset origins, intended use, and ethical considerations. Complementary tools, such as Model Cards and Data Statements, operationalize similar documentation standards for model transparency and accountability (Bender & Friedman, 2018; Gebru et al., 2021).

Between 2021 and 2025, the conversation evolved toward data justice, linking ethical data representation with human rights and social inclusion (Royal Society, 2024; Kazim et al., 2021). Scholars and regulators have emphasized that ethical data governance should ensure meaningful consent, privacy protection, and representation of marginalized voices (NIST, 2022). Bias mitigation strategies increasingly incorporate *causal modeling* to generate fairer synthetic data and reduce structural inequities (Gonzalez-Sendino et al., 2024). By 2025, data ethics frameworks integrate both technical mechanisms and procedural norms—data audits, stewardship protocols, and participatory governance—to ensure that data not only power AI systems but also reflect societal diversity and moral responsibility (AI Now Institute, 2023; Pham et al., 2025). In essence, ethical data representation serves as the moral and epistemic foundation of equitable AI. It ensures that datasets respect human dignity, minimize harm, and promote inclusivity—establishing trust at the root of the AI lifecycle (Bhambri, 2025).

1.1.9. Regulatory and Policy Compliance

Regulatory and policy compliance in AI refers to adherence to legal frameworks, standards, and institutional policies governing algorithmic systems' fairness, transparency, and accountability (Bhambri & Starostka-Patyk, 2025). After 2016, following publicized algorithmic failures (Angwin et al., 2016; Dastin, 2018), policymakers began recognizing AI as a domain requiring dedicated oversight. Early initiatives such as the EU's General Data Protection Regulation (GDPR) (2018) introduced the "right to explanation," laying groundwork for algorithmic accountability (Wachter, Mittelstadt, & Russell, 2017). Subsequent global efforts expanded this regulatory momentum, with the EU Artificial Intelligence Act (drafted 2021–2024) categorizing AI systems by risk level and mandating transparency, human oversight, and auditability (Royal Society, 2024; DRCF, 2021).

In the United States, the NIST AI Risk Management Framework (2022) and NIST SP 1270 established guidelines for bias identification and governance integration. These frameworks define compliance not as box-checking, but as risk-based management emphasizing accountability across the entire AI lifecycle (NIST, 2022). Internationally, organizations such as OECD, UNESCO, and ISO released ethical AI principles aligning fairness, explainability, and human rights (AI Now Institute, 2023). Scholars like Cech (2021) and Lam et al. (2024) highlighted the need for *algorithmic assurance*—independent validation of whether systems meet policy standards through audit trails, certification, and documentation.

Research between 2022 and 2025 underscores that effective regulation requires collaboration between technologists, ethicists, and lawmakers (Royal Society, 2024). Case studies in sectors like finance and healthcare demonstrate that compliance frameworks not only reduce legal risk but also foster public trust (Liu et al., 2022; Ferrara, 2023). However, literature also warns against "ethics washing," where organizations use superficial compliance language to avoid deeper accountability (AI Now Institute, 2023). The emerging consensus calls for dynamic regulation—adaptive, evidence-based, and supported by continuous algorithmic auditing (Vecchione et al., 2021; Kazim et al., 2021).

By 2025, regulatory and policy compliance evolved from fragmented ethical guidelines to a coherent, enforceable governance ecosystem. It now encompasses multi-level alignment between legal norms, technical standards, and institutional practices, ensuring that AI systems respect human rights and societal values across jurisdictions (Bhambri & Hamad, 2025).

1.1.10. Human-Centered AI Design

Human-centered AI design emphasizes embedding human values, cognition, and participation throughout the AI lifecycle—from problem definition to evaluation. Rooted in participatory design and human-computer interaction, this approach argues that equitable AI cannot exist without integrating human judgment and empathy into algorithmic decision-making (Lepri et al., 2018; Veale et al., 2018). Between 2015 and 2020, researchers began reframing AI as *collaborative intelligence* rather than autonomous replacement (O'Neil, 2016; Crawford & Paglen, 2019). The goal was to enhance human capability, not to automate moral or social reasoning.

Human-centered design promotes *explainability, usability,* and *co-design* by engaging stakeholders—developers, users, domain experts, and impacted communities—in iterative feedback loops (Veale & Binns, 2017; Vecchione et al., 2021). Scholars such as Raji et al. (2020) and Suresh & Guttag (2019) emphasized that ethical design requires attention to context and lived experience. Participatory evaluation frameworks and *Algorithmic Impact Assessments (AIAs)* have been proposed to integrate human feedback into risk identification before deployment (DRCF, 2021; Royal Society, 2024).

Between 2021 and 2025, human-centered AI became a cornerstone of organizational ethics strategies (AI Now Institute, 2023). Advances in *interactive bias mitigation* and *human-in-the-loop auditing* enabled continuous collaboration between algorithms and human reviewers (Wang et al., 2021; Lam et al., 2024). Educational AI research further demonstrated that incorporating empathy, cultural sensitivity, and accessibility principles enhances system trust and fairness (Pham et al., 2025). Regulatory frameworks such as the EU AI Act and NIST standards now explicitly reference human oversight as a core compliance requirement (NIST, 2022; Royal Society, 2024).

By 2025, human-centered AI design is understood as a comprehensive paradigm integrating usability, ethics, and social impact. It ensures that algorithmic systems complement human reasoning rather than replace it, thus fostering accountability, inclusivity, and resilience in AI ecosystems. In ethical and equitable AI, *the human remains both the beneficiary and the governor of technology* (Bhambri & Kant, 2025a).

1.1.11. Continuous Monitoring and Impact Assessment

Continuous monitoring and impact assessment represent the dynamic phase of ethical AI governance, ensuring that fairness, transparency, and accountability are maintained after deployment (Geetha et al., 2024). While early discussions of algorithmic ethics (Barocas & Selbst, 2016; O'Neil, 2016) focused on model design, research between 2018 and 2025 has emphasized that responsible AI must be monitored throughout its lifecycle (Lepri et al., 2018; NIST, 2022). Algorithms evolve as they interact with new data, users, and contexts; thus, their ethical performance can drift over time, producing unanticipated harms (Kazim et al., 2021; Raji et al., 2020).

Continuous assessment integrates *technical auditing* with *social evaluation*. Technically, it involves tracking input distributions, model outputs, and performance disparities across demographic or contextual groups (Gonzalez-Sendino et al., 2024). Socially, it requires structured feedback from affected communities and stakeholders (Veale & Binns, 2017). The combined approach allows organizations to detect and mitigate bias dynamically rather than retroactively. Recent tools such as AI fairness dashboards, model impact trackers, and risk-scoring APIs enable real-time bias visualization and compliance reporting (NIST, 2022; Royal Society, 2024).

From 2021 onward, researchers proposed Algorithmic Impact Assessments (AIAs)—pre- and post-deployment evaluations documenting potential harms, mitigation plans, and human-rights implications (DRCF, 2021; Lam et al., 2024). Governments in Canada, the EU, and the U.S. have adopted pilot frameworks requiring AIAs for high-risk systems in healthcare, finance, and employment. These assessments operate in parallel with regulatory requirements such as the EU AI Act and NIST AI Risk Management Framework, both of which mandate *ongoing* monitoring and re-certification (NIST, 2022; Royal Society, 2024).

Scholars emphasize that continuous monitoring demands *organizational culture change*—embedding ethics officers, audit teams, and documentation protocols into daily operations (AI Now Institute, 2023). Automation alone cannot ensure ethical outcomes; humans must interpret audit signals, contextualize anomalies, and decide corrective actions (Pham et al., 2025). Moreover, transparency reports and impact statements build public trust and satisfy governance requirements, transforming ethical oversight from a reactive task into a proactive process (Ferrara, 2023).

By 2025, continuous monitoring has evolved into a living governance process—an iterative loop of measurement, reflection, and improvement (Sharma & Bhambri, 2024). It operationalizes fairness not as a static property but as an ongoing negotiation between technology and society. Through adaptive audits and participatory assessments, organizations can ensure that AI systems remain aligned with ethical principles and public values across time, context, and scale (Kazim et al., 2021; Royal Society, 2024).

2. LITERATURE REVIEW

Table 1.

S. No.	Reference (APA Style)	Findings / Contributions	Research Gaps	Tools / Software / Frameworks
1	Barocas & Selbst (2016)	Identified that algorithmic bias arises from historical and institutional inequalities embedded in data.	Lacked empirical validation across domains.	Conceptual analysis; no software used.
2	O'Neil (2016)	Highlighted real-world harms of opaque algorithms ("Weapons of Math Destruction").	Limited technical remedies suggested.	Case analysis; ethical impact mapping.
3	Angwin et al. (2016)	Demonstrated racial bias in criminal-risk scoring (COMPAS).	No open methodology for auditing replication.	ProPublica dataset audit scripts (Python).
4	Wachter, Mittelstadt & Russell (2017)	Proposed GDPR "right to explanation" as legal basis for AI accountability.	Ambiguity in technical implementation.	GDPR regulatory framework.
5	Veale & Binns (2017)	Introduced human-centered legal interpretability for AI decisions.	Needed integration with developer tools.	Interpretability APIs (prototype).
6	Buolamwini & Gebru (2018)	Exposed gender and racial disparities in facial-recognition systems.	Limited dataset diversity.	IBM Watson Vision, Python OpenCV.
7	Lepri et al. (2018)	Proposed fairness-aware data governance for social good.	Minimal empirical metrics for fairness.	Data governance prototypes; R scripts.
8	Crawford & Paglen (2019)	Critiqued datasets as cultural artifacts reflecting bias.	Did not offer auditing solutions.	Visual analysis; ImageNet audit.
9	Suresh & Guttag (2019)	Developed taxonomy of bias sources (measurement, aggregation, evaluation).	No cross-domain validation.	FairML, SciKit-Learn.
10	Raji et al. (2020)	Proposed practical algorithmic auditing framework within corporations.	Lacked standardized audit reporting.	Audit pipeline (Jupyter, Python).

continued on following page

Table 1. Continued

S. No.	Reference (APA Style)	Findings / Contributions	Research Gaps	Tools / Software / Frameworks
11	AI Now Institute (2020)	Emphasized social accountability in AI oversight institutions.	Weak link to technical standards.	Institutional audit frameworks.
12	Cech (2021)	Analyzed ethics-washing in corporate AI policies.	Empirical testing absent.	Policy content analysis tools (NVivo).
13	Gebru et al. (2021)	Introduced *Datasheets for Datasets* to enhance dataset transparency.	Needed industry-wide adoption.	Template documentation tool (Google Docs).
14	Kazim et al. (2021)	Mapped governance frameworks for algorithmic fairness.	Insufficient evaluation metrics.	ISO/IEC AI management standards.
15	DRCF (2021)	Proposed cross-regulator algorithmic audit guidelines in the UK.	Limited cross-sector testing.	Policy prototype toolkit.
16	NIST (2022)	Released *AI Risk Management Framework* emphasizing bias mitigation and accountability.	Still voluntary; lacks enforcement.	NIST AI RMF, SP 1270.
17	Liu et al. (2022)	Studied AI compliance in healthcare; emphasized interpretability.	Focused on developed-country contexts.	TensorFlow Explainability Library.
18	AI Now Institute (2023)	Advocated for continuous algorithmic monitoring post-deployment.	Implementation pathways unclear.	Monitoring dashboards.
19	Ferrara (2023)	Evaluated misinformation detection algorithms and ethical concerns.	Need longitudinal bias tracking.	Python NLP toolkits (spaCy).
20	Gonzalez-Sendino et al. (2024)	Proposed causal-model-based bias mitigation for dynamic data.	Scalability challenges.	DoWhy, CausalNex libraries.
21	Royal Society (2024)	Published global report linking AI governance to societal trust.	Limited focus on Global South.	Policy mapping framework.
22	Lam et al. (2024)	Explored human-in-the-loop fairness correction in ML workflows.	High human resource cost.	Human-AI audit loops (Python+R).
23	Pham et al. (2025)	Proposed participatory ethical-AI co-design involving communities.	Limited quantitative evaluation.	Co-design survey platforms.
24	Wang et al. (2021)	Suggested dynamic feedback auditing for bias correction.	Lack of longitudinal performance data.	MLflow monitoring framework.
25	Vecchione et al. (2021)	Presented framework combining technical and ethical assurance.	Needed regulatory validation.	Fairness indicators (TensorFlow toolkit).

3. APPLICATIONS FOR ENSURING ETHICAL AND EQUITABLE AI-FROM BIAS MITIGATION TO ALGORITHMIC AUDITING

3.1. The Practical Need for Ethical AI Applications

Between 2015 and 2025, the integration of Artificial Intelligence (AI) into socially critical domains—such as healthcare, education, finance, and criminal justice—raised serious ethical questions regarding fairness, transparency, and accountability. Scholars like Barocas and Selbst (2016) and O'Neil (2016) demonstrated that machine learning systems often inherit or amplify societal inequalities due to biased data, unbalanced model training, or opaque decision-making. Consequently, the practical need for ethical and equitable AI applications has grown alongside the technical advancement of the field.

These applications aim to operationalize ethical principles—such as fairness, transparency, and human oversight—into measurable and actionable processes. The literature shows a progressive transition from identifying algorithmic bias (2015–2018) to embedding accountability through algorithmic auditing, fairness metrics, and policy compliance tools (2019–2025) (Kazim et al., 2021; NIST, 2022; Royal Society, 2024). The following sections analyze how ethical AI principles have been transformed into concrete applications across multiple sectors.

3.2. Lessons from High-Profile Algorithmic Failures

While the adoption of ethical AI tools is progressing, examining well-documented failures provides critical, ground-truth lessons on the stakes of inadequate oversight. These cases illustrate how bias, opacity, and flawed governance can lead to tangible harm, reinforcing the necessity of the frameworks discussed in this chapter.

- **Facial Recognition in Law Enforcement:** Landmark audits, most notably the "Gender Shades" study (Buolamwini & Gebru, 2018), demonstrated significantly higher error rates for darker-skinned and female faces in commercial facial analysis systems. When deployed for surveillance or suspect identification, these biased systems risked perpetuating and automating racial discrimination, leading to municipal bans and moratoriums across several U.S. cities and igniting global debate on appropriate use.
- **Automated Hiring and Recruitment:** Amazon's experimental recruitment tool, trained on a decade of résumés submitted to the company, taught itself to penalize applications containing words like "women's" (e.g., "women's chess club captain") because it inferred a historical bias against female candidates in the tech industry (Dastin, 2018). This case is a canonical example of

how models codify past inequities and the dangers of automating high-stakes decisions without continuous bias auditing.

- **Welfare Fraud Detection:** The Dutch government's use of the SyRi algorithm to predict welfare fraud was ruled a violation of human rights by a district court in 2020. The system's opaque risk-scoring of entire neighborhoods, based on data like ethnicity and financial history, created a "digital welfare prison" for low-income and immigrant communities, demonstrating the severe social consequences of unaccountable predictive systems in the public sector (Eubanks, 2018).
- **Key Implications:** These failures underscore common themes: the inadequacy of post-hoc technical fixes alone, the critical need for transparency and external auditability, and the imperative to center the rights and experiences of impacted communities in the design and evaluation process. They serve as a stark reminder that ethical AI is not an optional enhancement but a fundamental requirement for responsible innovation.

3.3. Bias Mitigation Applications in Real-World Systems

Bias mitigation is among the earliest and most widely adopted ethical AI applications. It involves detecting, quantifying, and reducing bias during model training or deployment.

a. Recruitment and Human Resources Systems

In recruitment, automated hiring systems were found to reinforce gender and racial disparities due to biased historical data (Dastin, 2018). As a result, companies began implementing bias-detection applications that use re-weighting algorithms and fairness constraints. For example, IBM's AI Fairness 360 Toolkit introduced statistical parity difference and disparate impact analysis to ensure equitable hiring outcomes (Raji et al., 2020). This toolkit, later extended through open-source integration with Python libraries, became foundational in bias monitoring for HR systems.

b. Healthcare and Clinical Decision Support

In healthcare, bias mitigation applications focus on diagnostic and predictive AI tools. Studies show that algorithms used for predicting patient outcomes often underrepresent minority populations, leading to unequal treatment recommendations (Liu et al., 2022). Consequently, hospitals began using algorithmic fairness frameworks such as TensorFlow Fairness Indicators and NIST SP 1270 guidance for continuous bias evaluation (NIST, 2022). These applications allow medical practitioners to visualize model fairness across demographic subgroups, supporting equitable healthcare delivery.

c. Education and Learning Analytics

AI-based assessment tools and learning management systems (LMS) also face bias concerns, particularly in grading and recommendation systems (Vigneshwari et al., 2024). Research in ethical educational AI introduced equity-focused recommender systems that adjust learning content delivery to minimize systemic bias (Pham et al., 2025). These applications combine *human-in-the-loop* feedback with statistical fairness corrections, ensuring that students from diverse backgrounds receive balanced academic opportunities.

3.4. Algorithmic Auditing and Compliance Applications

Algorithmic auditing applications represent a major shift from theoretical ethics to applied governance. Auditing ensures that AI models comply with fairness, privacy, and transparency standards both pre- and post-deployment (Raji et al., 2020; Vecchione et al., 2021).

a. Internal Corporate Auditing Platforms

Several technology firms introduced internal AI audit dashboards that automatically track model performance metrics and fairness indicators. For instance, Google's Model Cards (Gebru et al., 2021) serve as documentation tools summarizing model purposes, limitations, and ethical risks. These internal systems standardize transparency reporting, creating a "compliance trail" for legal or regulatory reviews.

b. Government and Policy Compliance Tools

Public-sector institutions have begun employing algorithmic auditing frameworks to assess AI systems used in welfare distribution, policing, and credit assessment (Royal Society, 2024). The UK's Digital Regulation Cooperation Forum (DRCF, 2021) developed prototype auditing guidelines that ensure AI applications conform to privacy and anti-discrimination laws. Similarly, Algorithmic Impact Assessments (AIAs) have been implemented in Canada and the EU to evaluate high-risk AI systems before deployment (Lam et al., 2024).

c. Ethical Certification and Risk Management Platforms

Emerging applications now link algorithmic auditing to certification. The NIST AI Risk Management Framework (2022) provides a structured tool for identifying, measuring, and mitigating risks related to bias and accountability. AI companies increasingly use such frameworks for ethical certification, allowing products to meet industry standards similar to ISO/IEC AI management systems (Kazim et al., 2021).

3.5. Data Ethics and Documentation Applications

The Dual Role of Open-Source Communities and Low-Resource Challenges: The open-source ecosystem, exemplified by platforms like Hugging Face and GitHub, has dramatically democratized access to AI models and fairness toolkits, accelerating innovation and peer review. However, it also presents challenges for ethical assurance, as community-shared models often lack rigorous documentation or audit trails, making bias difficult to trace (Kouroupis, K. & Efthymiou, I., 2026). Simultaneously, the application of ethical AI in low-resource settings faces distinct hurdles: scarce and non-representative data, limited compute infrastructure, and a shortage of local technical expertise. Innovations in frugal AI—such as efficient model architectures, synthetic data generation for underrepresented groups, and mobile-first fairness diagnostics, are emerging as critical responses. These efforts highlight that equitable AI must be not only technically sound but also globally accessible and adaptable to diverse socio-economic realities.

Transparent and ethical data collection is the backbone of equitable AI systems. Several applications and frameworks have emerged to standardize dataset documentation and prevent representational bias.

a. Datasheets for Datasets and Data Statements

Gebru et al. (2021) proposed Datasheets for Datasets, an application-driven documentation template that describes dataset origin, composition, collection methods, and potential biases. This model inspired similar tools such as Data Cards (Bender & Friedman, 2018), which integrate directly with model training pipelines to ensure traceability.

b. Open Data Auditing Tools

Applications such as Audit-AI, Fairlearn, and EthicalML emerged between 2020 and 2025 to automate the evaluation of dataset fairness and labeling transparency (Gonzalez-Sendino et al., 2024). These software packages assess label balance, demographic coverage, and sampling diversity, enabling developers to quantify and document fairness before deployment.

c. Data Provenance and Lineage Systems

With the expansion of large-scale generative AI, provenance-tracking tools like IBM's Watson Data Governance Suite and Google Cloud's AI Governance Toolkit were adopted for managing ethical lineage (AI Now Institute, 2023). These applications ensure that all datasets are traceable, labeled ethically, and compliant with global data-protection laws such as GDPR and the upcoming EU AI Act (Royal Society, 2024).

3.6. Explainability and Transparency Applications

Explainable AI (XAI) applications allow users and regulators to understand the rationale behind algorithmic outputs.

a. Model Explainability Frameworks

Since 2017, researchers have developed applications like LIME (Local Interpretable Model-Agnostic Explanations) and SHAP (SHapley Additive exPlanations) to visualize decision-making in complex models (Wachter et al., 2017). These tools quantify feature importance and highlight which variables most affect predictions—critical for ensuring fairness in sectors like finance and insurance.

b. Accountability Dashboards and Visualization Tools

Accountability dashboards emerged as practical transparency applications, integrating explainability modules into live systems. Examples include Microsoft's InterpretML and Google's What-If Tool, both of which help developers simulate different fairness interventions and visualize their impact on outcomes (Lepri et al., 2018).

c. Public Transparency Portals

Governments have also applied transparency applications to improve public trust. The AI Transparency Portal by the European Commission (2023) provides public summaries of AI systems used in critical services, including decision criteria and risk levels. Such tools operationalize the "right to explanation" under GDPR (Wachter et al., 2017).

3.7. Human-Centered and Participatory Design Applications

Human-centered AI applications prioritize human oversight, empathy, and inclusivity in system design.

a. Human-in-the-Loop (HITL) Auditing Systems

HITL applications allow humans to continuously evaluate algorithmic behavior. Lam et al. (2024) and Wang et al. (2021) describe auditing frameworks where humans review model outputs periodically to identify unintended bias or ethical concerns. These systems integrate annotation feedback directly into retraining cycles, creating adaptive and fair AI workflows.

b. Participatory Co-Design in Civic AI

Pham et al. (2025) demonstrated participatory co-design applications in community governance, where stakeholders collaboratively define AI ethics guidelines. Tools like Ethical Design Assistants enable non-experts to review and contribute to model-building decisions, promoting inclusivity in AI development.

c. Accessibility and Inclusive Design

Human-centered AI also applies to accessibility innovations. Recent applications use fairness-aware natural language processing (NLP) models to improve assistive technologies for users with disabilities or non-native language backgrounds (AI Now Institute, 2023). These inclusive design tools ensure ethical alignment between technology and human diversity.

3.8. Continuous Monitoring and Post-Deployment Impact Applications

AI systems evolve with new data and contexts; hence, ethical compliance must extend beyond deployment.

a. Real-Time Fairness Tracking

Continuous monitoring tools such as MLflow, TensorBoard Fairness Plugin, and Google Vertex AI Model Monitoring allow organizations to track bias drift over time (NIST, 2022). These applications compare live data against baseline fairness metrics, triggering alerts when disparities exceed predefined thresholds.

b. Automated Algorithmic Impact Assessments (AIAs)

AIAs have evolved into semi-automated platforms that collect system metadata, log potential risks, and generate compliance reports (DRCF, 2021; Royal Society, 2024). These applications help organizations document accountability throughout the AI lifecycle.

c. Feedback and Redress Mechanisms

Post-deployment feedback applications enable users to report unfair or incorrect outcomes. Ethical AI platforms increasingly incorporate user-feedback loops to address harm, aligning with the EU's AI Act (2024) requirements for ongoing human oversight.

3.9. A Practitioner's Toolkit for Implementing Ethical AI

Translating ethical principles into operational practice requires structured, actionable steps. The following phased toolkit synthesizes insights from the literature to guide organizations through the AI lifecycle, integrating technical measures with governance protocols.

Table 2.

Phase	Key Actions	Example Tools & Frameworks
Scoping & Design	Conduct stakeholder mapping; Draft an initial Algorithmic Impact Assessment (AIA); Define fairness goals contextually with domain experts.	Participatory design workshops; DRCF (2021) AIA template; NIST AI RMF Scoping Guide.
Data Governance & Documentation	Create Datasheets for Datasets; Audit data for representation, label quality, and historical bias; Establish data provenance trails.	Datasheets (Gebru et al., 2021); Fairlearn, Audit-AI; IBM Watson Data Governance.
Model Development & Training	Apply in-processing bias constraints (e.g., adversarial debiasing); Establish explainability baselines; Test for disparate impact across subgroups.	IBM AIF360; TensorFlow Fairness Indicators; SHAP, LIME; Aequitas toolkit.
Pre-Deployment Auditing	Perform an internal technical and procedural audit; Prepare and publish Model Cards; Validate against regulatory checklists (e.g., EU AI Act).	Internal audit checklist (Raji et al., 2020); Google Model Cards; Regulatory compliance scanners.

continued on following page

Table 2. Continued

Phase	Key Actions	Example Tools & Frameworks
Deployment & Continuous Monitoring	Launch with user-facing transparency notices; Implement real-time fairness and performance dashboards; Set up drift detection alerts.	MLflow, TensorBoard Fairness Plugin; Google Vertex AI Model Monitoring; Amazon SageMaker Clarify.
Governance, Feedback & Redress	Establish clear user feedback and appeal channels; Schedule periodic independent (external) audits; Maintain an immutable audit trail for accountability.	Ethics committee charters; Third-party audit protocols (e.g., IEEE P7003); Blockchain-based audit logs.

This toolkit is iterative, not linear. Findings from later phases (e.g., monitoring alerts) should feed back into model retraining and governance policy updates.

3.10. Sectoral Applications of Ethical and Equitable AI

a. Law Enforcement and Criminal Justice

Bias auditing tools are being used to reassess predictive policing and risk-assessment algorithms (Angwin et al., 2016). Governments now apply algorithmic accountability dashboards to evaluate fairness in sentencing and parole recommendations (Royal Society, 2024).

b. Financial and Credit Scoring Systems

Financial institutions apply fairness testing frameworks to credit models to prevent discriminatory lending. Applications such as FairScore and Explainable Credit AI provide transparency to regulators and customers (Vecchione et al., 2021).

c. Social Welfare and Public Policy

Governments employ ethical audit software to ensure welfare-distribution AI does not discriminate against vulnerable populations. Canada's Directive on Automated Decision-Making (2022) mandates impact assessments before deploying AI for benefits eligibility.

d. Environmental and Sustainability Applications

Recent work connects algorithmic equity with environmental ethics. Tools like EthicalAI for Climate assess how data-driven climate interventions impact marginalized communities (Royal Society, 2024).

3.11. Ethical AI and Environmental Justice

The pursuit of equitable AI is inextricably linked to environmental sustainability, forming a nexus of climate ethics. The development and deployment of large-scale models incur substantial carbon footprints, with training a single large language model potentially emitting over 500,000 lbs of CO_2 equivalent (Strubell et al., 2019). This environmental cost is disproportionately borne by marginalized communities least responsible for climate change. Furthermore, the lifecycle of AI hardware—from mineral extraction for chips to e-waste disposal—raises serious labor and environmental justice concerns. Future ethical ecosystems must, therefore, integrate sustainability scorecards, prioritizing energy-efficient architectures (e.g., sparse models), federated learning to reduce data transfer loads, and procurement policies that consider the full ecological and social cost of AI systems, aligning technological progress with planetary boundaries (AI Now Institute, 2020; Royal Society, 2024).

3.12. Toward Responsible AI in Practice

From 2015 to 2025, applications in bias mitigation, data ethics, algorithmic auditing, and regulatory compliance have transformed ethical AI from an abstract concept into a measurable, enforceable practice. Early work revealed the dangers of unregulated algorithms (O'Neil, 2016; Barocas & Selbst, 2016), prompting waves of practical interventions—open-source fairness toolkits, explainability frameworks, and post-deployment auditing dashboards (Gebru et al., 2021; NIST, 2022).

By 2025, ethical AI applications extend across the AI lifecycle:

1. Pre-deployment: bias detection, fairness-aware design, and data documentation.
2. Deployment: explainability and transparency dashboards.
3. Post-deployment: continuous auditing, human oversight, and adaptive fairness correction.

Together, these applications operationalize ethical and equitable AI principles, bridging technical innovation with societal accountability. They reflect a paradigm where fairness, inclusivity, and responsibility are not afterthoughts—but integral design features of modern intelligent systems.

4. FUTURE TRENDS FOR ENSURING ETHICAL AND EQUITABLE AI-FROM BIAS MITIGATION TO ALGORITHMIC AUDITING

The future of ethical and equitable artificial intelligence (AI) lies in a systemic transformation that integrates transparency, accountability, and inclusivity into every stage of the AI lifecycle. From 2015 to 2025, research evolved from identifying algorithmic bias (Barocas & Selbst, 2016; Angwin et al., 2016) to developing frameworks for accountability, fairness metrics, and human-in-the-loop decision-making (Raji et al., 2020; NIST, 2022; Pham et al., 2025). The coming decade will likely witness the institutionalization of these frameworks through multi-stakeholder governance, standardized auditing, and adaptive AI ethics infrastructures. Rather than viewing ethical AI as a reactive compliance requirement, future systems will embody proactive equity assurance integrated into design, deployment, and evaluation (Kazim et al., 2021; AI Now Institute, 2023). The future of ethical AI will be defined by systemic integration, moving from isolated tools to cohesive, adaptive ecosystems. The following interconnected trajectories will shape the next decade:

- **From Static to Dynamic Governance:** Continuous compliance automation will replace periodic audits. Real-time fairness dashboards, integrated with CI/CD pipelines, will enable automatic drift detection and mitigation, making ethical monitoring a perpetual feature of live systems.
- **Expanding the Epistemic Foundation of Fairness:** Fairness will evolve beyond statistical group parity toward intersectional and causal fairness metrics. This shift requires integrating sociological frameworks and community-defined harm models into technical assessments, acknowledging that fairness is context-dependent and multi-dimensional.
- **Institutionalization of Algorithmic Auditing:** Third-party algorithmic auditing will mature into a certified profession, with standardized protocols (akin to financial auditing). Regulatory "sandboxes" will allow for testing new models under supervision, and liability frameworks will clarify chains of accountability for AI harms.
- **Human-in-the-Loop as a Governance Architecture:** Human oversight will evolve from a checkpoint to a distributed governance layer. Participatory audit committees comprising domain experts, ethicists, and community advocates will be embedded in organizational structures, co-guiding model development and deployment.
- **Global Harmonization with Localized Implementation:** While transnational standards (e.g., OECD AI Principles, ISO/IEC norms) will provide a baseline, effective regulation will require local adaptation. This will spur the

development of region-specific audit tools, multilingual explainability interfaces, and fairness benchmarks reflective of diverse populations.

- **Convergence of Ethics, Privacy, and Sustainability:** Federated learning and differential privacy will become default ethical imperatives for model training, balancing data utility with individual rights. Sustainability metrics (carbon, water, energy use) will become mandatory components of AI impact assessments, linking algorithmic ethics to climate justice.

5. CONCLUSION

Between 2015 and 2025, the discourse on ethical and equitable AI has matured from narrow bias-detection methods to comprehensive governance ecosystems. The ten interlinked domains—bias mitigation, fairness metrics, algorithmic auditing, transparency, accountability, data ethics, regulatory compliance, human-centered design, and continuous monitoring—collectively re-define responsible AI as a socio-technical practice rather than a purely computational problem.

The literature (Barocas & Selbst, 2016; Gebru et al., 2021; NIST, 2022; Royal Society, 2024) consistently shows that achieving equity in AI requires synergy between technical innovation and institutional ethics. Bias mitigation alone cannot guarantee justice without transparent data governance, participatory oversight, and legally enforced accountability. Likewise, continuous auditing and monitoring ensure that fairness persists amid algorithmic drift and societal change. By integrating human-centered design and regulatory rigor, ethical AI systems can evolve adaptively while preserving public trust.

Ultimately, ensuring ethical and equitable AI means embedding reflection, responsibility, and responsiveness into every stage of the AI lifecycle—from data collection to post-deployment auditing. The 2015–2025 research landscape demonstrates that sustainable AI ethics is neither a fixed code of conduct nor a checklist, but a continuous dialogue between humans, technology, and governance that defines how intelligence—artificial or otherwise—serves humanity.

Future trajectories for ensuring ethical and equitable AI underscore a profound transformation—from static compliance models to dynamic, adaptive, and participatory ecosystems. Bias mitigation, algorithmic auditing, and governance will converge into continuous, data-driven, and contextually aware systems that uphold fairness, accountability, and human dignity. As interdisciplinary collaboration deepens and regulatory landscapes mature, the dream of truly equitable AI—transparent, inclusive, and just—will shift from aspiration to operational standard.

The journey toward ensuring ethical and equitable artificial intelligence (AI) has evolved significantly between 2015 and 2025, reflecting a paradigm shift from conceptual awareness of bias to practical frameworks for accountability, transparency, and fairness. Early discussions primarily identified algorithmic harms embedded within datasets and decision systems (Barocas & Selbst, 2016; Angwin et al., 2016), while contemporary research has moved toward institutionalizing algorithmic auditing and bias-mitigation practices (Raji et al., 2020; NIST, 2022). The transformation represents a critical turning point in technological ethics—AI is no longer assessed merely by performance accuracy but by its ability to serve all individuals equitably and to uphold democratic values in digital governance. The convergence of ethical design principles, algorithmic transparency, and participatory oversight marks the maturation of AI ethics into a structured discipline that directly shapes policy, regulation, and societal trust.

A central conclusion emerging from the literature is that bias mitigation must extend beyond static detection to become a continuous, adaptive process. Early mitigation techniques treated fairness as a post-hoc adjustment—correcting outcomes only after bias was discovered (Buolamwini & Gebru, 2018). However, scholars such as Suresh and Guttag (2019) and Gonzalez-Sendino et al. (2024) emphasize that bias evolves dynamically as datasets shift and user demographics change. Thus, the next phase of equitable AI requires continuous monitoring embedded directly into the machine-learning lifecycle. Automated fairness dashboards, drift-detection algorithms, and real-time audit trails will ensure that equity remains stable as systems learn and adapt over time. The future therefore lies not in eliminating bias once, but in institutionalizing fairness as an enduring design parameter that evolves with context and data.

Another core insight is that algorithmic auditing will become a standardized institutional practice, akin to financial or environmental audits (Raji et al., 2020; DRCF, 2021). The growing adoption of frameworks such as the NIST AI Risk Management Framework (2022) and the proposed EU AI Act demonstrates an international shift toward structured accountability. The literature converges on the notion that voluntary ethical codes are insufficient for ensuring justice in high-impact AI applications; instead, mandatory audits and certification mechanisms must validate both process and outcome fairness (Kazim et al., 2021). Algorithmic auditing, when performed by independent third parties, provides the necessary transparency to evaluate whether AI systems operate consistently with human rights principles. Future developments will likely include professional accreditation for AI auditors, standardized audit templates, and shared repositories of bias-detection benchmarks. Through these efforts, ethical compliance will transition from corporate self-regulation to public accountability.

A recurring conclusion across the reviewed studies is that fairness itself is context-dependent and multi-dimensional. Statistical parity or equalized odds may satisfy mathematical fairness, but they cannot capture complex social realities (Lepri et al., 2018; Cech, 2021). The integration of intersectional fairness models, which recognize overlapping identities such as race, gender, and disability, will redefine how AI systems evaluate equity. Future research must therefore combine quantitative metrics with qualitative, sociotechnical approaches that include lived experience and community impact assessments (Crawford & Paglen, 2019). Fairness will be reframed not as a universal number but as an ongoing negotiation between technical design and ethical interpretation. This evolution demands collaboration between engineers, ethicists, policymakers, and affected communities to co-construct definitions of fairness appropriate to each domain (Bhambri, 2024).

The conclusion also highlights the indispensable role of human-in-the-loop governance in ensuring that algorithmic decisions remain contextually accountable (Lam et al., 2024; Pham et al., 2025). While automation offers efficiency, it risks eroding moral responsibility if humans are excluded from oversight. Incorporating human reviewers into every critical decision point ensures that AI remains subordinate to human judgment and values. Participatory co-design—where end-users, communities, and ethicists collaboratively shape system objectives—will become the foundation of equitable AI. This human-centered paradigm strengthens interpretability and fosters public trust, which is essential for sustainable AI adoption in education, healthcare, and governance. The literature indicates that participation transforms ethics from an abstract ideal into a practical, lived process embedded in technological systems (Royal Society, 2024).

Global harmonization of AI regulation represents another decisive trend. As AI transcends national boundaries, fragmented governance creates ethical inconsistencies and compliance loopholes (Kazim et al., 2021; Royal Society, 2024). By aligning international standards—such as the OECD AI Principles, NIST AI RMF, and UNESCO's AI Ethics Recommendations—stakeholders can foster interoperability and accountability. The establishment of transnational auditing protocols will help developing nations participate in global AI governance rather than remain passive recipients of imported technology. Inclusive governance ensures that ethical norms reflect pluralistic worldviews rather than being dictated by a few dominant economies. Such collaboration may also reduce digital inequities by ensuring equitable data access, multilingual model development, and fairness across diverse populations.

A profound theme that runs through the literature is the fusion of explainability with ethical assurance. As Wachter, Mittelstadt, and Russell (2017) and Veale and Binns (2017) observed, interpretability is not merely a technical challenge but a moral necessity for accountability. Future AI systems will embed explainable AI (XAI) mechanisms directly within their ethical governance frameworks, producing contextualized explanations for regulators, developers, and users alike. Rather than treating transparency as optional documentation, organizations will integrate it into user interfaces and audit dashboards, enabling stakeholders to trace decision logic in real time. This synergy between interpretability and ethics transforms abstract accountability into operational visibility—empowering both oversight and informed consent.

Another central pillar in the conclusion is ethical data stewardship. The pioneering *Datasheets for Datasets* framework (Gebru et al., 2021) inspired a movement toward full lifecycle accountability, emphasizing provenance tracking, consent documentation, and bias labeling. Data is not ethically neutral—it embodies cultural and social power relations (Crawford & Paglen, 2019). Future data governance must therefore adopt traceable lineage mechanisms, potentially supported by blockchain technology, to verify dataset integrity and consent compliance (Ferrara, 2023). Ethical data management extends to active maintenance—periodically reviewing datasets for representational imbalances or outdated features. Equitable AI depends not only on how algorithms learn but also on what they learn from and whom they represent.

The future also links ethical AI to environmental and social sustainability, expanding the moral scope beyond human fairness. The AI Now Institute (2020) and Royal Society (2024) stress that the computational and environmental costs of large AI models must be included in ethical evaluation. Sustainable AI will assess not only accuracy but also carbon footprint, labor practices, and social impact. This perspective promotes responsible innovation aligned with the United Nations Sustainable Development Goals, ensuring that progress in AI does not exacerbate global inequalities or ecological degradation. The integration of ethics with sustainability reframes accountability from an individual or organizational responsibility to a collective planetary obligation.

In practical domains such as healthcare, education, and public administration, the application of ethical-AI frameworks is rapidly becoming a regulatory expectation. Studies show that explainable and bias-aware AI models in healthcare improve diagnostic reliability and patient trust (Liu et al., 2022), while fairness audits in education technology ensure equitable grading and learning outcomes (AI Now Institute, 2023). Public agencies adopting algorithmic decision-making tools will increasingly rely on algorithmic impact assessments to detect and correct systemic bias (Wachter et al., 2017). These cross-sectoral implementations demonstrate that ethical AI is no longer theoretical—it is a functional criterion for institutional legitimacy. Consequently, equitable AI is emerging as both a moral and competitive imperative in global digital economies.

Technological evolution will also drive the automation of ethical compliance. Future AI pipelines will incorporate built-in fairness metrics, explainability engines, and automated audit logging tools. Open-source frameworks such as IBM AIF360, Microsoft FairLearn, and TensorFlow Fairness Indicators are precursors to comprehensive compliance toolchains capable of generating real-time ethical reports (Vecchione et al., 2021). These systems will reduce the burden on human auditors while increasing transparency and reproducibility. Ethical compliance automation will thus bridge the gap between regulatory mandates and practical implementation, ensuring that fairness, accountability, and transparency are consistently monitored without manual intervention.

Yet, even as technology advances, the literature warns that ethical AI cannot rely solely on code. Socio-technical co-design—embedding community voices into development—remains essential to counter technocratic biases (Pham et al., 2025). Ethical governance thrives when users are treated as co-creators rather than data sources. Citizen audit workshops, participatory evaluations, and open-access datasets enable communities to shape technologies that impact their lives. This democratization of AI governance aligns ethics with empowerment, fostering long-term public legitimacy. The emphasis on participation also addresses epistemic justice, acknowledging that marginalized voices often hold crucial insights into algorithmic harms invisible to designers or regulators.

The convergence of law, ethics, and technology further underscores the growing institutionalization of accountability. By 2030, most jurisdictions are expected to mandate algorithmic impact assessments and establish legal liability for biased or opaque AI outcomes (DRCF, 2021; Wachter et al., 2017). These legal frameworks will clarify accountability chains from developers to deployers, establishing "algorithmic fiduciary duties" analogous to those in finance or medicine. Institutional mechanisms such as AI ombudspersons and regulatory sandboxes will mediate conflicts and test emerging ethical practices under real-world conditions. As legal systems evolve, ethical AI will transition from voluntary good practice to a binding civic responsibility.

Underlying all these transformations is the need for education and capacity-building. A sustained culture of ethical AI requires experts who understand both technical design and moral reasoning. The literature repeatedly calls for interdisciplinary curricula combining computer science, philosophy, sociology, and law (Kazim et al., 2021). Certification programs for AI auditors and ethicists will standardize professional competence, while organizational training will cultivate ethical reflexivity among developers. Such educational initiatives ensure that equity and accountability are internalized cultural norms rather than external enforcement measures. Building ethical capacity is therefore as crucial as building ethical algorithms.

Finally, privacy-preserving technologies such as federated learning and differential privacy will become foundational to equitable AI. These methods align with ethical imperatives by reducing centralized data dependencies and respecting user autonomy (NIST, 2022). Federated fairness learning allows models to train across distributed data sources without compromising individual rights. Combined with continuous algorithmic auditing ecosystems (AI Now Institute, 2023), these innovations promise to sustain fairness dynamically and transparently. The future of equitable AI thus lies in interconnected infrastructures—technological, institutional, and cultural—that work in harmony to uphold justice.

In synthesis, the decade from 2015 to 2025 has redefined how humanity conceptualizes fairness and accountability in intelligent systems. The trajectory from recognizing bias to institutionalizing audits reflects a collective moral awakening within the technological community. Ethical and equitable AI is no longer an optional afterthought; it is an operational necessity that defines the credibility of the digital age. The literature consistently underscores that achieving this vision demands integration across disciplines, sustained global cooperation, and public engagement. Future AI ecosystems will function as living ethical organisms—capable of self-reflection, adaptation, and mutual accountability. When combined with transparent governance, explainable logic, and participatory inclusion, these systems have the potential to reinforce democratic values rather than undermine them. In this emerging paradigm, fairness is not simply an algorithmic output but a societal commitment—one that must be renewed with every dataset collected, every model trained, and every decision automated. Ensuring ethical and equitable AI, therefore, is not the end of a journey but the foundation of a responsible technological future.

Power, Labor, and the Unfinished Project of Ethical AI: A truly equitable AI ecosystem must also confront the often-overlooked power dynamics and labor conditions within its own supply chain. The definition of "fairness," the selection of audit metrics, and the control over ethical standards are themselves exercises of power, often concentrated within a small number of corporations and nations. Furthermore, the "hidden labor" of data annotation and content moderation—work that is frequently outsourced, underpaid, and psychologically taxing—forms the essential yet marginalized foundation of AI systems (Gray & Suri, 2019). Ignoring these political economic dimensions risks reducing ethical AI to a technical façade that fails to address root causes of inequity. Therefore, the path forward demands a deliberately inclusive and democratic approach. It requires:

- **Epistemic Justice:** Centering knowledge from the Global South and marginalized communities in defining harms and solutions.
- **Labor Equity:** Ensuring fair wages and working conditions for all contributors in the AI data lifecycle.
- **Transparent Power Sharing:** Creating governance structures that distribute oversight authority among developers, regulators, and civil society.

Call to Action: The journey from bias detection to algorithmic auditing marks significant progress, but the work is incomplete. We call upon researchers, practitioners, and policymakers to move beyond compliance toward co-creation. Let us build AI ecosystems where fairness is dynamically maintained, accountability is legally enforced, and the benefits of innovation are justly shared. The goal is not merely to prevent harm but to actively foster a technological future that amplifies human dignity, equity, and democratic participation.

REFERENCES

AI Now Institute. (2023). *Algorithmic accountability: Moving beyond audits*. AI Now Institute.

Anand, A. J., & Bhambri, P. (Eds.). (2025). *Handbook of AI-driven threat detection and prevention: A holistic approach to security*. CRC Press., DOI: 10.1201/9781003521020

Angwin, J., Larson, J., Mattu, S., & Kirchner, L. (2016). *Machine bias: There's software used across the country to predict future criminals. And it's biased against blacks*. ProPublica.

Bajdor, P., & Bhambri, P. (2024). The principles of sustainability. In Bhambri, P., & Bajdor, P. (Eds.), *Handbook of technological sustainability: Innovation and environmental awareness* (pp. 3–17). CRC Press., DOI: 10.1201/9781003475989-2

Barocas, S., & Selbst, A. D. (2016). Big data's disparate impact. *California Law Review, 104*(3), 671–732.

Benjamin, R. (2019). *Race After Technology: Abolitionist Tools for the New Jim Code*. Polity Press.

Bhambri, P. (2024). Enhancing cybersecurity with SuperHyperSoft computing approaches. In Smarandache, F., & Majumder, P. (Eds.), *Modern SuperHyperSoft computing trends in science and technology* (pp. 127–148). IGI Global., DOI: 10.4018/979-8-3693-6875-6.ch005

Bhambri, P. (2025). Understanding AI and machine learning in security. In Bhambri, P., & Anand, A. J. (Eds.), *Handbook of AI-driven threat detection and prevention: A holistic approach to security* (pp. 1–17). CRC Press., DOI: 10.1201/9781003521020-1

Bhambri, P., & Bajdor, P. (2025). AI-powered predictive analysis for proactive cyber defense. In Bhambri, P., & Anand, A. J. (Eds.), *Handbook of AI-driven threat detection and prevention: A holistic approach to security* (pp. 308–321). CRC Press., DOI: 10.1201/9781003521020-18

Bhambri, P., & Hamad, A. (2025). Quantum computing and AI synergies: Strengthening cybersecurity resilience. In Bhambri, P., & Anand, A. J. (Eds.), *Handbook of AI-driven threat detection and prevention: A holistic approach to security* (pp. 337–352). CRC Press., DOI: 10.1201/9781003521020-20

Bhambri, P., & Kant, S. (2025a, February 1–2). A taxonomy of bias in machine learning: Classification, sources, and implications for ethical AI [Paper presentation]. Lincoln-SPAST Global Sustainability Programme (SGS-24), 1st International Conference of L-GPR Program, Lincoln University College, Malaysia. SPAST Proceedings, 1(2). https://spast.org/techrep/article/view/5387/724

Bhambri, P., & Kant, S. (2025b, April 26–27). Ethical AI systems: A comprehensive framework for bias mitigation and fairness in machine learning [Paper presentation]. Lincoln-SPAST Global Sustainability Programme (SGS-24), 2nd International Conference of L-GPR Program, Lincoln University College, Malaysia. SPAST Proceedings, 1(2). https://spast.org/techrep/article/view/5385/698

Bhambri, P., & Pawełoszek, I. (2025). Deep learning techniques for intrusion detection in critical infrastructure. In Bhambri, P., & Anand, A. J. (Eds.), *Handbook of AI-driven threat detection and prevention: A holistic approach to security* (pp. 322–336). CRC Press., DOI: 10.1201/9781003521020-19

Bhambri, P., & Starostka-Patyk, M. (2025). Integrating AI with blockchain for decentralized security and threat prevention. In Bhambri, P., & Anand, A. J. (Eds.), *Handbook of AI-driven threat detection and prevention: A holistic approach to security* (pp. 353–370). CRC Press., DOI: 10.1201/9781003521020-21

Brookings Institution. (2019). *Algorithmic bias detection and mitigation: Best practices and policies to reduce consumer harms*. Brookings Institution.

Buolamwini, J., & Gebru, T. (2018). Gender shades: Intersectional accuracy disparities in commercial gender classification. *Proceedings of Machine Learning Research*, *81*, 77–91.

Caton, S., & Haas, C. (2020). Fairness in machine learning: A survey. *ACM Computing Surveys*, *54*(6), 1–35. DOI: 10.1145/3457607

Cech, F. (2021). Mechanisms for algorithmic accountability through the lens of actor and forum agency. *Government Information Quarterly*, *38*(4), 101615. DOI: 10.1016/j.giq.2021.101615

Crawford, K., & Paglen, T. (2019). Excavating AI: The politics of images in machine learning datasets. https://excavating.ai

. Dastin, J. (2018, October 11). Amazon scraps secret AI recruiting tool that showed bias against women. Reuters.

DRCF / UK regulators. (2021). Auditing algorithms: The existing landscape, role of regulators and future outlook. UK Government. https://www.gov.uk/government/publications

Eubanks, V. (2018). *Automating Inequality: How High-Tech Tools Profile, Police, and Punish the Poor*. St. Martin's Press.

Ferrara, E. (2023). Fairness and bias in artificial intelligence: A brief survey of sources, impacts, and mitigation strategies. *Scientific Reports, 13*, 21854. DOI: 10.1038/s41598-023-48938-0

Gebru, T., Morgenstern, J., Vecchione, B., Vaughan, J. W., Wallach, H., Daumé, H.III, & Crawford, K. (2021). Datasheets for datasets. *Communications of the ACM, 64*(12), 86–92. DOI: 10.1145/3458723

Geetha, K., Vigneshwari, J., Bhambri, P., & Thangam, A. (2024). Sustainable solutions for global waste challenges: Integrating technology in disposal and treatment methods. In Bhambri, P., & Bajdor, P. (Eds.), *Handbook of technological sustainability: Innovation and environmental awareness* (pp. 46–56). CRC Press., DOI: 10.1201/9781003475989-5

Gonzalez-Sendino, R., Martínez-Rego, D., Alvares-Chillida, G., & Fontenla-Romero, Ó. (2024). Mitigating bias in artificial intelligence: Fair data generation using causal models. *Expert Systems with Applications, 238*, 122064. DOI: 10.1016/j.eswa.2023.122064

Goodman, B., & Flaxman, S. (2017). European Union regulations on algorithmic decision-making and a "right to explanation". *AI Magazine, 38*(3), 50–57. DOI: 10.1609/aimag.v38i3.2741

Gray, M. L., & Suri, S. (2019). *Ghost Work: How to Stop Silicon Valley from Building a New Global Underclass*. Houghton Mifflin Harcourt.

Hacohen, U. Y. (2023). User-based algorithmic auditing. *SSRN*. https://ssrn.com/abstract=4564776 DOI: 10.2139/ssrn.4540163

Hardt, M., Price, E., & Srebro, N. (2016). Equality of opportunity in supervised learning. *Advances in Neural Information Processing Systems*, •••, 29.

Hort, M., Chen, Z., Zhang, J. M., Harman, M., & Sarro, F. (2022). Bias mitigation for machine learning classifiers: A comprehensive survey. *ACM Computing Surveys, 55*(13s), 1–38. DOI: 10.1145/3583475

Kazim, E., Dennehy, D., & Koshiyama, A. (2021). Systematizing audit in algorithmic recruitment. *Journal of Intelligence Studies in Business, 11*(1), 26–41. PMID: 34564294

Kleinberg, J., Mullainathan, S., & Raghavan, M. (2017). Inherent trade-offs in the fair determination of risk scores. *Proceedings of the 8th Innovations in Theoretical Computer Science Conference (ITCS)*, 43:1–43:23.

. Kossow, N., & Transparency International. (2021). Algorithmic transparency and accountability: Tools for practitioners. Transparency International Knowledge Hub.

Kouroupis, K., & Efthymiou, I. (Eds.). (2026). *Legal Challenges of AI Across Interdisciplinary Sectors*. IGI Global Scientific Publishing., DOI: 10.4018/979-8-3373-5452-1

Kroll, J. A., Huey, J., Barocas, S., Felten, E. W., Reidenberg, J. R., Robinson, D. G., & Yu, H. (2017). Accountable algorithms. *University of Pennsylvania Law Review*, *165*, 633–705.

Lam, K., O'Neill, E., Candon, P., & Crow, D. (2024). A framework for assurance audits of algorithmic systems. *Proceedings of the 2024 ACM Conference on Fairness, Accountability, and Transparency*, 1564–1575. DOI: 10.1145/3630106.3658957

Lepri, B., Oliver, N., Letouzé, E., Pentland, A., & Vinck, P. (2018). Fair, transparent, and accountable algorithmic decision-making processes. *Philosophical Transactions. Series A, Mathematical, Physical, and Engineering Sciences*, *376*(2128), 20170353. DOI: 10.1098/rsta.2017.0353 PMID: 30082300

Li, F., Li, Y., & Li, F. (2023). Evaluating and mitigating bias in machine learning models: A practical review. npj. *Digital Medicine*, *6*, 170. DOI: 10.1038/s41746-023-00921-9

Liu, X., Glocker, B., McCradden, M. M., Ghassemi, M., Denniston, A. K., & Oakden-Rayner, L. (2022). The medical algorithmic audit. *The Lancet. Digital Health*, *4*(5), e384–e397. DOI: 10.1016/S2589-7500(22)00003-6 PMID: 35396183

Mehrabi, N., Morstatter, F., Saxena, N., Lerman, K., & Galstyan, A. (2021). A survey on bias and fairness in machine learning. *ACM Computing Surveys*, *54*(6), 1–35. DOI: 10.1145/3457607

Mohamed, S., Png, M. T., & Isaac, W. (2020). Decolonial AI: Decolonial Theory as Sociotechnical Foresight in Artificial Intelligence. *Philosophy & Technology*, *33*(4), 659–684. DOI: 10.1007/s13347-020-00405-8

Mohana Sundari, V., Ganeshkumar, M., & Bhambri, P. (2024). Environmental stewardship in the digital age: A technological blueprint. In Bhambri, P., & Bajdor, P. (Eds.), *Handbook of technological sustainability: Innovation and environmental awareness* (pp. 57–67). CRC Press., DOI: 10.1201/9781003475989-6

NIST. (2022). Towards a standard for identifying and managing bias in artificial intelligence (NIST Special Publication 1270). National Institute of Standards and Technology. DOI: 10.6028/NIST.SP.1270

O'Neil, C. (2016). *Weapons of math destruction: How big data increases inequality and threatens democracy*. Crown.

Pasquale, F. (2015). *The black box society: The secret algorithms that control money and information*. Harvard University Press. DOI: 10.4159/harvard.9780674736061

Pham, N., Kuo, R., & Huang, Y. (2025). Fairness for machine learning software in education: Mapping and review. *Computers & Education*, *220*, 105120. DOI: 10.1016/j.compedu.2024.105120

Raji, I. D., & Buolamwini, J. (2019). Actionable auditing: Investigating the impact of publicly naming biased performance results. Proceedings of the 2019 AAAI/ACM Conference on AI, Ethics, and Society, 429–435. DOI: 10.1145/3306618.3314244

Raji, I. D., Smart, A., White, R. N., Mitchell, M., Gebru, T., Hutchinson, B., Smith-Loud, J., Theron, D., & Barnes, P. (2020). Closing the AI accountability gap: Defining an end-to-end framework for internal algorithmic auditing. *Proceedings of the 2020 Conference on Fairness, Accountability, and Transparency*, 33–44. DOI: 10.1145/3351095.3372873

Rana, R., & Bhambri, P. (2024). Ethical considerations in artificial intelligence for environmental solutions: Striking a balance for sustainable innovation. In Bhambri, P., & Bajdor, P. (Eds.), *Handbook of technological sustainability: Innovation and environmental awareness* (pp. 389–396). CRC Press., DOI: 10.1201/9781003475989-31

Royal Society. (2024). Towards algorithm auditing: Managing legal, ethical and technical challenges. *Royal Society Open Science*, *11*(1), 231209. DOI: 10.1098/rsos.231209

Sandvig, C., Hamilton, K., Karahalios, K., & Langbort, C. (2014). *Auditing algorithms: Research methods for detecting discrimination on internet platforms*. Data & Society Research Institute.

Selbst, A. D. (2019). Negligence and AI's human users. *Boston University Law Review. Boston University. School of Law*, *100*, 1315–1344.

Selbst, A. D., & Barocas, S. (2018). The intuitive appeal of explainable machines. *Fordham Law Review*, *87*, 1085–1139.

Selbst, A. D., Boyd, D., Friedler, S. A., Venkatasubramanian, S., & Vertesi, J. (2019). Fairness and abstraction in sociotechnical systems. Proceedings of the ACM on Human-Computer Interaction, 3(CSCW), 1–36. DOI: 10.1145/3287560.3287598

Sharma, R., & Bhambri, P. (2024). Digital duplicity and the disintegration of trust: A quantitative inquiry into the impact of deep fakes on media sustainability and societal equilibrium. In Bhambri, P., & Bajdor, P. (Eds.), *Handbook of technological sustainability: Innovation and environmental awareness* (pp. 273–291). CRC Press., DOI: 10.1201/9781003475989-24

Strubell, E., Ganesh, A., & McCallum, A. (2019). Energy and Policy Considerations for Deep Learning in NLP. *Proceedings of the 57th Annual Meeting of the Association for Computational Linguistics*, 3645–3650. DOI: 10.18653/v1/P19-1355

Suresh, H., & Guttag, J. V. (2021). A framework for understanding sources of harm throughout the machine learning life cycle. *Proceedings of the 2021 ACM Conference on Fairness, Accountability, and Transparency*, 806–807. DOI: 10.1145/3465416.3483305

Thirumalaiyammal, B., Steffi, P. F., & Bhambri, P. (2024). Green horizons: Navigating environmental challenges through technological innovation. In Bhambri, P., & Bajdor, P. (Eds.), *Handbook of technological sustainability: Innovation and environmental awareness* (pp. 292–304). CRC Press., DOI: 10.1201/9781003475989-25

Veale, M., & Binns, R. (2017). Fairer machine learning in the real world: Mitigating discrimination without collecting sensitive data. *Big Data & Society*, *4*(2). Advance online publication. DOI: 10.1177/2053951717743530

Veale, M., Van Kleek, M., & Binns, R. (2018). Fairness and accountability design needs for algorithmic support in high-stakes public sector decision-making. *Proceedings of the 2018 CHI Conference on Human Factors in Computing Systems*, 1–14. DOI: 10.1145/3173574.3174014

Vecchione, B., Levy, K., & Barocas, S. (2021). Algorithmic auditing and social justice: Lessons from the field. *Proceedings of the 2021 ACM Conference on Fairness, Accountability, and Transparency*, 662–663. https://doi.org/DOI: 10.1145/3442188.3451338

Vigneshwari, J., Senthamizh Pavai, P., Maria Suganthi, L., & Bhambri, P. (2024). Eco-ethics in the digital age: Tackling environmental challenges through technology. In Bhambri, P., & Bajdor, P. (Eds.), *Handbook of technological sustainability: Innovation and environmental awareness* (pp. 201–213). CRC Press., DOI: 10.1201/9781003475989-18

Wachter, S., & Mittelstadt, B. (2021). A right to reasonable inferences: Re-thinking data protection law in the age of big data and AI. *Columbia Business Law Review*, *2021*(2), 494–620.

Wachter, S., Mittelstadt, B., & Russell, C. (2017). Counterfactual explanations without opening the black box: Automated decisions and the GDPR. *SSRN*, *31*(2), 841–887. DOI: 10.2139/ssrn.3063289

Wang, H., Wang, Y., & Wang, Y. (2021). An interactive approach to bias mitigation in machine learning. *Proceedings of the 2021 CHI Conference on Human Factors in Computing Systems*, 1–14. DOI: 10.1109/ICCICC53683.2021.9811333

Chapter 3
Ethical Considerations in Generative AI:
Navigating Uncertainty and Responsible Innovation

Azamat Ali
https://orcid.org/0000-0003-2931-2166
Lovely Professional University, India

ABSTRACT

Generative AI systems are fast changing the research, industry and society. These systems are used to generate new content, decisions and knowledge with minimum human involvement. This possibility brings to the fore serious ethical issues. This chapter explores some of the most important ethical aspects of the establishment and application of generative AI. Its interests lie in areas like prejudice, openness, responsibility, confidentiality and abuse. The chapter brings out the uncertainty surrounding the model behaviour, outcome, and long term social impact. It talks about the inadequacy of the current ethical frameworks in generative systems. Another concept that is discussed in the chapter is responsible innovation as a guiding technique. Responsible innovation focuses on foresight, human control and inclusive governance. The risks may be mitigated by incorporating ethical thinking in design and deployment. The chapter also intends to guide the researchers, developers, and policymakers to make informed and ethical decisions in the face of uncertainty.

1. INTRODUCTION

Generative artificial intelligence is computational systems that are capable of generating new text, images, audio, video and code. These systems are trained on

DOI: 10.4018/979-8-2600-0000-7.ch003

massive datasets and produce outputs that are similar to those created by human beings. Language models, image generators and music composition systems are usually common examples (Brindha et al., 2025). Over the last few years, generative AI has developed at an extremely fast pace because of the enhancement of machine learning algorithms, the increased power of different computers, and the accessibility of large amounts of data. Consequently, these technologies are currently in extensive application in the fields of education, healthcare, business, media and scientific research (Lv, 2023). The increased power of generative AI has altered the process of creation and exchange of information. The automated content generation can enhance efficiency, minimize the costs, and facilitate creativity (Liu & Kim, 2025). As an illustration, generative AI may help researchers write reports, educators create learning materials, and professionals in software development. Such advantages have stimulated organizations and governments to spend a lot on AI advancement. Nonetheless, the ability to create AI generative models is both powerful and problematic in terms of complex ethical considerations (Efthymiou, 2026; Sengar et al., 2025).

A significant ethical issue is associated with the use of data. The AI systems deployed as generative systems are trained with large data sets that tend to contain personal, copyrighted or sensitive data. In most instances, people do not have the knowledge that their information has been gathered or even used to train. Here the privacy and consent question and the ownership of the data are questioned. Mismanagement of data may result in infringement on the rights of people, as well as loss of community confidence (Banh & Strobel, 2023). The second important concern is prejudice and equity. Training data can be indicative of the prevailing social inequity, cultural bias, or historical discrimination. Generative AI systems will be able to reproduce these biases or even enhance them when such data is used. This may lead to the unfair treatment of some population and support dangerous narratives. The problem of bias is difficult to resolve since it involves a thorough selection of data and its assessment, as well as continuous monitoring (Routray et al., 2024).

Misinformation and misuse are also another concern of generative AI. It is easier to make fake or misleading because of the possibilities to create realistic text, pictures, or videos. Deepfakes, computer generated propaganda, and fake news may affect the masses and undermine democracy. These dangers emphasize the moral duty of the developers and users to block malicious applications (Mallikarjuna & Chittemsetty, 2024). Other ethical issues are accountability and transparency. Numerous generative AI frameworks are complicated systems that cannot be easily interpreted. Users do not even know how outputs are created and why a decision is made this way or that. This is not explainable thus making it difficult to attribute blame when mistakes or injuries take place. It also restricts the possibility of the victims to question or doubt AI generated results (Al-Kfairy et al., 2024).

Generative AI has frequently been implemented faster than ethical standards and regulations have been created. This becomes a source of doubt to the policymakers, organizations, and the society in general. The demand grows toward the responsible solutions that strike the right balance between the innovation and the ethical protection. These methods need cooperation between the scientists, programmers, policymakers, and citizens (Farina et al., 2025). The ethical issues of generative AI are thus critical to understand to make it an ethical process as well as to make sure that the benefits of generative AI are fairly and responsibly shared.

1.1 Research Gap

Although this field of study is increasingly expanding, there are still a number of significant research gaps in the ethics of generative artificial intelligence. The largest part of the available literature like this research deals with the conceptual and theoretical debates of the ethical principles which include fairness, transparency, accountability, and responsibility. There exists little empirical data that looks at the manifestations of these ethical issues in actual applications of generative AI. Consequently, the real implications of uncertainty, bias, and abuse in various sectors are hardly known. The other significant gap is situated in the domain specific analysis. Most of the existing literature views generative AI as a technology with a broad brush, without paying much attention to the fact that ethical risks may be significant in such sensitive areas as healthcare, education, law, or governance. These regions can experience special ethical implications that have to be addressed with specific structures and resolutions. Also, current systems of governance and ethics tend to remain fixed, whereas the generative AI systems are changing at a very high rate. This puts a distance between the development of technology and moral control. Besides, the fact that cultural and regional distinctions have limited attention on ethical values and regulatory abilities has been neglected. The vast majority of ethical frameworks are international or westernized and this might not have a complete representation of the various social situations. The gaps should be tackled in future studies by the use of empirical studies, industry based research, and integrating inclusive governance strategies that will keep up with the current technological change.

2. CONCEPTUAL FOUNDATIONS OF AI ETHICS

2.1 Fairness in AI Systems

Fairness is one of the fundamental principles of AI ethics and can be defined as just and fair treatment and social groups of people. The related issue of fairness that is considered in artificial intelligence systems is mainly the prevention of discrimination and the elimination of the results, which disproportionately harm certain populations. Since AI models are normally trained with prior data, they are prone to persisting and exaggerating current social biases of gender, race, age, or socioeconomic statuses. The irresponsible application of such data, i.e., without critically analysing every detail about the past, can result in the reproduction or strengthening of discriminatory tendencies existing in society (Baumann & Loi, 2023). Fairness is a crucial aspect that has to be thought of in the lifecycle of the AI system. An aspect of diversity and representativeness is necessary in data collection because it is a crucial phase. In case datasets are biased or inaccurate on specific groups, models may just work unequally, and yield biased results. Model design choices also have an impact on fairness. Algorithms can be made efficient or accurate often at odds to the goals of equity. Because of this, systems which seem to be successful technically can still produce unjustly or exclusionary results unless ethical factors are also asked to be incorporated into their design (Chen et al., 2023).

Another factor that is critical in the promotion of fairness is evaluation practices. The conventional performance indicators tend to be inadequate to identify the differences between various demographic groups. Hence, evaluation measures that are conscious of fairness should be included in order to examine how the outcomes change with the population. These steps assist in highlighting disparate effects and minimizing the threat of unfair treatment (De Cremer et al., 2024). There is however no one universal concept in the definition of fairness. It can be described in various ways with respect to the context, like equal opportunity, equal treatment, or equal outcomes. These interpretations can be in conflict and they may lead to moral complications whereby enhancing justice in one aspect may jeopardize the other. It therefore follows that context sensitive judgment and ethical reflection is needed to ascertain the idea of fairness. Gen itself can be described as one that respects fairness issues even with generative AI systems whose results can affect the culture and social norms. The development of ethical AI should then consider the values and interests of which people are taken into consideration, and it is critical to apply responsible and inclusive development methods (Alvarez et al., 2024).

2.2 Transparency and Explainability

Transparency is the degree to which AI systems and their operations make sense to users, programmers and other interested persons. Ethically, transparency helps in trust, making informed decisions and accountability. In cases where the critical decisions or social implications are affected by AI systems, users must be able to find out straightforward information pertaining to the functioning of these systems and the derivation of outputs. Lack of transparency makes it hard to detect any errors, bias, or negative impact (Balasubramaniam et al., 2023). The concept of explainability is more closely associated with transparency and deals with the possibility to interpret and understand AI generated choices or outputs. A variety of developed AI systems, deep learning models, are extremely complex and imprecise to comprehend. Such complexity has ethical implications particularly in the high risk areas like healthcare, finance, and criminal justice. In such regions, the use of AI in the decision making process can significantly affect the right and opportunity of people (Balasubramaniam et al., 2022).

Weak explainability may create undue unfairness and legitimacy. The affected individuals may fail to comprehend the reason why a given outcome happened and how to oppose it. Such ignorance undermines procedural justice and lowers the confidence in institutions that use AI systems. The use of AI products as objective or authoritative raises ethical concerns when there is not enough justification (Cetinkaya & Krämer, 2025). Ethical transparency may not necessarily mean that technical disclosure is complete. It, on the contrary, focuses on meaningful explanations, which are reasonable to various stakeholders. Detailed technical knowledge can be demanded by developers and simple explanations are not always simple by the user who wants to know the capabilities and limitations of the system. Policymakers and regulators might want to obtain data on risks, security and accountability systems. Layered transparency enables information to be shared in various levels without jeopardizing intellectual property and system security (Andrada et al., 2023).

2.3 Accountability and Responsibility

Accountability in AI ethics is the capacity to ascribe responsibility of the design, implementation and impact of AI systems. Ethical accountability will make sure that the harms, which AI creates, can be detected, solved, and avoided. In case the accountability mechanisms are weak or ill-structured, then organizations or individuals might evade accountability by placing blame on the negative outcomes on the automated processes as opposed to human decisions. This decentralization of loss of trust and reduces the ability to effectively address ethical malpractices (Novelli et al., 2024). Many stakeholders, such as the data providers, system designers, organizations, and end users, are involved in the development and use of AI systems. This mutual participation presents some difficult ethical issues. In case of biased outputs, technical errors and misuse, it is always hard to determine who should be blamed. The data providers can affect result by selecting data and the developers affect the system behaviour by design decisions. The manner in which systems are implemented and monitored is decided by the organization, and the user impacts the results of the interaction with the AI tools. Appropriate governance frameworks are thus necessary to establish roles, responsibilities and accountability throughout the AI lifecycle (Lechterman, 2024).

Ethical frameworks highlight the institutional and human responsibility in the AI systems. The developers have a role of guaranteeing the quality of data, minimizing bias and safety in the system design and training. When implementing AI systems, organizations should pay attention to performance monitoring, risk management, and ethical standards and legal requirements (Kouroupis & Efthymiou, 2026). These involve putting up internal control systems and equipping employees accordingly in terms of training. Informed and proper use is also the responsibility of end users, especially in a professional or high stakes situation (Ali, 2025). There should also be proper redress and correction mechanisms in accountability. People who are victims of AI driven results must be able to appeal decisions and find solutions. This requires the existence of easy procedures of complaints, accountable auditing, and the availability of detailed documentation of system behaviour. Periodic appraisals and impact analysis aid in finding new threats and steer the rectifying measures. With a higher level of accountability structures, AI systems can stay in connection with the human values and social responsibility, and ethical failures will be dealt with instead of being hidden (Camilleri, 2024).

2.4 Human Autonomy and Control

Human autonomy is one of the main ethical concepts, which underline the agency, free will, and self-determination. When discussing the issue of autonomy in terms of AI systems, it is more appropriate to assert the degree to which humans remain useful and significant in decision making and action related interventions caused or created by artificial intelligence. Ethical AI design is not aimed at substituting human judgment, but to assist and enhance it, which makes humans stay as the centre of the decision making process (Kamila & Jasrotia, 2025). AI systems are becoming an increasingly influential determinant of human behaviour because of the recommendation, automated decisions, and biased access to information. Whereas such capabilities are felt to increase efficiency and convenience, they also create the risk of manipulation and over dependency. A high reliance on automated products can decrease critical thinking and decrease human decision making to machine made decisions. These dynamics may lead to the loss of the personal autonomy and deterioration of the ability to make own judgments, especially when users blindly follow the AI based recommendations. Also, there are additional ethical issues in cases where people do not know about AI systems that are being used to make decisions or do not have any meaningful choices to pursue other than using AI. Autonomy is challenged in these situations since the user is unable to make knowledgeable decisions regarding their interaction with technology (Prunkl, 2024).

Transparency is thus very important in protecting autonomy. Users should be told when the AI systems are working, the way they operate on the high level, and what outcomes their work may produce. The autonomy of man is directly connected with the principle of informed consent and meaningful human supervision. Human-in-the-loop solutions are intended to make sure that the ultimate decisions are not handed over to the automated system, particularly in high stakes or sensitive systems including healthcare, money, or criminal justice. Ethical AI systems are also based on human accountability and responsibility as opposed to giving priority to complete autonomy of AI systems. The role of the human species should not be to be an inactive recipient of the automated results of the use of AI but to be an active decision-maker who views AI products critically (Fossa, 2025). All in all, in order to guarantee preservation of human autonomy it is necessary to design system intentionally to support human control, transparency, and responsibility in the context of AI aided decision making.

3. SOURCES OF UNCERTAINTY IN GENERATIVE AI SYSTEMS

3.1 Technical Uncertainty in Data and Model Design

Technical uncertainty denotes the constraints, inconsistencies, and uncertainty in the development of AI that occur during the collection and model development and training of the system. The systems of generative AI rely on very large data sets which frequently come in diverse and generally unchallenged data collection. These datasets are consequent of inaccuracies, omission or obsolete data, inappropriate tags, or irregular structure. These problems with quality of data diminish the applicability of training data and cause uncertainty in model results. Due to the magnitude and intricacy of contemporary data sets, they can prove to be quite challenging to identify, and even the slightest defect may greatly affect the actions of models when they are learning (Raglin, 2024). The systematic bias of generative AI production is also caused by uncertainty of data. Models can use the training data to generate content that is similar to the imbalances in some groups, perspectives, and topics. It may result in biased or incomplete representations, imbued in complicated data paths and not immediately apparent on system diagnosis. The generative AI systems being scalable further make it harder to determine and remedy such latent biases when making output more uncertain about the reasonableness and credibility of such outputs (Akerkar, 2026).

The model architecture and design decisions also have some level of uncertainty as well. Generative AI models are usually implemented in probabilistic modes where they produce an output destination dependent on the probability of occurrence as opposed to deterministic algorithms. This results in the same inputs giving different outputs with multiple system runs or following system updates. On the one hand, such variability allows being creative and flexible, but, on the other hand, it complicates the consistent evaluation and performance guarantees. Such changes as model fine tuning, updates to the training data, or a change in optimization strategies may only exacerbate uncertainty over time (Tomczak, 2024). The other important source of technical uncertainty is the difficulty of generalization. Generative AI systems are trained using past information, though they frequently apply in active and new situations. Models should not give a reliable or accurate result when models are faced with unfamiliar topics, the presence of vague prompts, or when a situation is in a rapid change. It is troubling especially in high stakes areas like healthcare, law and public policy where wrong decisions can have severe repercussions. These environments involve direct challenges of the assumption related to accuracy, reliability, and safety due to technical uncertainty (Uddin et al., 2025).

3.2 Social and Contextual Uncertainty

The combination of social and contextual uncertainty is the result of interaction between the generative AI system and social environment where the system is applied. There is no working autonomous system of AI. Their impacts are framed by cultural practices, institutional practices and human behaviours. Simultaneously, AI systems have the potential to affect such social environments in a manner that cannot be easily predicted or regulated. Owing to such reciprocal relationship, the social effect of generative AI remains unclear in many cases (Parker et al., 2024). User interpretation is one of the key causes of social uncertainty. Various users may interpret the same AI generated content in a different way. These differences are based on individual background, education, cultural values and expectations. Consequently, the outputs of AI seem to be neutral or harmless in one setting and lead to confusion, abuse, or injury in a different setting. Certain users can trust the information generated by AI excessively and suppose that it is objective or authoritative. Such blind faith causes making of bad decisions, misinformation, or unsuitable dependence on automated systems (Nordström, 2022).

Communication patterns and norms can also be manipulated by generative AI systems. The incessant exposure to AI generated texts, images, or suggestions can potentially influence the way individuals use language, demonstrate creativity or make a decision. The effects are usually of gradual built up and hard to quantify. Since they are coming out as time goes by, the same may be ignored until it is already everywhere. There is further social uncertainty when implementing the generative AI system on scale. Even minor design decisions, like framing and prioritization of outputs, might generate significant and unintended outcomes in the society (Solaiman et al., 2023). Another source of uncertainty is institutional context. Without knowing all their limitations, organizations can use generative AI systems. Lack of time, financial benefits, or a lack of training may cause the use of automated outputs to be inadequate. Besides this, variations in regulations, governance models, and cultural perception of different areas influence the use and perception of AI systems. Such differences make it hard to predict the consistent outcomes in various social environments (Birkstedt et al., 2023). The ethical aspect is indicated by social and contextual uncertainty and the inability to predict long term and downstream effects of AI systems. Generative AI technologies cannot be entirely controlled by users with regard to how they can adapt, modify, or misuse them. Such uncertainty demands continuity, stakeholder involvement, and adaptive forms of governance technologies to discover and act on arising risks as they arise (Al-Emran et al., 2025).

3.3 Epistemic Uncertainty and Knowledge Limitations

Epistemic uncertainty is associated with restrictions in knowledge and understanding and justification of AI generated products. The output of generative AI systems seems both coherent and confident despite the lack of information or misinformation supporting the content. This poses ethical dilemmas of truth, reliability and authority (Cuzzolin & Sultana, 2024). Epistemic uncertainty through the absence of immediate basis in well-established knowledge is one of the sources of uncertainty. Generative AI models are trained to learn statistical trends, but not factual knowledge. They lack enlightenment and understanding of meaning. Consequently, they can come up with realistic yet incorrect information. In the case of the content being fluent, users might be unable to differentiate the correct and incorrect outputs. The other epistemic issue is explainability. Generative AI systems can be opaque models, and they are not very interpretable. The user and researchers might not comprehend clearly why a particular output was made. This ignorance decreases the trust in the system and makes it difficult to consider the accuracy or bias. It restricts the power to refine models on the basis of explicit causal cognition as well (Cruz-Aguilar, 2025).

Responsibility and trust are also subject to epistemic uncertainty. In decision making, using AI generated content it is not clear what weight is supposed to be assigned to automated outputs. Human judgment can be compromised by over-reliance and potential benefits can be curtailed by under reliance. These risks have to be equalised by paying special attention to the epistemic limits of the system (Sahebi & Formosa, 2025). Epistemic uncertainty is further aggravated by the fast changing nature of generative AI. Models and capabilities are created at a rate that allows no thorough assessment to take place. There is little ethical knowledge regarding the long term impacts, emerging behaviours and cumulative effects. The difference between technological advancement and ethical literacy is clear, which is why it is necessary to be careful in use and to learn every day. Combined, the technical, social, and epistemic uncertainties will explain why generative AI is especially problematic in terms of ethics. These uncertainties are not the transient shortcomings but the structural characteristics of complicated AI systems. To ensure responsible innovation and wise governance, it is important to identify and take care of them (Alvarado, 2023).

4. BIAS, FAIRNESS, AND SOCIAL IMPACT

4.1 Algorithmic Bias in Generative AI

The algorithmic bias is described as the disadvantageous and systematic behavior of AI systems to pursue specific groups, views, or results instead of others. In generative AI, prejudice is most frequently caused by the training data with which the models are built. Big data is a mirror of the actual social order, culture, and the inequalities of the past. Learning on this data, generative AI systems can recreate biased patterns in the text, images, or other pieces, which the systems produce (Zhou et al., 2024). Training data bias: It happens when certain groups, languages or perspectives are over-represented whereas others are under-represented or absent. To illustrate the example, materials that belong to the prevalent cultures or are widely used languages are more plentiful over the web and will thus tend to be found in the training datasets. Consequently, AI systems have the potential to be more effective when it comes to these populations and discriminate less represented people. These imbalances may instead of diminishing social inequalities strengthen them (Jain & Jain, 2024).

The design and development choices may also result in algorithmic bias. Decisions on the selection of data, labeling, model structure and optimization objectives affect the behavior of models. Considerations of fairness can be hampered when effectiveness or business interests have the upper hand. Furthermore, the diversity of data can also be limited by technical limitations and the lack of computational resources (Vartiainen et al., 2025). Generative AI systems have multiple implicit and unintentional bias. Users are not likely to doubt biased outputs because the outputs created by an AI seem neutral or authoritative. To mitigate the adverse effects of algorithmic bias, it is necessary to monitor it all the time, consider hiring multidisciplinary development teams, and embrace ethical introspection to minimize its negative effects (Banh & Strobel, 2023; Feuerriegel et al., 2024).

4.2 Representational Harms and Cultural Implications

Representational harm takes place when AI systems portray individuals or societal groups in false, stereotypical, or demeaning manners. Generative AI systems, especially, are shaping the meaning of representation since the systems generate narratives, images, and descriptions that add to the social sense and cognition. The distortion and the bias of these representations may harm human dignity, identity formation, and social inclusion (Denia, 2025). Since the effects of generative AI can be subtle but expansive because they tend to seem authoritative or neutral. Representational harm can occur in text based generative systems when the language used attributes some negative characteristic, role, or behaviour to specific social groups. These connections have the power to strengthen negative stereotypes and legitimize discriminatory beliefs. Likewise, in the generation of images, stereotypical looks or roles attributed to particular groups could be recreated by use of recurrent visual patterns. In the long term, these types of outputs may lead to social stigmatization and support the existing power structure. Although such representations are not aimed to convey a particular comment to a certain individual, the overall impact of such representation can influence the collective attitude and reinforce inequality (Choi & Zu, 2025).

The issue of cultural context has a vital role in representational context. Most generative AI algorithms are trained using datasets that are collected worldwide with preference to major languages and cultures. Due to this fact, such systems might not be able to detect local meanings, values, and traditions. This drawback might result in cultural oversimplification or misrepresentation. Cultures of minorities and indigenous knowledge systems are especially at risk because the minority views might be underrepresented or misconstrued, or even not recorded at all in the training data. This may lead to cultural erasure and distortion, which may have a negative impact on cultural diversity and social recognition (Karell et al., 2025). The harms of representation cannot be easily quantified since they are subjective and have a social interpretation. This is because what is viewed as being neutral or acceptable by one group can be harmful or offensive in another group. This complication renders the evaluation of ethics difficult and situational. The solution to representational harm is thus the involvement with the communities that have been affected and interdisciplinary methods that involve technical, social, and cultural knowledge. The creators and programmers need to deeply consider whose opinion goes to AI systems and whose are omitted. Generative AI design has ethical responsibility that includes incessant evaluation of the representational influence, and a dedication to inclusive and respectful representation (Pires Ting et al., 2024).

4.3 Unequal Social Impacts and Power Asymmetries

Generative AI does not have an equal social impact. There are unequal impacts on benefits and risks by different individuals and communities. Such inequalities are determined by access to technology, digital literacy, economic resources and social power. When AI systems enhance existing disparities, the issue of ethics emerges (Capraro et al., 2024). Largely, marginalized populations tend to be the most at risk due to biased or wrongful AI results. The lack of resources can diminish their possibility to oppose the negative representations or choices. Meanwhile, such communities can be a disadvantage of AI innovation because of the absence of infrastructure or participation in the design processes. There are power asymmetries between the developers and the users of the technology as well. Big organizations dictate data, models and deployment. Users might know little or have no control over the impact of the generation AI systems on them. The result of such an imbalance is the minimization of accountability and restricting meaningful consent (Danaher, 2024). Labour, education, and creative sectors can also be impacted by the generative AI. Automation can oust some classes of work and open up new prospects of employment. Without a careful consideration of these changes, they can increase economic inequality. The long term social consequences and distributional justice should be discussed in terms of ethical analysis. The inclusion and participation is necessary to address the imbalanced impacts. The collaboration between policymakers, developers and civil society is needed to make sure that generative AI can be used to serve the interests of society in general. The ethical systems must be focused on the social responsibility and the overall welfare. Ethical development of AI can be in better position to guarantee fairness and social justice by understanding power dynamics and social context (Shafik, 2024).

5. TRANSPARENCY, EXPLAINABILITY, AND ACCOUNTABILITY

5.1 Challenges of Transparency in Generative AI Systems

Transparency is defined as the openness of AI systems, how the systems are, how they perform and what data they consume. Transparency is especially difficult to achieve in the case of generative AI. These systems are extremely complicated and they are trained on huge datasets. Its size and technical complexity ensure that users, regulators and even professionals cannot entirely know their working mechanisms. Consequently, the full transparency is not possible in reality in most cases (Uddagiri & Isunuri, 2024). Training data is one of the greatest obstacles to transparency. Studies Generative AI systems are usually trained using large and diverse datasets that are neither fully documented. These data collections can consist of publicly accessible information, licensed assets, and user generated information. The sources of the data are not clear in most instances. Such ambiguity leads to such ethical issues as consent, copyright, and fair representation. People, the data of whom have been utilized, might not know that their content has been used to train the AI systems. This fact may negatively impact trust and casts doubts on the ownership of data and its responsibility (Al-Kfairy et al., 2024).

Model opacity is another issue of major concern. Generative AI systems are based on deep neural networks and their millions or even billions of parameters. These inner architectures cannot be analyzed or described in a way that is comprehensible to humans. This causes the fact that it is not always clear how given inputs result in given outputs. This unaccountability cannot help audit the behaviour of a system, determine the causes of error, or find bias. The insufficient interpretability also makes the quality of holding accountable more difficult in the event that dangerous or wrong outputs have been generated (Franzoni, 2023). Privacy is also curtailed by the considerations of commerce and security. To defend intellectual property or to curb the abuse of a system, organizations can restrict the information sharing. Although these are legitimate concerns, decreased transparency may undermine confidence and conscious amenity. Users might not know what the systems which they work with can and cannot do and this is likely to result in the risk of abuse or overdependence. Ethical transparency should then have a moderate treatment. Complete technical disclosure does not necessarily occur, however, there has to be effective communication. Mutually agreeable information should be made accessible to the developers regarding the capabilities and limitations of the system as well as the risks that become known. This transparency may help people have informed utilization, develop trust, and ensure responsible use of generative AI systems (Taeihagh, 2025).

5.2 Explainability of Generative Outputs

Explainability is a concept that describes the possibility to know and communicate the way AI systems generate their output. Explainability is particularly problematic in generative AI since such systems are not making decisions in any conventional sense. Instead, they are created through probabilistic a algorithm that is based on learned statistical patterns as opposed to explicit rules or logic. Consequently, in most cases, it is not evident why a certain output was produced rather than a great number of other alternative outcomes (Uddin, 2025). The absence of clear causal reasoning is also one big problem of explainability. Generative AI systems do not have transparent lines of reasoning which can be easily traced or justified. They produce text, images or other outputs, which are calculated according to complicated inside presentations that show associations among the exploration data. This complicates the process of explaining one sentence, image or reply that comes out at the end of output. Lack of fit understanding of causal explanations hinders the users and developers ability to completely comprehend system behaviour. Explainability is also a factor of specifications of various audiences. Technical descriptions can be relevant to the developers or system designers but can be too complex to the end users. Majority of the users need a straight forward and practical explanation that would enable them know the reliability of an output, the degree of uncertainty and how to utilize the output in a proper manner. In instances where explanations have not been given, the users will misinterpret the AI generated content or trust too much its accuracy and objectiveness (Schneider, 2024).

Lack of explainability is a big ethical issue where high stakes are involved. Whenever generative AI engines are deployed in aiding the decision making process in the field of healthcare, law, or education, ambiguity in their outputs may have an impact on rights, opportunities, and well-being. Where users are not able to comprehend or query AI generated content, it is hard to refute errors or prejudice. Such ambiguity may undermine procedural fairness and accountability in the decision making (Goktas, 2024). Ethical methods of explainability pay attention to the creation of meaningful and useful information as opposed to complete technical disclosure. This involves the communication of the system aim, the boundaries, as well as the degree of trust that its products are of. Even in a case of inadequate explanation, design strategies can aid in responsible use. These measures are labeling content generated by AI, stating uncertainty, and providing direction to the user. These actions may assist the user to understand the results in the right way and eliminate the possibilities of abuse (ZadZiabari & Tabatabaei, 2025).

5.3 Accountability across the AI Lifecycle

Accountability is a process of responsibility in terms of creating AI systems, deploying and influencing them. Accountability in generative AI is complicated by the fact that there are numerous actors who are responsible. These are data providers, model developers, platform operators and end users (Elia et al., 2025). Diffusion of responsibility is one of the accountability issues. In the event that the harmful outputs take place it might not be evident who is to blame. Developers can claim it is the users who are abusing the system and the users may use the authority of the system. Such uncertainties undermine ethical and legal responsibility. The issue of accountability should be taken throughout the AI lifecycle. Data quality, bias mitigation and safety are the areas of responsibility of the developers during design and training. Organizations need to have a check of system performance and risk management during deployment. Feedback, correction and redress mechanisms are necessary after deployment (Novelli et al., 2024). Accountability is achieved through documentation and governance tools. Now, model cards, data statements and impact assessments aid in understanding the intent and limitations of systems. Supervision and constant improvement is facilitated by auditing and evaluation processes. Well-defined reporting lines will enable users to hold concerns and demand solutions (Cheong, 2024). Accountability is also influenced by legal and regulatory systems. The regulations may establish a set of responsibility, norms, and accountability regarding the use of AI. Nevertheless, regulation should keep up with generative AI that is in a state of change. Ethical responsibility is not just complying with the law but it focuses on social responsibility (Janssen, 2025). With increased transparency, explainability, and accountability, the ethical risks of generative AI can be more effectively managed by the stakeholders. These principles also assist in ensuring human values, supervision, and accountability of AI systems throughout their lifecycle.

6. RESPONSIBLE INNOVATION AND GOVERNANCE FRAMEWORKS

6.1 Ethical Design Strategies for Generative AI

Ethical design plans are concerned with implementing ethical values directly into generative AI system development. This method appreciates the fact that ethics must not be a peripheral issue. Rather, ethics issues should be incorporated throughout the process of design and development. The designers have been advised to foresee the possible harms early and mitigate ethical risks prior to the widespread deployment of systems (Chellakkannu, 2024). Value sensitive design is one of the useful ethical design approaches. The proposed strategy will focus on determining the topical human values of fairness, privacy, accountability, and safety and converting them into specific technical requirements. To illustrate, inclusiveness can be used in making decisions regarding data selection and data representation. The decision can be made in the design of choices that will include different groups and viewpoints in training data. Besides that, model constraints, content moderation, and safety filters can be used to minimize harmful or unsafe outputs (Hermann & Puntoni, 2025).

One more potential approach is the risk based design. It is a method that considers the generative AI systems based on the extent of harm they can do. There are applications that are more prone to ethical or social risks that should have more rigorous protection, increased testing, and increased attention. Risk based design aids in the allocation of resources and focus in areas where damage may happen the most. It also helps in responsible deployment since high risk systems are given proper oversight. Constant testing and iteration is also part of ethical design. The generative AI systems must be put into test on different situations and users so that undesired results or unexpected cases can be detected. Feedback mechanisms also give developers an opportunity to learn, through the real world, and subsequently enhance system performance with time. Ethical design acknowledges that no system is ideal but focuses on constant enhancement as opposed to solution based once only answers (Kim et al., 2025). It is important to note that interdisciplinary cooperation is a key component of the design of ethics. The contributions of ethicists, social scientists, legal experts, and affected communities can be able to expose ethical risk areas that cannot be identified by the technical teams only. These views assist in making sure that AI systems generated are powerful and socially accountable and ready to conform to human ideals. The cooperation between disciplines enhances the quality and validity of AI systems in terms of ethics (Farina et al., 2025).

6.2 Human Oversight and Participatory Approaches

One of the major components of responsible innovation in generative AI is human oversight. Monitoring will hold human beings accountable to make decisions and results by AI systems. This principle will address the danger of being over automated and losing control over humans (Sedkaoui & Benaichouba, 2024). Oversight is usually ensured through human in the loop. The products of AI are checked, checked, or modified by humans in these systems before they can be used. This is particularly crucial in sensitive areas like healthcare, education and in the administration of the populace. Human control can be used to identify mistakes, biases and inappropriate material. Other than individual management, participatory methods focus on the presence of wider stakeholders. The users, the communities it affects and the civil society organizations will be able to offer valuable views of the system impact. The involvement can assist in the identification of social dangers and aid in legitimacy and trust (Gunasekara et al., 2025). Some of the participatory design techniques are consultations, user testing and public engagement programs. These approaches enable the various voices to shape system objectives and limitations. They also assist in aligning the AI development with the values and expectations of society. Effective oversight requires both training and education. The users and professionals should be aware of the abilities and the limitations of generative AI. As long as there is no proper knowledge, the oversight is only a symbol but not a meaningful one. Responsible innovation hence involves investment in digital literacy and ethical awareness (Bodini, 2024).

6.3 Regulatory and Policy Approaches

The regulatory frameworks are vital towards the regulation of generative AI. Regulation gives official guidelines and norms that stipulate what good practice is, and what safeguards the interests of the people. Nevertheless, it is difficult to regulate generative AI because of a rapid change in technology and its global implementation (Huang et al., 2024). Risk based regulation is also one of the regulatory approaches. This model categorizes AI systems based on their possible impact and places corresponding demands. Low risk apps can be given limited attention whereas the high risk systems have strict compliance. This is a type that balances innovation and protection. The other strategy is to establish transparency and accountability requirements. Laws can necessitate the documentation of training data, purpose of the system and known limitations. Oversight and enforcement are facilitated with reporting and audit requirements (Taeihagh, 2025). Transnational coordination is becoming more significant. Generative AI systems tend to be cross border and legal structures tend to be national. The inconsistency in regulation may produce gaps or discrepancies. The global standards and collaborative systems assist in the advancement of common morality (Ferrari et al., 2025). Soft law tools like guidelines, codes of conduct and best practices are also policy approaches. Such tools are flexible and are able to adjust with the rapid change in technology. They have no legal binding but affect the organizational behaviour and uphold ethical standards. The effective regulation depends on the constant communication between policymakers, industry, researchers and the society. The existence of static regulations can be obsolete and therefore responsiveness and adaptive governance is necessary (Cheong et al., 2025).

6.4 Ethical Governance Models and Institutional Responsibility

Models of ethical governance offer organizationally and systematically structured methods of coordinating responsibility. These models establish the roles, processes, and mechanisms of decision making about ethical risks in generative AI. Organizational ethics governance is one of the models. This involves ethics committees, review boards and internal guidelines governing AI projects. These frameworks guarantee ethical thinking and responsibility in organizations. They also offer pre-deployment forums to resolve ethical issues (Dabis & Csáki, 2024). The other model is focused on lifecycle governance. Ethical responsibility is implemented at all the levels of AI development, such as collecting data, or post-deployment surveillance. This method acknowledges the fact that ethical risks change as time goes by and should be addressed on a continuous basis. Transparency, public accountability, and democratic oversight are the topics of the models of the public sector governance. Governments have a role in developing standards, supporting research and securing societal values. Innovation and the trust of the people in the government should be reconciled through ethical governance of the government (Raza et al., 2025).

The models of collaborative governance include partnerships between the state institutions, private organizations and civil society. These models understand that the ethical issues of generative AI cannot be handled by only one actor. Collaborative accountability promotes more thorough and all-embracing governance (Ali & Aysan, 2025). Redress and learning also need to be in place in ethical governance. Harms should act and learn after failures when the harms take place and when the institutions are involved. This helps in the continuous improvement and commitment to ethics. Generative AI should be governed by responsible innovation and governance frameworks that offer necessary recommendations. The strategies of ethical design, human regulation, and institutional governance can help societies manage uncertainty better and encourage ethical and sustainable AI development (Ghose et al., 2024).

7. IMPLICATIONS

The generative AI ethical analysis discussed in this chapter has a number of notable implications in the research, practice, and policy. To researchers, the findings indicate that they need to consider ethical thinking in technical development. Ethical issues like bias, uncertainty, and explainability ought to be regarded as research priorities but not as secondary concerns. This means that there is a transition toward interdisciplinary studies in the form of integration of computer science and social sciences, ethics, and law. Researchers are urged to find ways of evaluating performance, as well as social impact and risk.

Practitioners and developers may find the implications to be the focus on responsibility throughout the AI lifecycle. The ethical implications of design choices are connected to the choice of data, model structure, and the deployment environment. Companies need to invest in ethical design, what is continually monitored and human monitored. There is also the need to have training programs that enhance ethical awareness and digital literacy of developers and users. These measures have the capacity to minimize damage, enhance responsibility, and create popular confidence in the generative AI systems.

The analysis highlights the need of risk based and adaptive governance structures to policy makers and regulators. The constantly changing nature of generative AI might not be approached with the help of static regulations. The policies must encourage transparency and accountability as well as safeguard basic rights without restraining innovation. There is a particular significance of international cooperation because of the global environment of the implementation of generative AI technologies.

At the societal level the implication is on equity and inclusion. Depending on its governance, generative AI may strengthen or cause social good, or create social harm. There should be inclusive participation, opening up and ethical governance hence. On the whole, the suggestions are that there should be a sense of collective responsibility and long term ethical engagement to make sure that generative AI is used in ways yielding beneficial and sustainable developments in innovation and sustainability.

8. CONCLUSION, LIMITATIONS, AND FUTURE SCOPE

8.1 Conclusion

The role of generative AI in social, economic, and institutional practices is increasing, and there are major ethical concerns. Some of the discussed key ethical issues in this chapter are uncertainty, bias, transparency, accountability, and governance. The generative AI systems are driven by the uncertainty about the technical, social, and epistemic aspects, rendering their behaviour unpredictable and challenging to control to some extent. These ambiguities demand some ethical deliberations when designing and using such systems. Main ethical values, such as fairness, human autonomy and responsibility, are required to lead to responsible development and use. The evaluation points out that discrimination and disproportionate social effects are still a looming danger especially to vulnerable and underrepresented groups. Generative AI systems can strengthen the status quo of inequalities and social evils without conscious effort. Explainability and transparency are also important to establish trust and facilitate forces capable of using it, and accountability to make sure that the consequences of AI driven changes are under the control of human participants and organizations. Governance systems and responsible innovation offer viable ways of dealing with these ethical dilemmas. Scientists, designers, and policymakers must incorporate ethical principles into the design of systems, regulation systems, and oversight to develop a form of generative AI that offers social value, increases trust, and encourages technological advancement that is sustainable and safe.

8.2 Limitations

There are a number of limitations in this chapter. One, the discussion is rather conceptual and lacks empirical studies of cases and quantitative analysis of generative AI systems. Second, the ethical issues are considered on a general level, which is not necessarily able to reflect the domain risks in such fields as healthcare, law, or education. Third, due to the rapid development of generative AI, certain governance and ethical issues can alter with time. Lastly, the ethical values and regulatory ability differ across cultures and regions are not discussed in detail, which can potentially restrict the generalizability of the analysis.

8.3 Future Scope

Future studies are supposed to concentrate on empirical researchers that discuss actual effects of generative AI in various fields and populations. More focus should be placed on assessing the bias, social harm and scale effects. Adaptive and international governance frameworks to respond to the rapid technological change should be given priority in the policies. It should also focus on participatory and inclusive design in the future that engages different stakeholders. The further development of ethical tools, standards, and evaluation techniques will facilitate sustainable innovation directions and make sure that generative AI will develop based on human values and social accountability.

REFERENCES

Akerkar, R. (2026). Blueprints for Intelligence: Building Robust AI Systems. In *Artificial Intelligence: Transcending Traditional Paradigms* (pp. 447-491). Cham: Springer Nature Switzerland. DOI: 10.1007/978-3-031-91084-5_12

Al-Emran, M., Abu-Hijleh, B., & Alsewari, A. A. (2025). Examining the impact of Generative AI on social sustainability by integrating the information system success model and technology-environmental, economic, and social sustainability theory. *Education and Information Technologies*, *30*(7), 9405–9426. DOI: 10.1007/s10639-024-13201-0

Al-Kfairy, M., Mustafa, D., Kshetri, N., Insiew, M., & Alfandi, O. (2024, September). Ethical challenges and solutions of generative AI: An interdisciplinary perspective. In *Informatics* (Vol. 11, No. 3, p. 58). Multidisciplinary Digital Publishing Institute. DOI: 10.3390/informatics11030058

Ali, A. (2025). Ethical Considerations and Bias in AI Applications in Healthcare. In *Artificial Intelligence in Healthcare Information Systems—Security and Privacy Challenges* (pp. 129–146). Springer Nature Switzerland., DOI: 10.1007/978-3-031-84404-1_7

Ali, H., & Aysan, A. F. (2025). Ethical dimensions of generative AI: A cross-domain analysis using machine learning structural topic modeling. *International Journal of Ethics and Systems*, *41*(1), 3–34. DOI: 10.1108/IJOES-04-2024-0112

Alvarado, R. (2023). AI as an epistemic technology. *Science and Engineering Ethics*, *29*(5), 32. DOI: 10.1007/s11948-023-00451-3 PMID: 37603120

Alvarez, J. M., Colmenarejo, A. B., Elobaid, A., Fabbrizzi, S., Fahimi, M., Ferrara, A., Ghodsi, S., Mougan, C., Papageorgiou, I., Reyero, P., Russo, M., Scott, K. M., State, L., Zhao, X., & Ruggieri, S. (2024). Policy advice and best practices on bias and fairness in AI. *Ethics and Information Technology*, *26*(2), 31. DOI: 10.1007/s10676-024-09746-w

Andrada, G., Clowes, R. W., & Smart, P. R. (2023). Varieties of transparency: Exploring agency within AI systems. *AI & Society*, *38*(4), 1321–1331. DOI: 10.1007/s00146-021-01326-6 PMID: 35035112

Balasubramaniam, N., Kauppinen, M., Hiekkanen, K., & Kujala, S. (2022, March). Transparency and explainability of AI systems: ethical guidelines in practice. In *International working conference on requirements engineering: foundation for software quality* (pp. 3-18). Cham: Springer International Publishing. DOI: 10.1007/978-3-030-98464-9_1

Balasubramaniam, N., Kauppinen, M., Rannisto, A., Hiekkanen, K., & Kujala, S. (2023). Transparency and explainability of AI systems: From ethical guidelines to requirements. *Information and Software Technology*, *159*, 107197. DOI: 10.1016/j.infsof.2023.107197

Banh, L., & Strobel, G. (2023). Generative artificial intelligence. *Electronic Markets*, *33*(1), 63. DOI: 10.1007/s12525-023-00680-1

Baumann, J., & Loi, M. (2023). Fairness and risk: An ethical argument for a group fairness definition insurers can use. *Philosophy & Technology*, *36*(3), 45. DOI: 10.1007/s13347-023-00624-9 PMID: 37346393

Birkstedt, T., Minkkinen, M., Tandon, A., & Mäntymäki, M. (2023). AI governance: Themes, knowledge gaps and future agendas. *Internet Research*, *33*(7), 133–167. DOI: 10.1108/INTR-01-2022-0042

Bodini, M. (2024). Generative Artificial Intelligence and Regulations: Can We Plan a Resilient Journey Toward the Safe Application of Generative Artificial Intelligence? *Societies (Basel, Switzerland)*, *14*(12), 268. DOI: 10.3390/soc14120268

Brindha, R., Pongiannan, R. K., Bharath, A., & Sanjeevi, V. K. S. M. (2025). Introduction to Multimodal Generative AI. In *Multimodal Generative AI* (pp. 1–36). Springer Nature Singapore., DOI: 10.1007/978-981-96-2355-6_1

Camilleri, M. A. (2024). Artificial intelligence governance: Ethical considerations and implications for social responsibility. *Expert Systems: International Journal of Knowledge Engineering and Neural Networks*, *41*(7), e13406. DOI: 10.1111/exsy.13406

Capraro, V., Lentsch, A., Acemoglu, D., Akgun, S., Akhmedova, A., Bilancini, E., Bonnefon, J.-F., Brañas-Garza, P., Butera, L., Douglas, K. M., Everett, J. A. C., Gigerenzer, G., Greenhow, C., Hashimoto, D. A., Holt-Lunstad, J., Jetten, J., Johnson, S., Kunz, W. H., Longoni, C., & Viale, R. (2024). The impact of generative artificial intelligence on socioeconomic inequalities and policy making. *PNAS Nexus*, *3*(6), pgae191. Advance online publication. DOI: 10.1093/pnasnexus/pgae191 PMID: 38864006

Cetinkaya, N. E., & Krämer, N. (2025). Between transparency and trust: Identifying key factors in AI system perception. *Behaviour & Information Technology*, •••, 1–15. DOI: 10.1080/0144929X.2025.2533358

Chellakkannu, S. (2024, April). Embracing generative AI in design: practical implementations. In *International Conference on Artificial Intelligence and its Applications in the Age of Digital Transformation* (pp. 189-206). Cham: Springer Nature Switzerland. DOI: 10.1007/978-3-031-71429-0_15

Chen, P., Wu, L., & Wang, L. (2023). AI fairness in data management and analytics: A review on challenges, methodologies and applications. *Applied Sciences (Basel, Switzerland), 13*(18), 10258. DOI: 10.3390/app131810258

Cheong, B. C. (2024). Transparency and accountability in AI systems: Safeguarding wellbeing in the age of algorithmic decision-making. *Frontiers in Human Dynamics, 6*, 1421273. DOI: 10.3389/fhumd.2024.1421273

Cheong, I., Caliskan, A., & Kohno, T. (2025). Safeguarding human values: Rethinking US law for generative AI's societal impacts. *AI and Ethics*, *5*(2), 1433–1459. DOI: 10.1007/s43681-024-00451-4 PMID: 40352581

Choi, I., & Zu, J. (2025). Generating language assessment content free from representational harms. *Language Testing*, *42*(4), 539–560. DOI: 10.1177/02655322251349560

Cruz-Aguilar, M. A. (2025). The epistemic revolution of AI: Reconfiguring the foundations of scientific knowledge. *AI & Society*, •••, 1–17. DOI: 10.1007/s00146-025-02658-3

Cuzzolin, F., & Sultana, M. (2024). *Epistemic Uncertainty in Artificial Intelligence*. Springer Nature Switzerland., DOI: 10.1007/978-3-031-57963-9

Dabis, A., & Csáki, C. (2024). AI and ethics: Investigating the first policy responses of higher education institutions to the challenge of generative AI. *Humanities & Social Sciences Communications*, *11*(1), 1–13. DOI: 10.1057/s41599-024-03526-z

Danaher, J. (2024). Generative AI and the future of equality norms. *Cognition*, *251*, 105906. DOI: 10.1016/j.cognition.2024.105906 PMID: 39067317

De Cremer, D., Narayanan, D., Nagpal, M., McGuire, J., & Schweitzer, S. (2024). AI Fairness in Action: A human-computer perspective on AI fairness in organizations and society. *International Journal of Human-Computer Interaction*, *40*(1), 1–3. DOI: 10.1080/10447318.2023.2273673

Denia, E. (2025). AI narratives model: Social perception of artificial intelligence. *Technovation*, *146*, 103266. DOI: 10.1016/j.technovation.2025.103266

Efthymiou, I. P. (2026). Ethical Considerations and Regulatory Standards for Adversarial Defense. In Nowroozi, E., Taheri, R., & Cordeiro, L. (Eds.), *Adversarial Example Detection and Mitigation Using Machine Learning*. Springer., DOI: 10.1007/978-3-031-99447-0_2

Elia, M., Ziethmann, P., Krumme, J., Schlögl-Flierl, K., & Bauer, B. (2025). Responsible AI, ethics, and the AI lifecycle: How to consider the human influence? *AI and Ethics*, *5*(4), 1–18. DOI: 10.1007/s43681-025-00666-z

Farina, M., Yu, X., & Lavazza, A. (2025). Ethical considerations and policy interventions concerning the impact of generative AI tools in the economy and in society. *AI and Ethics*, *5*(1), 737–745. DOI: 10.1007/s43681-023-00405-2

Ferrari, F., Van Dijck, J., & Van den Bosch, A. (2025). Observe, inspect, modify: Three conditions for generative AI governance. *New Media & Society*, *27*(5), 2788–2806. DOI: 10.1177/14614448231214811

Feuerriegel, S., Hartmann, J., Janiesch, C., & Zschech, P. (2024). Generative ai. *Business & Information Systems Engineering*, *66*(1), 111–126. DOI: 10.1007/s12599-023-00834-7

Fossa, F. (2025). Artificial intelligence and human autonomy: The case of driving automation. *AI & Society*, *40*(3), 1851–1862. DOI: 10.1007/s00146-024-01955-7

Franzoni, V. (2023, June). From black box to glass box: advancing transparency in artificial intelligence systems for ethical and trustworthy AI. In *International Conference on Computational Science and Its Applications* (pp. 118-130). Cham: Springer Nature Switzerland. DOI: 10.1007/978-3-031-37114-1_9

Ghose, A., Ali, S. A., & Deshmukh, S. (2024). Navigating the Legal and Ethical Framework for Generative AI: Fostering Responsible Global Governance. In *Exploring the Ethical Implications of Generative AI* (pp. 168-184). IGI Global Scientific Publishing. DOI: DOI: 10.4018/979-8-3693-1565-1.ch010

Goktas, P. (2024). Ethics, transparency, and explainability in generative ai decision-making systems: A comprehensive bibliometric study. *Journal of Decision Systems*, •••, 1–29. DOI: 10.1080/12460125.2024.2410042

Gunasekara, L., El-Haber, N., Nagpal, S., Moraliyage, H., Issadeen, Z., Manic, M., & De Silva, D. (2025). A systematic review of responsible artificial intelligence principles and practice. *Applied System Innovation*, *8*(4), 97. DOI: 10.3390/asi8040097

Hermann, E., & Puntoni, S. (2025). Generative AI in marketing and principles for ethical design and deployment. *Journal of Public Policy & Marketing*, *44*(3), 332–349. DOI: 10.1177/07439156241309874

Huang, K., Joshi, A., Dun, S., & Hamilton, N. (2024). AI regulations. In *Generative AI security: theories and practices* (pp. 61–98). Springer Nature Switzerland., DOI: 10.1007/978-3-031-54252-7_3

Jain, R., & Jain, A. (2024, July). Generative AI in writing research papers: A new type of algorithmic bias and uncertainty in scholarly work. In *Intelligent Systems Conference* (pp. 656-669). Cham: Springer Nature Switzerland. DOI: 10.1007/978-3-031-66329-1_42

Janssen, M. (2025). Responsible governance of generative AI: Conceptualizing GenAI as complex adaptive systems. *Policy and Society*, *44*(1), 38–51. DOI: 10.1093/polsoc/puae040

Kamila, M. K., & Jasrotia, S. S. (2025). Ethical issues in the development of artificial intelligence: Recognizing the risks. *International Journal of Ethics and Systems*, *41*(1), 45–63. DOI: 10.1108/IJOES-05-2023-0107

Karell, D., Sachs, J., & Barrett, R. (2025). Synthetic duality: A framework for analyzing generative artificial intelligence's representation of social reality. *Poetics*, *108*, 101966. DOI: 10.1016/j.poetic.2024.101966

Kim, J. J., Um, R. S., Lee, J. W., & Ajilore, O. (2025). Generative AI can fabricate advanced scientific visualizations: Ethical implications and strategic mitigation framework. *AI and Ethics*, *5*(5), 4481–4493. DOI: 10.1007/s43681-024-00439-0

Kouroupis, K., & Efthymiou, I. (Eds.). (2026). *Legal Challenges of AI Across Interdisciplinary Sectors*. IGI Global Scientific Publishing., DOI: 10.4018/979-8-3373-5452-1

Lechterman, T. M. (2024). *The concept of accountability in AI ethics and governance. The Oxford Handbook of AI Governance*. Oxford University Press.

Liu, W., & Kim, H. G. (2025). The visual communication using generative artificial intelligence in the context of new media. *Scientific Reports*, *15*(1), 11577. DOI: 10.1038/s41598-025-96869-9 PMID: 40185848

Lv, Z. (2023). Generative artificial intelligence in the metaverse era. *Cognitive Robotics*, *3*, 208–217. DOI: 10.1016/j.cogr.2023.06.001

Mallikarjuna, B., & Chittemsetty, P. (2024). Generative artificial intelligence: Fundamentals and evolution. In Generative, A. I. (Ed.), *Current trends and applications* (pp. 3–17). Springer Nature Singapore., DOI: 10.1007/978-981-97-8460-8_1

Nordström, M. (2022). AI under great uncertainty: Implications and decision strategies for public policy. *AI & Society, 37*(4), 1703–1714. DOI: 10.1007/s00146-021-01263-4 PMID: 34511737

Novelli, C., Taddeo, M., & Floridi, L. (2024). Accountability in artificial intelligence: What it is and how it works. *AI & Society, 39*(4), 1871–1882. DOI: 10.1007/s00146-023-01635-y

Parker, L., Hayes, J., Loper, A. J., & Karakas, A. L. (2024). Navigating the unknown: anticipating concerns and gaps in generative AI research. In *General Aspects of Applying Generative AI in Higher Education: Opportunities and Challenges* (pp. 3-20). Cham: Springer Nature Switzerland. DOI: 10.1007/978-3-031-65691-0_1

Pires Ting, C., Filippovna, V., & Araujo, M. M. (2024). AI and the Visualization of Paradise: Cultural Paradigms, Aesthetic Evolution, and Cognitive Exploration. *Logica Universalis, 18*(4), 505–519. DOI: 10.1007/s11787-024-00365-z

Prunkl, C. (2024). Human autonomy at risk? An analysis of the challenges from AI. *Minds and Machines, 34*(3), 26. DOI: 10.1007/s11023-024-09665-1

Raglin, A. (2024, June). Leveraging generative AI concepts with uncertainty of information concept. In *International Conference on Human-Computer Interaction* (pp. 230-238). Cham: Springer Nature Switzerland. DOI: 10.1007/978-3-031-60611-3_17

Raza, S., Qureshi, R., Zahid, A., Kamawal, S., Sadak, F., Fioresi, J., . . . Shoman, M. (2025). Who is responsible? the data, models, users or regulations? a comprehensive survey on responsible generative ai for a sustainable future. *arXiv preprint arXiv:2502.08650*. https://doi.org//arXiv.2502.08650DOI: 10.48550

Routray, S. K., Jha, M. K., Sharmila, K. P., Javali, A., Pappa, M., & Singh, M. (2024, October). Generative Artificial Intelligence: Principles, Potentials and Challenges. In *2024 2nd International Conference on Self Sustainable Artificial Intelligence Systems (ICSSAS)* (pp. 211-216). IEEE. DOI: 10.1109/ICSSAS64001.2024.10760992

Sahebi, S., & Formosa, P. (2025). The AI-mediated communication dilemma: Epistemic trust, social media, and the challenge of generative artificial intelligence. *Synthese, 205*(3), 1–24. DOI: 10.1007/s11229-025-04963-2

Schneider, J. (2024). Explainable Generative AI (GenXAI): A survey, conceptualization, and research agenda. *Artificial Intelligence Review, 57*(11), 289. DOI: 10.1007/s10462-024-10916-x

Sedkaoui, S., & Benaichouba, R. (2024). Generative AI as a transformative force for innovation: A review of opportunities, applications and challenges. *European Journal of Innovation Management*. Advance online publication. DOI: 10.1108/EJIM-02-2024-0129

Sengar, S. S., Hasan, A. B., Kumar, S., & Carroll, F. (2025). Generative artificial intelligence: A systematic review and applications. *Multimedia Tools and Applications*, *84*(21), 23661–23700. DOI: 10.1007/s11042-024-20016-1

Shafik, W. (2024). Generative AI for social good and sustainable development. In *Generative AI: current trends and applications* (pp. 185–217). Springer Nature Singapore., DOI: 10.1007/978-981-97-8460-8_10

Solaiman, I., Talat, Z., Agnew, W., Ahmad, L., Baker, D., Blodgett, S. L., Subramonian, A. (2023). Evaluating the social impact of generative ai systems in systems and society. *arXiv preprint arXiv:2306.05949*. https://doi.org/10.48550/arXiv.2306.05949

Taeihagh, A. (2025). Governance of generative AI. *Policy and Society*, *44*(1), 1–22. DOI: 10.1093/polsoc/puaf001

Tomczak, J. M. (2024). Why deep generative modeling? In *Deep generative modeling* (pp. 1–13). Springer International Publishing., DOI: 10.1007/978-3-031-64087-2_1

Uddagiri, C., & Isunuri, B. V. (2024). Ethical and privacy challenges of generative AI. In *Generative AI: Current Trends and Applications* (pp. 219-244). Singapore: Springer Nature Singapore. DOI: 10.1007/978-981-97-8460-8_11

Uddin, M., Arfeen, S. U., Alanazi, F., Hussain, S., Mazhar, T., & Arafatur Rahman, M. (2025). A Critical Analysis of Generative AI: Challenges, Opportunities, and Future Research Directions. *Archives of Computational Methods in Engineering*, •••, 1–31. DOI: 10.1007/s11831-025-10355-z

Uddin, M. Z. (2025). Trustworthy AI and explainability. In *Trustworthy Multimodal Intelligent Systems for Independent Living* (pp. 111–137). Springer Nature Switzerland., DOI: 10.1007/978-3-031-97359-8_4

Vartiainen, H., Kahila, J., Tedre, M., López-Pernas, S., & Pope, N. (2025). Enhancing children's understanding of algorithmic biases in and with text-to-image generative AI. *New Media & Society*, *27*(9), 5342–5368. DOI: 10.1177/14614448241252820

ZadZiabari. A., & Tabatabaei, A. (2025). Ethics and Regulations in Generative AI. In *Application of Generative AI in Healthcare Systems* (pp. 197-211). Cham: Springer Nature Switzerland. DOI: 10.1007/978-3-031-82963-5_8

Zhou, M., Abhishek, V., Derdenger, T., Kim, J., & Srinivasan, K. (2024). Bias in generative AI. *arXiv preprint arXiv:2403.02726*. https://doi.org//arXiv.2403 .02726DOI: 10.48550

Chapter 4
Behavioral Law and the Governance of Artificial Intelligence:
Bridging Ethics and Regulation

Iris-Panagiota Efthymiou
https://orcid.org/0000-0001-9656-8378
Regent College London, UK

ABSTRACT

This chapter per the authors, examines the role of behavioural law in shaping AI governance, integrating ethical principles with regulatory frameworks. Drawing on behavioural economics, it analyses how cognitive biases and human-AI interactions create ethical challenges in AI deployment. The chapter aims to develop a hybrid model that bridges fairness, transparency, and accountability with enforceable regulations to mitigate risks like algorithmic bias and privacy breaches. Through case studies on data privacy, autonomous systems, and decision-making algorithms, it demonstrates how behavioural tools, nudges, defaults, and transparency mandates enhance regulatory efficacy while promoting innovation. It advocates for empirical research to refine these approaches, proposing adaptive frameworks that incorporate stakeholder input and global standards to foster trust and equity in AI ecosystems. By blending theory and practice, this chapter per the authors, provides policymakers, developers, and scholars with actionable strategies for sustainable AI governance in a technology-driven world.

DOI: 10.4018/979-8-2600-0000-7.ch004

1. INTRODUCTION

Artificial Intelligence (AI) has two meanings. One refers to machines that can perform tasks requiring intelligence, simulating intelligence, and behaving like humans. The second meaning, relevant to many regulatory frameworks, is an intelligent machine that guides human decisions (Kantorowicz-Reznichenko, E., & Kantorowicz, J., 2025). Regulatory frameworks addressing AI must differentiate between these two meanings. AI can control or influence human decision-making at different levels: perception, evaluation, and action or choice (Efthymiou, I. & Egleton, T. E., 2025). One can encounter many types of AI. For instance, Damage Detection Systems predict damages in a construction project or recommend books according to a person's interest, and Intelligent Traffic Systems synchronize traffic lights based on real-time traffic detection. Human decisions in these examples are far from being autonomous. Likewise, many frameworks, including the United Nations Educational, Scientific and Cultural Organization (UNESCO)/OECD (Organisation for Economic Co-operation and Development) AI Act and the European Commission's AI Act, address the second meaning of AI. By guiding human decisions, AI can have a profound impact with ethical aspects on different dimensions, such as social and political. It represents a shift requiring the law to balance more diverse and varied values (Kouroupis, K., 2022).

Guiding human decisions can involve several cognitive phenomena (behaviours, actions, thoughts, or processes) and unique cognitive characteristics. Mental processes of these phenomena can differ from simple act until complex reasoning or problem solving. When legally and technically relevant, however, these mental processes associated with AI governance are often understood through the lens of the Behavioural Law and Economics (BLE) theory (Mitchell, G., 2014; Newman, J. C., 2020).

The need to establish a scientific understanding of any mental process arises to formulate certain regulations, such as guiding AI development according to people's feedback and cognition. BLE proposes several models that encompass different theories, offering distinctive explanations about human behaviour and cognitive decision-making processes (Rachlinski, J. J., 2011). The application of BLE to guidelines, principles, or other existing proposals on technology governance leads to multi-faceted literature focusing specifically on standards, data, testing, and auditability of transparency. The necessity arises to channel the understanding about the interface of AI and the mind through BLE in order to derive from different guiding frameworks pertaining to AI (Ayres, I., t al., 2025; Epstein, R. A., 2025; Ooi, K. L., 2025).

2. THEORETICAL FOUNDATIONS OF BEHAVIOURAL LAW

Behavioural Law is an emerging field at the intersection of Behavioural Science, Law and Governance (Gofen et al., 2021; Bodig and Ződi, 2024; Ewert et al., 2021). It studies how cognitive biases affect human interaction with technology and influence decision-making models. It has much to offer in the context of Artificial Intelligence. As these new technologies penetrate every part of life, the question of how to govern them effectively has grown more urgent. Behavioural Law is well placed to address this challenge. Far from being a hazard to personal freedom, it can bolster autonomy by increasing the predictability of human behaviour and making freedom from unforeseen influences better attainable. Behavioural Law enables the unbundling of the three traditional regulatory objectives of promoting autonomy, safety and efficiency. It also permits a mapping of regulatory responses onto prevailing theoretical approaches, thereby clarifying the options and making their analysis more systematic.

The age of rapid digital transformation has brought serious attention to how people interact with artificial intelligence. It has led to an increasingly urgent discourse on whether and how Artificial Intelligence should be regulated. There are many unanswered questions that need to be assessed, such as how to define Artificial Intelligence, how to gauge the risk or harm associated with an algorithm, how to map the nature and frequency of societal exposure to algorithms, how to evaluate their impact, and how to develop an actionable governance framework. To make sense of the existing literature on Artificial Intelligence regulation, it is useful to map the variety of approaches that are emerging (Smuha, N. A., 2025). Within this emerging literature, some theoretical perspectives recur that can usefully be summarised.

Cognitive Psychology explores how cognitive processes influence human perception and behaviour (Wallsten, 2024; Ke et al., 2025). The Heuristics and Biases literature identifies systematic tendencies to rely on simplifying rules that lead to predictable deviations from normative preferences (Summerfield & Parpart, 2022; Hunt et al., 2024; Ayaa et al.2022; Moseley & Thomann, 2021). Behavioural Economics focuses on anomalous phenomena in commercial contexts that cannot be accounted for by traditional expected-utility theory and investigates implications for economic modelling (Challoumis, 2024; Umapathy T., 2024). Normative or Descriptive Social Norms explore how individuals are influenced by information about the actions of others, and how norms change over time in response to the actions of opinion leaders, including through cascading effects. Each of these perspectives has systematic implications for technological design and governance, and each is particularly relevant in the case of Artificial Intelligence (Gervais, 2023).

3. BEHAVIOURAL INSIGHTS IN ARTIFICIAL INTELLIGENCE GOVERNANCE

Artificial intelligence (AI) enhances human decision-making capacity but also challenges regulatory frameworks. The distinction between AI systems generating decisions and artefacts accompanying human decision processes constitutes a significant conceptual hurdle (Herath Pathirannehelage, S., et al., 2025; Bao Y., et al., 2023; Garbuio M. & Lin N., 2021). The potential regulatory tension lies within AI-generated recommendations that may relieve human users of accountability for the decision made. Information overload and cognitive biases have long challenged humans in deep, complex, or substantial decision-making. These phenomena become further complicated with the augmented AI dimension. The proliferation of information tends to facilitate decision-making but also has the opposite effect, particularly with complex technical data. Humans still perform better than AI in complex narrative decision-making domains, where the broader context is essential. AI systems operating in these domains risk exacerbating the situation, complicating their governance, particularly regarding liability and responsibility when deployed across different regulations.

Literature Review

Behavioural Law and the Governance of Artificial Intelligence: Bridging Ethics and Regulation

The convergence of behavioural law and artificial intelligence governance is becoming increasingly vital in today's rapidly evolving technological landscape. As artificial intelligence systems significantly affect societal behaviours and decision-making processes, their impact cannot be underestimated. Recent studies underscore the necessity of aligning ethical principles with enforceable regulations to ensure the responsible, transparent, and trustworthy implementation of AI technologies across various sectors, including healthcare, finance, and education. This integration calls for a collaborative effort among policymakers, tech developers, and ethicists to navigate the complexities of AI's influence on our daily lives.

Key Challenges in AI Governance

- **Privacy and Behavioural Risks:** The utilization of artificial intelligence by government entities for the prediction of human behaviours, as well as the automation of various decision-making processes, presents not only considerable but also significant privacy issues that warrant serious attention

(Efthymiou, I. P., 2025). Alongside these concerns is the potential for widespread modification of societal behaviours, which could significantly alter the very fabric and dynamics of our communities. Therefore, in order to ensure effective and responsible governance in this evolving landscape, it is absolutely essential to implement comprehensive regulations that strike a careful balance between fostering innovation and safeguarding the inherent rights and privacy of citizens. This balancing act becomes particularly crucial within the contexts of extensive surveillance measures and increasingly complex data-driven decision processes, which impact lives on multiple personal and societal levels. It is imperative that these regulations do not stifle progress but instead guide the ethical use of technology, ensuring that citizens maintain control over their personal data while benefiting from advancements that can improve public services and overall quality of life (Saura J., et al., 2022; Zhang J. & Zhang Z., 2023; Rassolov I. & Chubukova S., 2022).

- **Ethics vs. Law:** Ethical guidelines often demonstrate a notable lack of enforceability, rendering them ineffective in numerous situations where they are intended to apply. Meanwhile, legal frameworks frequently struggle to adapt and keep pace with the rapid technological advancements we witness today, leading to a significant mismatch between necessary regulation and swift innovation. Therefore, a comprehensive and synergistic approach becomes essential; ethical principles must actively and assertively guide the creation of regulations that are not only flexible but also binding in nature. Such regulations should be specifically designed to thoroughly address and confront the myriad of risks inherent in the development and deployment of artificial intelligence technologies that we are seeing more and more frequently. This integration of ethics and law is crucial for fostering and maintaining a responsible technological environment that benefits society as a whole and ensures safety and accountability in technological advancements (Carrillo M., 2020; Larsson S,, 2020; Camilleri M., 2023).

Frameworks and Approaches

- **Dynamic and Multidisciplinary Models:** Modern frameworks emphasize the critical need for dynamic feedback mechanisms that facilitate continuous input and adaptation, fostering collaboration among various stakeholders involved in the process. Additionally, these frameworks advocate for the seamless integration of diverse perspectives, including legal, technological, and market viewpoints, which are vital for comprehensive understanding. Such models encourage the promotion of ongoing ethical evaluations and thorough risk analyses, as well as the implementation of adaptive regulatory measures.

This approach is essential to ensure alignment with the rapid advancements taking place in the realm of artificial intelligence, creating an environment where innovation can thrive while adhering to ethical standards (Liu Y., et al., 2024; Wang J., et al., 2024; Xue L. & Pang Z., 2022; Rodríguez N., et al., 2023).

- **Value Mapping and Socio-Technical Assemblages:** Understanding the intricate mechanisms through which various values are generated and effectively integrated within artificial intelligence systems is of utmost importance in today's rapidly evolving technological landscape. Regulatory frameworks should thoughtfully consider the contextual, dynamic, and multifaceted characteristics that are inherently present both in legal norms and in the advanced technologies associated with AI, thereby transcending traditional, rigid regulations. This approach would favour governance that prioritizes adaptable and process-oriented methodologies, allowing for more effective oversight and alignment with modern AI practices and ethical standards (Gordon G., et al., 2022; Judge B., et al., 2024).

Regulatory Innovations

- **Risk-Based and Sector-Specific Regulation:** The European Union's proactive adoption of a comprehensive risk-based approach, alongside the implementation of innovative regulatory sandboxes, as well as the establishment of detailed sector-specific guidelines, are progressively gaining recognition as exemplary practices across multiple domains (European Union, 2024). These thoughtfully devised strategies are specifically designed to promote transparency, enhance accountability, and ensure equity, all while fostering and encouraging a culture of ongoing innovation within diverse sectors of the economy (Rodríguez N., et al., 2023; Wang J., et al., 2024; Camilleri M., 2023).
- **International and Human Rights Perspectives:** The expansion of human rights obligations to fully encompass private entities, in tandem with drawing valuable lessons from the intricate historical context of internet governance, is regarded as absolutely essential for the establishment of a comprehensive, robust, and effective framework aimed at global AI governance. This approach recognises the critical importance of integrating human rights into the practices of both public and private sectors, ensuring that advancements in technology align with fundamental human dignity and societal values (Koniakou V., 2022; Ienca M., et al., 2021).

Connecting behavioural law, ethics, and regulation within the rapidly evolving realm of artificial intelligence governance necessitates the establishment of adaptive, multidisciplinary frameworks that seamlessly combine ethical principles with enforceable legal norms tailored to address the unique challenges posed by AI technologies. Scholarly discourse underscores the importance of dynamic oversight, risk-based regulatory measures, and the imperative for both globally applicable and contextually specific strategies that can effectively address the complexities of AI systems. These strategies are essential to ensure that AI technologies remain reliable, transparent, and in harmony with the ever-evolving societal values and expectations. The intersection of law, ethics, and regulation is critical in fostering an environment where innovation can thrive while protecting individuals and communities from potential harms and ethical dilemmas arising from advanced artificial intelligence applications (Chatzivasileiou, D., et al. 2024).

Bridging behavioural law, ethics, and regulation in AI governance requires adaptive, multidisciplinary frameworks that integrate ethical values with enforceable legal standards. The literature emphasizes dynamic oversight, risk-based regulation, and the need for both global and context-specific approaches to ensure AI systems are trustworthy, transparent, and aligned with societal values (Kumar, D., & Suthar, N., 2025; Ghosh, A., et al., 2025).

4. ETHICAL CONSIDERATIONS IN AI REGULATION

Regulatory safeguards against artificial intelligence's potential harms must not compromise ethical considerations (Mittelstadt B., 2019). However, it is crucial that these safeguards do not compromise the essential ethical considerations that are necessary to uphold. The use of AI can generate significant harm in various ways, whether it be through the complete replacement of human decision-making processes or by assisting in decisions that remain uncertain, ambiguous, or fraught with complexity. It is essential to recognize that while technical advancements can provide solutions, they also often raise new ethical concerns even as they address existing ones (Morley J., et al., 2021). Within the extensive domain of artificial intelligence, the paramount significance of ethical evaluations can be articulated through several fundamental principles that underpin responsible development.

Key principles for ethical evaluation in the sphere of Artificial Intelligence governance include the foundational values of human dignity and autonomy, which emphasize the importance of preserving human agency (Blanchard, A., et al., 2025; Law, R., et al., 2025). It is essential that you, as individuals, preserve the ultimate authority to make decisions.. Automated systems should assist you, but they should never replace your decision-making role entirely.

Next, accountability and responsibility are crucial. It is essential to establish clear and transparent responsibility chains. This clarity ensures that you know exactly who is liable whenever AI systems impact individuals or influence public policy decisions. This can lead to increased trust and integrity in the deployment of AI technologies.

Fairness and justice stand as critical pillars as well. It is vital to actively prevent discrimination in the data and algorithms utilized, addressing any bias that may arise in data, models, and their resulting outcomes. Striving for equality in these areas contributes to a more just society.

Furthermore, transparency and explainability are necessary for effective governance. People must have the ability and means to understand AI decisions in clear, practical terms; the technical complexities of AI should not obscure the decision-making process.

In terms of privacy and data protection, it is important to respect the rights associated with personal data. This involves applying strict and consistent rules concerning data collection, usage, and obtaining consent from individuals whose data is being utilized.

Safety, robustness, and reliability are also of utmost importance. AI must demonstrate consistent performance, actively avoid causing harm, and function safely within real-world environments, where unpredictable conditions may exist.

Proportionality and necessity should be guiding principles. Employ AI only in situations where it contributes tangible value to the tasks at hand. Excessive automation should be avoided, ensuring that human involvement remains central and essential.

Human oversight and control are fundamental to ethical AI governance. We must maintain meaningful supervision over AI systems, steering clear of developments that would effectively lock humans out of critical decision-making processes.

Lastly, beneficence, or serving human well-being, is a principle that underpins ethical AI practices. AI should actively support social good, promote public interest, and enhance the overall quality of life for individuals within society.

To ensure ethical governance, ongoing evaluation and adaptation are necessary. This includes continuous monitoring and auditing of AI systems as they evolve, thereby ensuring alignment with ethical guidelines and societal expectations.

The principles outlined encompass beneficence, which underscores the importance of ensuring that artificial intelligence systems yield significant advantages for users; non-maleficence, which stresses the necessity of preventing any detrimental impacts that could result from the implementation of AI; justice, which is concerned with guaranteeing equitable access to services facilitated by AI while simultaneously working to alleviate any prospective negative outcomes; and autonomy, which emphasizes the importance of maintaining or promoting user self-determination amid the swiftly changing technological environment. It is crucial to consider these ethical aspects as we traverse the intricate relationship between technological progress and the welfare of individuals and the broader society

Regulatory design, along with the individual deployment of diverse systems, plays critically important roles in shaping and determining how the distribution of these ethical benefits and harms unfolds and plays out in our complex society. The complexity, opacity, and inherent uncertainties associated with AI systems only serve to further compound, complicate, and amplify these ongoing concerns (Pistilli G., et al., 2023) . Such uncertainty not only complicates the challenging task of assessing competing designs or policies but also significantly heightens the essential need for rigorous experimentation. Without practical trials and thorough evaluations, it becomes exceedingly difficult and challenging to arrive at informed and sound conclusions. This ongoing situation poses additional hurdles, especially when multiple countervailing considerations remain in play, each requiring careful balancing to effectively navigate these intricate ethical dilemmas that our society faces today. In light of these complexities, the importance of responsible regulatory frameworks cannot be overstated, as they are crucial for guiding the development and deployment of AI technologies in ways that promote ethical outcomes for all members of society.

5. REGULATORY FRAMEWORKS FOR AI: A COMPARATIVE PERSPECTIVE

The regulatory landscape for artificial intelligence (AI) draws on diverse frameworks (precautionary, risk-based, data-centric, and rights-based) each with distinct strengths, weaknesses, and relevance to foundational behavioural insights. Precautionary regimes emphasise scientific uncertainty, emphasising oversight and governance without deployment (De Gregorio, G. & Dunn, P., 2022; Alanoca S., et al., 2025). Data-centric frameworks further prioritise rapid deployment by simplifying scrutiny to dataset connection, overextending prohibition beyond health and safety threats (Adepoju, D. A., & Adepoju, A. G; Patel, 2024). Rights-based perspectives define fundamental attributes of human dignity underpinning systems operating on definitive data, though with attendant overbroad scope (Edeji O. C., 2025; DALY, E. & MAY, J. R.)

A precautionary framework is thus advanced as a viable avenue for further consideration. Such an approach enables time-sensitive regulatory elucidation of socio-technical systems and operation-based activity knows no technological distinction, accentuating the enduring tension across diverse regimes. Regulatory authorities project a hesitance to restrict AI adoption. Preparing supplementary, adaptive regulation alongside initial legal articulation embraces behavioural principles while addressing additional dimensions. Attention is directed to the UK approach to high-risk AI, juxtaposed against the US White House Office of Science and Technology Policy's expansive Pixels Framework (Gervais D. J., 2023).

6. COMPLIANCE, ENFORCEMENT, AND ACCOUNTABILITY MECHANISMS

AI systems have become increasingly ubiquitous in contemporary society, leading to the emergence of various mechanisms aimed at facilitating the complex articulation of ethical charters, legal tools, and technical documentation governing the evolution of artificial intelligence in ways that are beneficial to a diverse array of stakeholders (Taj I. and Zaman N., 2022; Chen J., et al., 2022). It has become paramount in our interconnected world to develop such frameworks that address the multifaceted challenges presented by AI advancements. Although numerous organizations have made strides in articulating guiding principles that have been formalized into ethical charters to frame the development of AI technologies and to promote their appropriate usage, these charters are intended to protect human stakeholders and uphold societal norms. However, despite these efforts, the operationalization of ethical charters as effective instruments for navigating the complexities of AI implementation remains frustratingly limited. Various regulatory initiatives are underway in an attempt to adapt existing legislation to better address the unique and often unforeseen technical realities associated with AI and to enhance the documentation of these systems.

The intersection of ethical compliance and legal compliance is exceedingly intricate; it should be noted that legal compliance does not necessarily equate to moral compliance. For instance, transparency in corporate ethics requires that organizations adhere to their legal disclosure duties; however, it also necessitates more broadly communicating practices that are perceived to be beneficial and in the best interest of both consumers and society at large (Efthymiou, I.-P., 2024). The ongoing controversy surrounding data scraping to train AI models serves as a vivid illustration of the divergent views that exist regarding legality, consent, and ethical considerations; while the practice of data scraping may be legally permissible under certain conditions, ethical implications may suggest otherwise, presenting a complex dilemma for stakeholders involved.

Moreover, assessing compliance necessitates a comprehensive understanding of AI system behaviours, such as memorization and inherent bias, which play a crucial role in influencing the associated legal and ethical dimensions. Consequently, the significance of fostering comprehensive technical documentation cannot be overstated; it extends well beyond mere compliance with existing laws and regulations. Rather, effective documentation serves to inform and support both ethical and legal frameworks, creating a more holistic approach to the governance of artificial intelligence technologies in the rapidly evolving digital landscape (Pistilli G., et al., 2023).

7. GOVERNANCE BY DESIGN: ALIGNING AI SYSTEMS WITH LAW AND ETHICS

Societal expectations and various policy measures mandate that AI systems must integrate a comprehensive array of legal and ethical compliance considerations from the very beginning stages of their development. Governance by design fundamentally involves incorporating essential constraints at every stage of the AI lifecycle, which encompasses procurement, development, deployment, as well as the eventual decommissioning process. Important elements crucial for successful implementation consist of interlocks specifically designed to effectively prevent any illegal or unethical submissions from being processed, alongside robust audit trails that meticulously document all actions taken during the entire lifecycle of the AI system.

Additionally, there must be verifiable records in place to reliably confirm and guarantee that adherence to applicable laws and ethical requirements is consistently maintained throughout the system's operation. These comprehensive practices ensure that AI technologies are not only effective in fulfilling their intended purposes but also remain responsible and trustworthy, aligning meticulously with the best practices established for societal norms and legal frameworks that govern the use of technology in today's rapidly evolving landscape (Joshi H., 2025; De Silva D. & Alahakoon D., 2022).

The risk-harm-redress framework offers extensive architectural guidance designed specifically for AI applications, ensuring that various aspects of risk management are thoroughly considered. Key components of this framework include the careful identification of potential hazards that may arise during the deployment and usage of AI technologies. There is also a crucial assessment of liability regarding actual harm that might result from these applications. Additionally, it encompasses the specification of effective mitigation measures to minimize risks. Beyond that, the framework emphasizes the establishment of comprehensive post-market monitoring obligations, which are essential to track the ongoing performance and safety of AI systems. Finally, it involves the formulation of timelines for the prompt remediation of any harm that does occur, ensuring that necessary corrective actions are taken swiftly and effectively (Morley J., et al., 2021).

8. RISK, HARM, AND REDRESS IN AI APPLICATIONS

AI applications necessitate a thorough and comprehensive understanding of the concepts of risk, potential harm, and the avenues for redress, which encompass a wide range of hazards that are associated with both the design and functionality of a given technological system or application (Efthymiou, I. P., & Egleton, T. E., 2023). Risk assessment is an essential process that involves identifying and analysing foreseeable hazards that may emerge during the development and deployment of AI technologies. On the other hand, harm assessment requires a meticulous examination to pinpoint how the designated technology can lead to negative consequences, whether directly affecting individuals or indirectly impacting society as a whole (Psomiadi, A., 2026).

This rigorous assessment framework is equally applicable to artificial intelligence systems; it is crucial to determine the risks involved and to evaluate the adverse effects that individuals or communities are likely to suffer or which may be caused by the operational use of AI. Moreover, it is equally important to explore any available mitigation strategies or remediation methods that can be employed to counteract these risks and effects, ensuring that AI technologies are developed and utilized responsibly and ethically (Walz A. & Firth-Butterfield K., 2019).

The market introduction of AI solutions must be accompanied by comprehensive enabling mechanisms that are designed to effectively monitor not only the nature of use but also the associated risks inherent in these technologies. A robust mechanism to alert policymakers on emerging hazards, potential harms, and instances of non-compliance is considered absolutely necessary (Sotiropoulos, L., 2025). The insights derived from these mechanisms are instrumental in fine-tuning regulatory measures related to the technology in question, enabling a more precise understanding of how these innovations impact society. Moreover, it plays a crucial role in determining how and when to include specific applications within the scope of governance activities. Ideally, such provisions ought to exhibit a degree of bounded flexibility, allowing for adjustments as new information arises and as the technological landscape evolves over time. This approach ensures that regulations remain relevant and effective in addressing the fast-paced changes driven by advancements in artificial intelligence.

9. INTERNATIONAL DIMENSIONS OF AI REGULATION

The proliferation of artificial intelligence (AI) technologies has raised regulatory questions with significant international dimensions (Walter Y., 2024). Differentiated regulatory frameworks across jurisdictions contribute to a fragmented compliance landscape, thereby hampering international innovation and trade opportunities (Sotiropoulos, L. D., 2023). Certain frameworks promote the restrictive consequences of wide-ranging regulation or an outright ban, whereas jurisdictions with an overly permissive regulatory approach pose risks to health and safety. There is also concern that supporting elements such as cross-border data transfers, retention and access requirements, or requirements for secure storage and algorithmic neutrality may be jeopardized. Different approaches in the regulation of AI lead to divergent technical and ethical standards for products and services and can be a source of inefficiency in international trade.

Cross-border data transfers are essential for the free flow of information and knowledge. Access to datasets and their processing constitutes the basis for the training of models, the annual costs of which can reach hundreds of millions of euros. Concerns are emerging that legally mandated storage and access obligations may be imposed at the same time, or that requests to obtain stored data pertaining to large language models may begin to be made in respect of information that has been publicly disclosed. Generation of AI-generated texts in one commercial language can constrain the digital development and competitiveness of products and services in a region with a different commercial language.

The regulation of artificial intelligence merits a co-ordinated approach. Relevant exchanges have been at the G7, G20, OECD, and UNESCO, advancing governing principles and establishing model mechanisms, a significant but limited step towards a wider accord (Teng, Z., et al., 2025). Existing bilateral arrangements on direct information exchange across specified products and technologies could be adapted to a broader range of selected AI services while facilitating international monitoring of cross-border implications. Agreement on a common deadline for completion of independent technical assessments for each class of AI service that conducts a similar function, regardless of size, would mitigate competitive squeezes arising from differential timetables (Hartmann, D., et al., 2025; Bringas Colmenarejo A., et al., 2022). The AI Act is a timely opportunity to foster dialogue between the EU and its partners and to enhance participatory cooperation on AI services, yet regulatory performance worldwide remains slow (Novelli, C., et al., 2025; Ebers, M. (2025).

10. CHALLENGES AND LIMITATIONS OF CURRENT REGIMES

Governing AI applications presents profound challenges. Conventional regulatory approaches often falter under competitive pressures, such as weakening investor protections, and find it difficult to keep pace with rapid technological change (Smuha N. A., et al., 2025; Walter Y, 2024; Ayub T., 2024; Marsal, A. P., & Hardyansah, R., 2025). Traditional doctrines of strict liability and other ex-ante safety strategies remain ill-suited to assess pervasive harms arising long after deployment. Even consolidated regulatory efforts cannot anticipate harm-mitigation methods where mechanistic harm loops allow for limitless re-harm, nor do they grapple with alternative institutional designs and assistance technologies that avoid the need to attract scarce attention.

Emerging algorithmic systems empower the production of high-quality documents, videos, and other media recordings. Yet even static content of replayed or shifted media continues to draw attention, increased uncertainty surrounding realistic mimicry of speech, and the re-emergence of hyper-harm scale presents urgent and novel governance challenges (Kouroupis, K. & Efthymiou, I. (Eds.), 2026). Such technologies generate simulated documents and other outputs that flow through domestic networks, constructed from the simulating design of formally distributed and licensed specifications that remain characteristic of proprietary trade secret designs and shape competitive behaviour in the algorithm-development marketplace. Missing-specification processes generate outputs without explicit prior specification of the desired content, allowing alteration of literal content beyond input attention (Rocco S., 2022).

Competing narrative descriptions about alternative decisions or reconsideration-desired recovery, for example, long predate such technologies. Clamp measures that theoretically prevent extraction cannot guard against or detect the necessary execution of program, knowledge-management, and content-deployment specifications that disclose the complementary structural-characterization knowledge needed to reproduce full backup emergent-organization support for profoundly larger not narrowly even internal and deployed alternative-mathematical-structure, recurrent, and off-mathematical-structure documents and catalogues (Lecoeur B., et al., 2023; Ullah A., et al., 2024; Zhong E., 2023).

Harm reproducers, including the original generator or haven-less mandatory-disclose support cascade program co-disclosed by common entry from afar, provide no route for restrictive safeguarding against acquisition and rely on semi-safeguarding means regarding material acquired under such mathematics, yet consideration of potentially extending re-harming conditions and attention-cultures indicates that earlier reach merely involves extending log-length attention over already generated material (and extensive emergence-control program remains) to comparable mass or along a range (both preventable under algorithm possibilities) despite any unchanged provision and completion of consideration (Anderljung M., et al., 2025; Firmin C., et al., 2025). Subjecting other-organization acquired support under distinctive and uncovered mathematics instead of mandatory common regimen, however, secures reach qualified merely by safeguarding against extending supplementary alternative encouraging-locus attention through alternative reflectives or situated segments, with securing alternative merely and extension trapping plugging known though actual extension-via or remaining-sequencing mere invite at least substantially more-conceivable en masse but after now less actively an option regarding widely preserved.

11. OPPORTUNITIES FOR INNOVATION WITHIN LEGAL BOUNDS

Regulatory approaches for artificial intelligence (AI) are evolving rapidly, and yet, behaviours, speed of change and genuinely innovative opportunities remain. Regulatory regimes for AI are proliferating yet often focus too narrow (Dunleavy, P., & Margetts, H., 2025). New opportunities for wide ranging innovation remain in many domains, within existing legal frameworks, restrictively or liberally defined. Options include scalable compliance frameworks, modular regulations on risk, and widespread opportunities for "responsible AI by design" (Krishnan, 2025; Swan, A., & Aguilar, Y., 2025). Wet-lab or economic simulation sandboxes in concept are joined by simulators and competitive game-theoretic sandboxes, where planners can simulate in thousands of parametric scenarios consequences of broad classes of formulation. Regulatory sandboxes increasingly combine earlier safety-actionable aspects of creative AI with multi-task or multi-agent approaches or environmental-variability/consumer-type variants of creative-agent simulation. Performance simulations broaden yet remain within supple structures as awareness builds of wide-ranging, pervasive, potentially unintended, and hard-to-correct cascading feedbacks across-the-board.

Regulatory opportunities address existing regulatory roadblocks as coalescing understanding of the many and profound fine-tuned feedbacks grows. Traditional regulation accompanying innovation is intrinsically reactive. The criterion of scaling rather than labelling extensively enables huge spread of concept within steadily bounding categories and directs immediate attention to extending existing general approaches of regulation. Regulatory opportunities within the law fence clear space for broad societal consideration of the large range of innovative, regulatory and societal directions potentially available.

A governance model for artificial intelligence that is informed by behavioural insights fosters an environment conducive to innovation whilst upholding stringent legal and ethical protections. Policy frameworks are enhanced when they integrate behavioural law as a proactive strategy in design, rather than viewing ethical considerations as a mere limitation imposed after implementation. AI systems have the capacity to incorporate behavioural insights, thereby facilitating informed decision-making, encouraging responsible digital conduct, and mitigating the risks of manipulation or coercion. This methodology aligns innovative efforts with the public interest, as opposed to prioritizing market-driven opportunism.

The formulation of well-defined accountability frameworks is essential for promoting innovation in a responsible way. Clarity in legal liability, risk distribution, and obligations promotes research and implementation without leading to regulatory stagnation. Regulatory sandboxes, which serve as controlled testing environments, along with tiered compliance strategies, enable AI systems to develop under oversight, offering valuable real-world data while safeguarding individuals and organizations from excessive harm.

Robust data governance policies serve as facilitators of innovation. Regulations pertaining to data quality, consent, access rights, and security contribute to fostering public trust and generating dependable datasets, thereby enhancing overall system performance. Secure data-sharing frameworks, combined with measures to safeguard privacy, enable collaboration among research institutions, governmental bodies, and industry stakeholders without compromising legal or ethical standards.

Innovation is enhanced when policy distinctly emphasizes human-centric system design. Artificial intelligence should serve to enhance human judgment rather than replace it, integrating substantial oversight, accountability, and the capacity for contestation within technological frameworks. The transformation of overarching ethical commitments into quantifiable standards—via audits, certification processes, and compliance metrics—effectively converts ethical principles into tools for operational governance.

Adaptive regulation is essential. Fixed regulation is unable to effectively address swift technological advancements or the changing effects on behaviour. Governance that is informed by behavioural insights fosters an environment of iterative learning, ongoing monitoring, and the evolution of policies. This approach enhances democratic legitimacy, maintains public trust, and safeguards human rights, all while facilitating technological advancement.

In conclusion, the advancement of innovation within legal parameters is contingent upon interdisciplinary cooperation. It is essential for legal scholars, behavioural scientists, engineers, policymakers, and ethicists to collaborate in order to identify ethical blind spots, mitigate systemic risks, and foster AI ecosystems that yield social benefits. This approach, which emphasizes collaboration and behavioural awareness, offers a productive avenue for the responsible development of artificial intelligence (Efthymiou–Egleton, T. W., et al. 2020).

12. POLICY RECOMMENDATIONS FOR BRIDGING ETHICS AND REGULATION

Ethical imperatives concerning artificial intelligence (AI) centre on beneficence, non-maleficence, justice, and autonomy, ensuring that the technology is deployed in ways that enhance social goods, minimize harm to individuals and society, distribute benefits and burdens fairly, and respect individual rights (Σιδηρόπουλος, Σ., 2024; B. de Laat, 2021). Given the relevance of trustworthy, responsible, and ethical AI to regulator deliberations, such considerations are a necessary foundation for the subsequent analysis of regulatory frameworks. The discussion of these ethical considerations then transitions to an exploration of the dimensions of justice, particularly distributional equity of data, explainability, and consent to use (Diareme, K. C. et al., 2022; Livieri, G., et al., 2024).

Recommendations for Policymakers, AI Developers, and Ethicists

Policymakers are encouraged to adopt regulations that are firmly rooted in nudge theory principles. This includes, but is not limited to, the introduction of user-friendly and simplified privacy consent forms (Psomiadi, A., et al, 2020). The goal of these initiatives is to promote greater user engagement while ensuring compliance with crucial legal standards, such as the General Data Protection Regulation (GDPR) (Rahman, P., 2025; Leiser, M., 2025; Parrilli, D. M. (2025). By implementing these measures, we can cultivate a climate of trust among users, which is essential. Importantly, this approach is designed to foster such trust without imposing any obstacles to the continued advancement and development of artificial intelligence technologies.

AI developers should establish thoughtful architectural design choices that incorporate robust default configurations specifically aimed at prioritizing ethical outcomes, such as sophisticated algorithms that work to effectively mitigate bias (Lund, B., et al., 2025). This proactive approach not only significantly minimizes user dependency but also actively fosters greater transparency within artificial intelligence systems. By emphasizing these ethical considerations in the design phase, developers can ensure a more responsible implementation of AI technologies that benefit society as a whole.

Ethicists strongly recommend the thoughtful implementation of empirical research aimed at rigorously evaluating behavioural interventions (Efthymiou, I. P., 2024). This approach is essential for ensuring that governance frameworks remain adaptable to the complexities of actual human-AI interactions. Moreover, these frameworks should effectively promote fair and internationally consistent standards that can be universally understood and applied across various contexts.

Policy must systematically and exhaustively translate ethical principles into explicit and enforceable standards that integrate strong accountability and liability frameworks to promote compliance and responsibility.

Regulation informed by behavioural insights should not only forecast the profound impact of artificial intelligence on human decision-making processes but also actively mitigate potential risks associated with misuse, abuse, or manipulation of these advanced technologies.

Transparency should go beyond mere technical disclosures; it must strive to offer coherent explanations that are readily accessible and comprehensible to all individuals, thereby enabling meaningful engagement with the presented information and empowering them to challenge any inaccuracies or misinterpretations (Radanliev, P., 2025).

The incorporation of human oversight, alongside stringent data governance practices, is crucial in establishing essential foundations for fostering public trust, ensuring equity, and satisfying legal compliance standards across diverse interconnected sectors.

Governance frameworks should remain strategically nimble over time, bolstered by innovative regulatory sandboxes that promote experimentation, continuous auditing efforts, and targeted initiatives for institutional capacity enhancement to address evolving challenges.

Lastly, public involvement, as well as international collaboration, are vital in maintaining legitimacy and boosting public trust while supporting a culture of responsible innovation practices that effectively address societal needs and ethical concerns.

13. CONCLUSION

Humanity currently finds itself amidst what many refer to as a momentous second industrial revolution, the significance of which cannot be overstated. However, it is critical to point out that the structures of governance we rely on and adhere to still remain firmly entrenched within the systems and frameworks that were originally designed to accommodate the very first industrial revolution. This presents a considerable and significant challenge, as the primary issue revolves around the advent of machines that exhibit the remarkable capability of substituting human behaviour across a multitude of diverse, dynamic, and non-homogeneous environments. These autonomous systems are meticulously engineered to interpret, process, and interact with users and their environments using not only ordinary language but also recognizable behavioural patterns. Yet, when it comes to the intricate and complex processes of decision-making, these systems still invoke essential elements of human behaviour that remain irreplaceable. This includes complex aspects such as personal preferences, deep-seated biases, and subjective judgments that, at this pivotal juncture in our technological evolution, cannot be entirely delegated to or left to machines alone.

At the very heart of both legal and ethical frameworks governing the use of AI technologies lie essential and underlying requirements for sustained human oversight. This oversight is crucial to ensure not only safety but also accountability, and to establish a robust ethical foundation in all AI-driven operations that we are increasingly witnessing today across various sectors. As governance frameworks begin to evolve gradually, there is an increasing emphasis on the concept of supervision, rather than outright control over these increasingly intelligent systems and algorithms. However, despite this notable shift toward oversight, the legal safeguards that are enshrined within these frameworks still necessitate a significant degree of human involvement and active oversight, which cannot be overlooked or underestimated. The intricate and multifaceted process of programming machines is intricately tied to various legal regimes and regulations that are deeply embedded in human supervision as well as ethical considerations. Thus, it ultimately revolves around human activity, decision-making, and cognitive processes, highlighting the importance of human judgment when dealing with sophisticated technologies.

This indicates that the governance of artificial intelligence is intrinsically linked to long-established governance frameworks that have been in place for several decades, which necessitates a collaborative approach between both spheres. The relationship between traditional and contemporary systems underscores the urgent requirement for a meticulous and comprehensive reassessment of current governance structures to adequately address the distinct and complex challenges presented by these advanced autonomous technologies. As we progress into this unprecedented domain, it is imperative that our governance structures adapt, evolve, and transform in response to the intricacies associated with the integration of artificial intelligence across diverse societal contexts. This underscores the vital importance of human oversight within this continually changing technological environment..

In conclusion, behavioural insights offer essential perspectives that can significantly enhance and improve the formulation of more effective regulations, default parameters, and an extensive and well-defined accountability framework. A thorough understanding of human behaviour enables the development of guidelines that are not only much more efficient but also far more accessible and user-friendly to end users. This deeper comprehension can lead to the creation of policies that resonate better with individuals, thereby facilitating compliance and improving overall satisfaction.

REFERENCES

Adepoju, D. A., & Adepoju, A. G. Establishing Ethical Frameworks for Scalable Data Engineering and Governance in AI-Driven Healthcare Systems.

Alanoca, S., Gur-Arieh, S., Zick, T., & Klyman, K. (2025, June). Comparing Apples to Oranges: A Taxonomy for Navigating the Global Landscape of AI Regulation. In *Proceedings of the 2025 ACM Conference on Fairness, Accountability, and Transparency* (pp. 914-937). DOI: 10.1145/3715275.3732059

Anderljung, M., Hazell, J., & von Knebel, M. (2025). Protecting society from AI misuse: When are restrictions on capabilities warranted? *AI & Society*, *40*(5), 3841–3857. DOI: 10.1007/s00146-024-02130-8

Ayaa, M. M., Peprah, W. K., Mensah, M. O., Owusu-Sekyere, A. B., & Daniel, B. (2022). Influence of heuristic techniques and biases in investment decision-making: A conceptual analysis and directions for future research. *International Journal of Academic Research in Business & Social Sciences*, *12*(5), 1252. DOI: 10.6007/IJARBSS/v12-i5/13339

Ayres, I., Bankman, J., & Hemel, D. J. (2025). (forthcoming). The Law and Economics of Guilt and Shame. *The University of Chicago Law Review. University of Chicago. Law School.*

Ayub, T. (2024). Regulating AI: Balancing Innovation, Ethics, and Public Policy A Critical Analysis of AI Governance and Policy Approaches. Ethics, and Public Policy A Critical Analysis of AI Governance and Policy Approaches (December 11, 2024).

Bao, Y., Gong, W., & Yang, K. (2023). A literature review of human–AI synergy in decision making: From the perspective of affordance actualization theory. *Systems*, *11*(9), 442. DOI: 10.3390/systems11090442

Blanchard, A., Thomas, C., & Taddeo, M. (2025). Ethical governance of artificial intelligence for defence: Normative tradeoffs for principle to practice guidance. *AI & Society*, *40*(1), 185–198. DOI: 10.1007/s00146-024-01866-7 PMID: 41640455

Bodig, M., & Ződi, Z. (2024). Is Code Destined to Supersede Law? Behavioural Control by Code and Legal Theory. Behavioural Control by Code and Legal Theory (June 20, 2024).

Bringas Colmenarejo, A., Nannini, L., Rieger, A., Scott, K. M., Zhao, X., Patro, G. K., & Kinder-Kurlanda, K. (2022, July). Fairness in agreement with European values: An interdisciplinary perspective on AI regulation. In *Proceedings of the 2022 AAAI/ACM Conference on AI, Ethics, and Society* (pp. 107-118). DOI: 10.1145/3514094.3534158

Camilleri, M. (2023). Artificial intelligence governance: Ethical considerations and implications for social responsibility. *Expert Systems: International Journal of Knowledge Engineering and Neural Networks, 41*(7), e13406. Advance online publication. DOI: 10.1111/exsy.13406

Carrillo, M. (2020). Artificial intelligence: From ethics to law. *Telecommunications Policy, 44*(6), 101937. DOI: 10.1016/j.telpol.2020.101937

Challoumis, C. (2024). Behavioral Economics Concepts and the QE Method. [IJMMU]. *International Journal of Multicultural and Multireligious Understanding, 11*(10), 166–212. DOI: 10.18415/ijmmu.v11i10.6138

Chatzivasileiou, D., Psomiadi, A., Efthymiou-Egleton, T. W., & Kassar, L. (2024). AI, International Relations & Religion. *Journal of Politics and Ethics in New Technologies and AI, 3*(1), e37109–e37109. DOI: 10.12681/jpentai.37109

Chen, J., Sun, J., & Wang, G. (2022). From unmanned systems to autonomous intelligent systems. *Engineering (Beijing), 12*, 16–19. DOI: 10.1016/j.eng.2021.10.007

DALY, E. & MAY, J. R. (). DIGNITY UNDER LAW: A GLOBAL HANDBOOK FOR CIVIL SOCIETY.

De Gregorio, G., & Dunn, P. (2022). The European risk-based approaches: Connecting constitutional dots in the digital age. *Common Market Law Review, 59*(2), 473–500. DOI: 10.54648/COLA2022032

de Laat, P. B. (2021). Companies committed to responsible AI: From principles towards implementation and regulation? *Philosophy & Technology, 34*(4), 1135–1193. DOI: 10.1007/s13347-021-00474-3 PMID: 34631392

De Silva, D., & Alahakoon, D. (2022). An artificial intelligence life cycle: From conception to production. *Patterns (New York, N.Y.), 3*(6), 100489. DOI: 10.1016/j.patter.2022.100489 PMID: 35755876

Diareme, K. C., Liapakis, A., & Efthymiou, I. P. (2022). Data, sentiment analysis, and examples during the COVID-19 pandemic. *Hellenic Archives of Psychiatry and Clinical Psychology in Behavioral Sciences, 33779*(2), 21–30. Advance online publication. DOI: 10.12681/hapscpbs.33779

Dunleavy, P., & Margetts, H. (2025). Data science, artificial intelligence and the third wave of digital era governance. *Public Policy and Administration*, *40*(2), 185–214. DOI: 10.1177/09520767231198737

Ebers, M. (2025). Truly risk-based regulation of artificial intelligence how to implement the EU's AI Act. *European Journal of Risk Regulation*, *16*(2), 684–703. DOI: 10.1017/err.2024.78

Edeji, O. C. (2025). Intersections of the Right to Education and Human Dignity in International Human Rights Law: A Purpose-Based Analysis. *Laws*, *14*(3), 33. DOI: 10.3390/laws14030033

Efthymiou, I. (2025). Diversity, Inclusion, and Equity in Gig Economies. In Doon, R. (Ed.), *Sustainability and Adaptability of Gig Economies in Global Business* (pp. 231–260). IGI Global Scientific Publishing., DOI: 10.4018/979-8-3693-9385-7.ch009

Efthymiou, I., & Egleton, T. E. (2025). The Impact of Deep Fakes in Markets and Economies. In Gupta, G., Bohara, S., Kovid, R., & Pandla, K. (Eds.), *Deepfakes and Their Impact on Business* (pp. 19–50). IGI Global Scientific Publishing., DOI: 10.4018/979-8-3693-6890-9.ch002

Efthymiou, I. P. (2024). Exploring the frontier: New and future trends in research. In B. Orlando (Ed.), Innovation capabilities and entrepreneurial opportunities of smart working (pp. 48–68). IGI Global Scientific Publishing. DOI: 10.4018/978-1-7998-8797-3.ch003

Efthymiou, I.-P. (2024). AI, ethics and the future of work: What's left for us? Independently published.

Efthymiou, I. P. (2025). The role of e-government and e-governance in modern societies. In Wongmahesak, K., Ahmad, J., & Mustanir, A. (Eds.), *Public governance practices in the age of AI* (pp. 45–68). IGI Global Scientific Publishing., DOI: 10.4018/979-8-3693-9286-7.ch003

Efthymiou, I. P., & Egleton, T. E. (2023). Artificial intelligence for sustainable smart cities. In *Handbook of research on applications of AI, digital twin, and internet of things for sustainable development* (pp. 1–11). IGI Global. DOI: 10.4018/978-1-6684-6821-0.ch001

Efthymiou–Egleton, T. W., Sidiropoulos, S., Spanos, E., & Stougiannou, E. (2020). Big Data and Democracy. *HAPSc Policy Briefs Series*, *1*(2), 18–25. DOI: 10.12681/hapscpbs.26473

Epstein, R. A. (2025). Nudges, preferences and competences: A critique of both neoclassical and behavioral economics. *Behavioural Public Policy*, *9*(3), 1–15. DOI: 10.1017/bpp.2024.54

European Union. (2024). Regulation (EU) 2024/1689 of the European Parliament and of the Council of 12 July 2024 laying down harmonised rules on artificial intelligence (Artificial Intelligence Act) and amending certain Union legislative acts. Official Journal of the European Union, L 234, 1–83. https://eur-lex.europa.eu/legal-content/EN/TXT/?uri=OJ:L:2024:234:TOC 23

Ewert, B., Loer, K., & Thomann, E. (2021). Beyond nudge: Advancing the state-of-the-art of behavioural public policy and administration. *Policy and Politics*, *49*(1), 3–23. DOI: 10.1332/030557320X15987279194319

Firmin, C., Lefevre, M., Langhoff, K., & Ruch, G. (2025). A line in the sand: Moving from surface improvement to foundational shifts to develop effective responses to extra-familial risks and harms. *British Journal of Social Work*, *55*(3), 1296–1316. DOI: 10.1093/bjsw/bcae193

Garbuio, M., & Lin, N. (2021). Innovative idea generation in problem finding: Abductive reasoning, cognitive impediments, and the promise of artificial intelligence. *Journal of Product Innovation Management*, *38*(6), 701–725. DOI: 10.1111/jpim.12602

Gervais, D. J. (2023). Towards an effective transnational regulation of AI. *AI & Society*, *38*(1), 391–410. DOI: 10.1007/s00146-021-01310-0 PMID: 34776652

Ghosh, A., Saini, A., & Barad, H. (2025). Artificial intelligence in governance: Recent trends, risks, challenges, innovative frameworks and future directions. *AI & Society*, *40*(7), 1–23. DOI: 10.1007/s00146-025-02312-y

Gofen, A., Moseley, A., Thomann, E., & Kent Weaver, R. (2021). Behavioural governance in the policy process: Introduction to the special issue. *Journal of European Public Policy*, *28*(5), 633–657. DOI: 10.1080/13501763.2021.1912153

Gordon, G., Rieder, B., & Sileno, G. (2022). On mapping values in AI Governance. *Computer Law & Security Review*, *46*, 105712. DOI: 10.1016/j.clsr.2022.105712

Hartmann, D., De Pereira, J. R. L., Streitbörger, C., & Berendt, B. (2025). Addressing the regulatory gap: Moving towards an EU AI audit ecosystem beyond the AI Act by including civil society. *AI and Ethics*, *5*(4), 3617–3638. DOI: 10.1007/s43681-024-00595-3

Herath Pathirannehelage, S., Shrestha, Y. R., & von Krogh, G. (2025). Design principles for artificial intelligence-augmented decision making: An action design research study. *European Journal of Information Systems*, *34*(2), 207–229. DOI: 10.1080/0960085X.2024.2330402

Hunt, A. K., Wang, J., Alizadeh, A., & Pucelj, M. (2024). Advancing a theoretical framework for exploring heuristics and biases within HR decision-making contexts. *Personnel Review*, *53*(7), 1823–1841. DOI: 10.1108/PR-03-2023-0192

Ienca, M., Fins, J., Jox, R., Jotterand, F., Voeneky, S., Andorno, R., Ball, T., Castelluccia, C., Chavarriaga, R., Chneiweiss, H., Ferretti, A., Friedrich, O., Hurst, S., Merkel, G., Molnár-Gábor, F., Rickli, J., Scheibner, J., Vayena, E., & Kellmeyer, R. (2021). Towards a Governance Framework for Brain Data. *Neuroethics*, *15*(2), 20. Advance online publication. DOI: 10.1007/s12152-022-09498-8

Joshi, H. (2025). AI Governance by Design for Agentic Systems: A Framework for Responsible Development and Deployment.

Judge, B., Nitzberg, M., & Russell, S. (2024). When code isn't law: Rethinking regulation for artificial intelligence. *Policy and Society*. Advance online publication. DOI: 10.1093/polsoc/puae020

Kantorowicz-Reznichenko, E., & Kantorowicz, J. (2025). Law & Economics at sixty: Mapping the field with bibliometric and machine learning tools. *Journal of Economic Surveys*, *39*(1), 201–245. DOI: 10.1111/joes.12613

Ke, L., Tong, S., Cheng, P., & Peng, K. (2025). Exploring the frontiers of llms in psychological applications: A comprehensive review. *Artificial Intelligence Review*, *58*(10), 305. DOI: 10.1007/s10462-025-11297-5

Koniakou, V. (2022). From the "rush to ethics" to the "race for governance" in Artificial Intelligence. *Information Systems Frontiers*, *25*(1), 71–102. DOI: 10.1007/s10796-022-10300-6

Kouroupis, K. (2022). The AI Act in light of the EU Digital Agenda: A critical approach. *Journal of Data Protection & Privacy*, *5*(3), 216–229. DOI: 10.69554/FQUZ5239

Kouroupis, K., & Efthymiou, I. (Eds.). (2026). *Legal Challenges of AI Across Interdisciplinary Sectors*. IGI Global Scientific Publishing., DOI: 10.4018/979-8-3373-5452-1

Krishnan, R. (2025). "FROM POLICY TO PRACTICE: RESEARCH DIRECTIONS FOR TRUSTWORTHY AND RESPONSIBLE AI "BY DESIGN"."

Kumar, D., & Suthar, N. (2025). Balancing Innovation and Regulation: Ethical Frameworks for AI Policy and Governance. In *Navigating the Intersection of AI Policy, Technology, and Governance* (pp. 171-200). IGI Global Scientific Publishing.

Larsson, S. (2020). On the Governance of Artificial Intelligence through Ethics Guidelines. *Asian Journal of Law and Society*, *7*(3), 437–451. DOI: 10.1017/als.2020.19

Law, R., Ye, H., & Lei, S. S. I. (2025). Ethical artificial intelligence (AI): Principles and practices. *International Journal of Contemporary Hospitality Management*, *37*(1), 279–295. DOI: 10.1108/IJCHM-04-2024-0482

Lecoeur, B., Mohsin, H., & Donaldson, A. F. (2023). Program reconditioning: Avoiding undefined behaviour when finding and reducing compiler bugs. *Proceedings of the ACM on Programming Languages, 7*(PLDI), 1801-1825. DOI: 10.1145/3591294

Leiser, M. (2025). *Dark Patterns, Deceptive Design, and the Law: AI's Hidden Influence on Our Digital Experience*. Bloomsbury Publishing. DOI: 10.5040/9781509987122

Liu, Y., Zheng, W., & Su, Y. (2024). Enhancing Ethical Governance of Artificial Intelligence Through Dynamic Feedback Mechanism. **, 105-121. DOI: 10.1007/978-3-031-57867-0_8

Livieri, G., Anastasopoulou, C., Briola, K., Nikolopoulou, P., & Efthymiou, I. P. (2024). Ethical issues arising from the use of AI in drug discovery. *Journal of Politics and Ethics in New Technologies and AI*, *3*(1), e37093–e37093. DOI: 10.12681/jpentai.37093

Lund, B., Orhan, Z., Mannuru, N. R., Bevara, R. V. K., Porter, B., Vinaih, M. K., & Bhaskara, P. (2025). Standards, frameworks, and legislation for artificial intelligence (AI) transparency. *AI and Ethics*, *5*(4), 1–17. DOI: 10.1007/s43681-025-00661-4

Marsal, A. P., & Hardyansah, R. (2025). Legal Implications of Digital Platform Models: Regulatory Approaches to Consumer Protection, Data, and Competition for Start-up Sustainability. *Bulletin of Science, Technology & Society*, *4*(2), 17–26.

Mitchell, G. (2014). Alternative behavioral law and economics. *The Oxford handbook of behavioral economics and the law*, 167.

Mittelstadt, B. (2019). Principles alone cannot guarantee ethical AI. *Nature Machine Intelligence*, *1*(11), 501–507. DOI: 10.1038/s42256-019-0114-4

Morley, J., Elhalal, A., Garcia, F., Kinsey, L., Mökander, J., & Floridi, L. (2021). Ethics as a service: A pragmatic operationalisation of AI ethics. *Minds and Machines*, *31*(2), 239–256. DOI: 10.1007/s11023-021-09563-w PMID: 34720418

Moseley, A., & Thomann, E. (2021). A behavioural model of heuristics and biases in frontline policy implementation. *Policy and Politics*, *49*(1), 49–67. DOI: 10.1332/030557320X15967973532891

Newman, J. C. (2020). Decision Points in AI Governance: Three Case Studies Explore Efforts to Operationalize AI Principles. UC Berkeley, Center for Long-Term Cybersecurity. https://cltc.berkeley.edu/wp-content/uploads/2020/05/Decision_Points_AI_Governance.pdf 19

Novelli, C., Hacker, P., Morley, J., Trondal, J., & Floridi, L. (2025). A robust governance for the AI Act: AI office, AI board, scientific panel, and national authorities. *European Journal of Risk Regulation, 16*(2), 566–590. DOI: 10.1017/err.2024.57

Ooi, K. L. (2025). Modern Behavioural Finance Theories. In *Demystifying Behavioral Finance: Foundational Theories to Contemporary Applications and Future Directions* (pp. 59–69). Springer Nature Singapore.

Parrilli, D. M. (2025). *Informational Privacy for Service Design: An Ethical Framework for Designing Privacy-Oriented Services*. Springer Nature. DOI: 10.1007/978-3-031-76926-9

Patel, K. (2024). Ethical reflections on data-centric AI: balancing benefits and risks. Available at *SSRN* 4993089. DOI: 10.2139/ssrn.4993089

Pistilli, G., Munoz Ferrandis, C., Jernite, Y., & Mitchell, M. (2023). *Stronger Together: on the Articulation of Ethical Charters*. Legal Tools, and Technical Documentation in ML.

Pistilli, G., Muñoz Ferrandis, C., Jernite, Y., & Mitchell, M. (2023, June). Stronger together: on the articulation of ethical charters, legal tools, and technical documentation in ML. In *Proceedings of the 2023 ACM Conference on Fairness, Accountability, and Transparency* (pp. 343-354). DOI: 10.1145/3593013.3594002

Psomiadi, A. (2026). Social Dynamics and AI-Driven Care for Older Adults. In *AI for Geriatric Care in an Aging Society: Ethical, Clinical, and Policy Challenges* (pp. 123-154). IGI Global Scientific Publishing.

Psomiadi, A., Diareme, K. C., & Chatzivasileiou, S. (2020). HAPSc Policy Briefs Series.

Rachlinski, J. J. (2011). The psychological foundations of behavioral law and economics. *U. Ill. L. Rev.*, 1675.

Radanliev, P. (2025). AI ethics: Integrating transparency, fairness, and privacy in AI development. *Applied Artificial Intelligence, 39*(1), 2463722. DOI: 10.1080/08839514.2025.2463722

Rahman, P. (2025). *Exploring ethics and trust in persuasive technologies* (Doctoral dissertation, University of British Columbia).

Rassolov, I., & Chubukova, S. (2022). Artificial Intelligence and Effective Governance: Legal Framework. *Kutafin Law Review*. DOI: 10.17803/2713-0525.2022.2.20.309-328

Rocco, S. (2022). *Implementing and managing Algorithmic Decision-Making in the public sector* (No. ex93w_v1). Center for Open Science.

Rodríguez, N., Ser, J., Coeckelbergh, M., De Prado, M., Herrera-Viedma, E., & Herrera, F. (2023). Connecting the Dots in Trustworthy Artificial Intelligence: From AI Principles, Ethics, and Key Requirements to Responsible AI Systems and Regulation. *Information Fusion, 99*, 101896. DOI: 10.1016/j.inffus.2023.101896

Saura, J., Ribeiro-Soriano, D., & Palacios-Marqués, D. (2022). Assessing behavioural data science privacy issues in government artificial intelligence deployment. *Government Information Quarterly, 39*(4), 101679. DOI: 10.1016/j.giq.2022.101679

Smuha, N. A. (2025). *The Cambridge Handbook of the Law*. Ethics and Policy of Artificial Intelligence.

Sotiropoulos, L. (2025). Shaping the Future of Shipping: Legal and Policy Impacts of EU Emissions Trading and Fuel EU Maritime Regulations. In *AI, Sustainability and Solidarity in Advancing European Values in a Globalized Age* (pp. 95–106). Springer Nature Switzerland. DOI: 10.1007/978-3-032-06508-7_8

Sotiropoulos, L. D. (2023). Addressing Smart Contracts in the Insurance Sector: Institutional Framework and Practical Aspects. *ENTHA, 20*, 22.

Summerfield, C., & Parpart, P. (2022). Normative principles for decision-making in natural environments. *Annual Review of Psychology, 73*(1), 53–77. DOI: 10.1146/annurev-psych-020821-104057 PMID: 34555286

Swan, A., & Aguilar, Y. (2025). RESPONSIBLE AI BY DESIGN, NOT FOR COMPLIANCE.

Taj, I., & Zaman, N. (2022). Towards industrial revolution 5.0 and explainable artificial intelligence: Challenges and opportunities. *International Journal of Computing and Digital Systems, 12*(1), 295–320. DOI: 10.12785/ijcds/120128

Teng, Z., Xia, H., & He, Y. (2025). Algorithmic Fairness and Digital Financial Stress: Evidence from AI-Driven E-Commerce Platforms in OECD Economies. *Journal of Theoretical and Applied Electronic Commerce Research*, *20*(3), 213. DOI: 10.3390/jtaer20030213

Ullah, A., Chan, T. C., & Xie, Z. Y. (2024). Improving CNC turning machine precision through vibration analysis for clamping error detection employing principal component analysis.

Umapathy, T. (2024). *Behavioural Economics: Understanding Human Decision-Making*. Academic Guru Publishing House.

Wallsten, T. S. (Ed.). (2024). *Cognitive processes in choice and decision behavior*. Taylor & Francis. DOI: 10.4324/9781003469544

Walter, Y. (2024). Managing the race to the moon: Global policy and governance in Artificial Intelligence regulation—A contemporary overview and an analysis of socioeconomic consequences. *Discover Artificial Intelligence*, *4*(1), 14. DOI: 10.1007/s44163-024-00109-4

Walz, A., & Firth-Butterfield, K. (2019). Implementing ethics into artificial intelligence: A contribution, from a legal perspective, to the development of an AI governance regime. *Duke L. & Tech. Rev.*, *18*, 176.

Wang, J., Huo, Y., Mahe, J., Ge, Z., Liu, Z., Wang, W., & Zhang, L. (2024). Developing an Ethical Regulatory Framework for Artificial Intelligence: Integrating Systematic Review, Thematic Analysis, and Multidisciplinary Theories. *IEEE Access : Practical Innovations, Open Solutions*, *12*, 179383–179395. DOI: 10.1109/ACCESS.2024.3501332

Xue, L., & Pang, Z. (2022). Ethical governance of artificial intelligence: An integrated analytical framework. *Journal of Digital Economy*. DOI: 10.1016/j.jdec.2022.08.003

Zhang, J., & Zhang, Z. (2023). Ethics and governance of trustworthy medical artificial intelligence. *BMC Medical Informatics and Decision Making*, *23*(1), 7. Advance online publication. DOI: 10.1186/s12911-023-02103-9 PMID: 36639799

Zhong, L. (2023). A survey of prevent and detect access control vulnerabilities. arXiv preprint arXiv:2304.10600. Σιδηρόπουλος, Σ. (2024). *Παγκόσμια υγεία και κοινωνία των πολιτών: Χρηματοδοτικά εργαλεία και πολιτικές*. Εκδόσεις ΤΖΙΟΛΑ.

Chapter 5
Reconsidering the Rules:
Legal and Ethical Frameworks for the Governance of Foundation Models

Kumar Priyanshu Ranjan
https://orcid.org/0009-0008-5109-0276
Christ University, India

ABSTRACT

The emergence of foundation models in the recent past has completely changed the environment of artificial-intelligence governance. Unlike the strictly application-specific systems, foundation models are general-purpose infrastructures implemented in various sectors and jurisdictions, which creates systemic and changing risks. The available legal systems, based on sector regulation, data-protection law, and general liability principles, are poorly adapted to the magnitude, obscurity, and accountability dispersal of these technologies. This chapter has performed an extensive research of the legal and ethical dilemmas posed by foundation models through both doctrinal and comparative studies. It promotes an organization, life cycle regulatory framework which balances the enforceable legal requirements and documented ethical requirements. Through the concepts of proportionality, accountability by design, transparency, human controls and strong system of governance mechanisms, the chapter suggests a consistent and practical answer to governance of foundation models.

DOI: 10.4018/979-8-2600-0000-7.ch005

INTRODUCTION: A REVIEW OF THE RULES IN THE AGE OF FOUNDATION MODELS

The recent breakthroughs in the field of artificial intelligence resulted in the creation of more powerful systems whose abilities go far further than those of previous generations of AI (Russell and Norvig, 2021; Floridi, 2023). One of them is foundation models which are big, pre-trained models built on large, heterogeneous datasets and have shown emergent capabilities that outperform traditional task-specific or so-called narrow AI systems (Bommasani et al., 2021). More specifically, large language models and multimodal systems can serve a wide variety of downstream tasks, such as text generation, software creation, medical decision support, and customer support, commonly with fewer task-specific fine-tuning (Brown et al., 2020; Bommasani et al., 2021).

It has been identified that one of the key features of foundation models is their general-purpose nature. One model can be implemented in several sectors and institutional settings, providing high flexibility at the same time that creates an increased degree of uncertainty and risk (OECD, 2019). In contrast to the previous AI models that had very specific task involvement, foundation models serve as a platform around which many applications are created. As a result, biases, design mistakes, or constraints at the model level can be reproduced in a whole ecosystem of downstream applications (Bender et al., 2021). Researchers are increasingly referring to this trend that has intensely altered the nature of artificial intelligence systems design and implementation (Bommasani et al., 2021; Veale and Zuiderveen Borgesius, 2021).

This revolution poses an immediate threat to the current regulatory authorities. Most of the current strategies of AI governance, such as industry-specific policies and early risk-based designs, had discrete purpose-constrained systems in mind (European Commission, 2021; OECD, 2019). The regulatory interventions were traditionally directed at identifiable applications like biometric surveillance, credit evaluation or automated job new-hires. By comparison, foundation models are not directly dependent on a specific function or industry. They are susceptible mainly to post-deployment adoption and context of use, making their ex ante classification schemes more and more ineffective (Veale & Zuiderveen Borgesius, 2021).

This challenge is further complicated by the fact that there has always been a discrepancy between technical advancement and regulatory ability. The lawmakers and regulatory agencies usually do not possess the expertise or factual access to assess the actual functioning of the large-scale AI systems (Cave et al., 2019; Yeung, 2018). Without the honesty of transparency and record keeping, legal rules are prone to be under-protective in that they do not protect real harms, or over-protective, thus limiting innovation in ways not commensurate. Consequently, even laws developed specifically to control AI might fail to apply well to foundation models (Floridi et al., 2018).

Although an emerging body of literature is now questioning the rule of foundation models, this scholarship is still divided into three main schools of thought. The former is the risk-based regulatory model, which includes the European Union AI Act, as well as OECD models, and aims at categorising AI systems based on their possible harm and imposing respective compliance requirements (European Commission, 2021; OECD, 2019). The focus of this model is on ex-ante categorisation, technical risk management and conformity assessment. Nevertheless, researchers have raised concerns about whether it is appropriate to foundational models, which due to their general-purpose character can be transferred between risk types as time moves forward (Veale & Zuiderveen Borgesius, 2021).

There is a second line of research that focuses on innovativeness-driven and adaptive governance, such as regulatory sandboxes, soft-law, and voluntary standards (Cave et al., 2019; Raji et al., 2020). This view appreciates flexibility and involvement in the industry, yet has been criticized as being prone to under-regulating very powerful systems and favoring major technology players.

The third tradition is a rights-based and ethical approach to governance based on human rights and fairness and dignity (Floridi et al., 2018; UNESCO, 2021). Although normatively binding, this form of structure is frequently lacking in imposing machinery and is ineffective in addressing abstract values into functional regulatory responsibilities.

This chapter breaks away on its own of each of the three paradigms. It argues that neither risk classification nor voluntary ethics nor innovation-first regulation can be effective to govern foundation models. Rather it promotes an integrated lifecycle based model of governance integrating legal enforceability, ethical grounding and procedural oversight, specifically in response to general purpose, systemic AI.

THE ARTIFICIAL INTELLIGENCE DEVELOPMENT AND THE FOUNDATION MODEL EMERGENCE

1. Historical Development of Artificial Intelligence

The history of artificial intelligence has been marked by a series of different paradigms each of which is indicative of the current assumptions of how intelligent behaviour might be computationally simulated. Symbolic AI and expert systems dominated early studies in the middle of the twentieth century, which attempted to create models of human cognition based on explicitly defined rules and logical representations (Russell and Norvig, 2021). These systems were good in narrow scopes, like computing taxes or using medical rules, but were fragile when faced with uncertainty or new cases.

Criticism of the purely symbolic methods along with episodes of minimal investment and interest often termed as AI winters, led to a methodological dedication to statistical and data-driven approaches (Nilsson, 2010). Since the 1990s machine learning has become the new paradigm, allowing systems to deduce patterns using the data instead of basing their logic on hand-written rules (Mitchell, 1997; Domingos, 2015). Despite the fact that these models saw a marked extension of the field of AI use, they were still highly task-specific and relied on hand-selected features.

Another change was brought about by the emergence of deep learning at the beginning of the 2010s. The growth of computation, existence of massive datasets, and the creation of neural network designs resulted in significant improvements in performance across image recognition and natural language processing (LeCun, Bengio, and Hinton, 2015). In their specific form, transformer-based architectures have greatly redefined language modelling models as they allow systems to respond to long-range dependencies and contextual relationships at scale (Vaswani et al., 2017). Due to this, AI systems were more often integrated into daily applications, such as voice assistants or medical imaging.

2. Characteristic Features of Foundation Models

Recently, a novel group of AI systems known as foundation models has come into existence in the field (Bommasani et al., 2021). Such models are commonly described by massively large neural networks that are trained on large and diverse data that can be text, images, audio, and source code. Instead of being trained on one task, foundation models can be trained to perform a very broad range of downstream tasks via fine-tuning or prompt-based adaptation.

The main characteristic of foundation models over the previous systems lies in their size and generality in functionality (Bommasani et al., 2021). Their magnitude and scope in training data allow the capabilities to be developed that are not explicitly programmed by developers. One model can conduct translation, summarisation, and creative generation and rudimentary reasoning with little task-specific adaptation. Therefore, more and more specialised AI applications are being constructed as layers over a small set of underlying foundation models, which are strengthening the concentration in the AI ecosystem (Liang et al., 2023).

Another characteristic trait is Opacity. The internal logic of foundation models is represented in high-dimensional parameter spaces that are not easily intuited (Burrell, 2016). Even developers can not be sure of the way these systems will behave in new situations. This is not interpretable and makes it hard to diagnose any mistakes, responsibility, or genuinely explain any particular output (Weller, 2019).

3. Regulatory Implications Foundation Models

The extent, generality, and obscurity of foundation models create regulatory issues that are not only more significant, but qualitatively distinct, than those relating to previous AI systems. Training datasets usually feature huge amounts of scraped or aggregated information, which provokes the questions of privacy, intellectual property, and consent (Taddeo and Floridi, 2018; European Data Protection Board, 2020). Furthermore, the computational resources needed to train and deploy such models present environmental and sustainability issues that are not frequently considered in the law (Strubell et al., 2019).

Unpredictability is also enhanced by generality. In some situations, foundation models generate results that seem to be coherent and authoritative but factually false, biased, or harmful, based on the context and prompting (Bender et al., 2021; Weidinger et al., 2022). Since these systems can be used in a variety of domains, it is the case that sector-specific regulatory regimes are unable to realize the entire range of risk potentiality. Additionally, the liability can be shared between developers, deployers, and users, which generates confusion in the conventional liability frameworks (Edwards and Veale, 2017).

These features imply that foundation models cannot be reasonably governed by means of an incremental adjustment to the current legal regimes only. Rather than that, they need regulatory assumptions to be re-evaluated in relation to risk assessment, accountability, and the connection between law and ethics. The subsequent sections expound on this until they take a look at the shortcomings of existing legal frameworks and examine governance methods that can address the systemic aspect of foundation models.

RISK, TRANSPARENCY, AND SYSTEMIC GOVERNANCE OF FOUNDATION MODELS

The foundation models introduce a qualitatively different risk profile to the one that is related to previous generations of task-specific artificial intelligence systems. Conventional AI solutions are normally associated with localised, context-specific, and function or industry-specific risks. In comparison, foundation models are general-purpose infrastructures, which can be used in multiple ways in almost all domains, including healthcare, finance, education, media, and public administration (Bommasani et al., 2021).

Since the same underlying model can be used to implement many applications, vulnerabilities in the model (such as discriminatory representations, hallucinations, or security flaws) can spread to a number of applications at the same time. This brings the danger of systemic damage, and not isolated cases of inaccuracy. Systemic harm This is the propagation of negative impacts across systems that are interdependently related, such that a failure in one system can cause cascading impacts across a larger socio-technical context (Floridi, 2023).

There are a few facets of systemic risk that are of particular importance to foundation models. First, it has the possibility of amplification. Training data or optimisation goals can also have biases, and they can be recreated in numerous downstream tasks, greatly increasing their effects on society. Second, systemic risk has a close connection to emergence and unpredictability. When trained on a specific task, foundation models can have behaviours that were not explicitly trained or thought to happen, particularly in new or high-stakes environments (Weidinger et al., 2022). These behaviours are mostly revealed when scalable deployment and interaction with different user groups take place.

Another issue has something to do with the infrastructural concentration. The growing use of few and powerful foundation models that in many cases are governed by a few private players casts doubt on single points of failure and market concentration, as well as disproportional power over information ecosystems (Floridi, 2023). The consequences of decisions made by these actors such as technical, commercial or strategic acts of decision-making can thus extend beyond personal deployments.

These characteristics make it difficult to take regulatory approaches based on the intended use that is primarily ex ante classification of risks. A model initially used in low risk situations can subsequently be generalized to high risk decision making negating the viability of fixed regulatory types. Furthermore, numerous risks related to foundation models are hidden, situation-specific, and can only be seen with time, and it is difficult to assume that everything is foreseeable and controllable.

1. Dynamic and Precautionary Approaches to Risk Governance

The rapid change in foundation models has caused an increasing amount of skepticism towards stasis as a form of regulation. Traditional regulatory methods are often based on an ex ante assessment. In the symbiotic relationship of regulation and foundation models, this would define, at the time of deployment, whether the foundation model is based on a purpose (function) or sector (the Industry that it is used in). Traditional models work reasonably well when regulating restricted, narrow applications in stable environments; however, they do not address the 'evolving' type of capabilities being created with the foundation models. An example of this is how changing a little, insignificant line of code could produce a risk, with time, that is many times greater than the original risk of harm associated with the small change.

As a result, researchers and policy-makers have become more vocal on the need for dynamic precautionary regulations around the development of AI risk management.

In the case of dynamic regulatory measures, the recognition that AI risk is constantly changing with the maintenance and consistent upgrading of the model(s) is of utmost importance. Dynamic Precautionary regulation affords policymakers the ability to apply regulation to the ongoing identification and evaluation of uncertainty, which continues to shape the development of Artificial Intelligence Systems through the implementation of Adaptive Governance Models (Raji et al., 2020).

An essential element associated with these types of models is extending compliance evaluation from a single moment in time to an ongoing relationship with AI, rather than treating deployment as a discrete event that concludes all regulatory activity regarding the model. Within the framework of dynamic governance, the emphasis is on monitoring the performance of AI models post-deployment to enable continued oversight of the entire AI lifecycle, periodic reevaluation, and processes for recognizing and mitigating the potential emergence of new risks from the use of an AI model.

Dynamic governance is positively supported by similar methods of continuous monitoring and evaluation implemented in other areas, including but not limited to, high-risk industries such as pharmaceuticals and aviation, as well as banks and insurance companies, whereby risk and uncertainty are managed through continuous monitoring and reassessment, instead of through a single static endorsement or approval of an AI model.

In regard to foundation models, mechanisms of post-deployment governance may include mandatory reporting of incidents, monitoring of ongoing performance across different demographic and contextual settings, and monitoring for bias, misuse, and security-issues through continual audits. The tools utilized to govern foundation models recognize and reflect that there are many serious negative consequences associated with using foundation models, such as discriminatory outcomes, creating falsehoods and unsafe recommendations, that are only realized when the models are used in large numbers and interact with a wide range of users.

Also, dynamic governance requires institutional elasticity whereby the control authorities retain the capacity to consult and revise the standards, to revise the guidance and to recalibrate the enforcement priorities with the introduction of new evidence or to the creation of technological advances. Technology specific and inflexible rules and regulations are at risk of becoming obsolete within a world characterized by swift and frequent innovations. In this light, adaptive instruments [regulatory guidance; technical standards; controlled experimentation environments], are now viewed as more and more desirable supplements to the binding legal rules.

Meanwhile, the precautionary measures cause the issues of proportionality and innovation. Critics claim that regulation that is made on the basis of uncertain results, can impose unwarranted regulatory costs or deter socially beneficial applications of the technology. It has been proposed by precaution that caution should be taken when the foundation model technologies establish some critical socio-technical infrastructures that can influence the basic rights, public confidence, and democratic processes on a mass scale.

2. Transparency and Documentation as Governance Tools

In a modern-day context, transparency is an essential part of conversations surrounding the governance of foundation models; it acts as both a normative standard and a mechanism for effectively regulating them. For the governing body, transparency for AI systems means more than "explaining" the output. The AI system's development process (including Training & Evaluating) is just as important as how it produces its outputs and how well it maintains that functionality over time.

As a result of the vast scale and complexity of foundation models' Technical Structures alongside the fact that much of the decision-making process cannot be understood by anyone other than the developers, the industry has turned its attention to using Documentation as a means of Transparent Regulation rather than requiring Comprehensive knowledge of each proprietary aspect of the Technical Structure of a foundation model in order to oversee it. (Source: Doshi-Velez & Kim, 2017).

Two most prominent documentation techniques mentioned in the literature are the use of model cards and building a datasheet for datasets. A model card contains structured information around what the model is intended for, how it is supposed to perform, its limitations, and any ethical implications. In contrast to a model card, a datasheet for a dataset describes how the data was created, its composition, how it was collected, as well as any potential bias that may exist within the training data set.

Documentation transparency is key to Governing Foundation Models since many of the risks associated with foundation models exist upstream: at the point of selecting data or designing a foundation model; thus, it is essential for developers to make known all relevant information related to their training processes, methods of evaluation, and any risks they know about in order for regulatory agencies to help identify risk as soon as possible, and help deployers make better-informed decisions about deploying their models.

Although documentation standards are considered appealing in concept, they have not yet been widely used. Empirical analyses show that many of the AI developers leading the field provide very little or highly selective information about the sources of their training data, the processes of fine-tuning, and the results of their evaluations. This limits the capacity of outside parties to perform meaningful risk assessments on these models (Liang et al., 2023). These limitations have led to calls for the establishment of minimum standardised transparency requirements for foundation models that have the potential for a broad and systemic impact.

3. Layered Transparency, Explainability, and Procedural Accountability

Foundation Model Governance Depends on Sufficiently Distinct and Consolidated Levels of Transparency for Stakeholders

All users of AI will have different transparency needs when it comes to AI systems. Regulators, Developers, Deployers, and End-users, each interact with AI in differing ways as such, each will require different levels and types of transparency. The provision of a standardised transparency requirement will not meet the various levels of transparency required by all.

In order to effectively oversee AI, regulatory authorities must have access to very detailed documentation on a foundation model, evaluation results, and documents that outline the foundation AI's risk management activities. Examples of the type of information necessary to support the above oversight activities could include: model cards, dataset documentation, Red Team test reports, and logs of AI operational data collected through post-deployment monitoring. Although the disclosure of these documents may raise issues with respect to proprietary or trade secret information, a Regulatory authority can implement Controlled Access mechanisms to provide them regulatory oversight, while still recognising the rights of AI models' developers to maintain their commercial interests.

Downstream Developers and System Integrators will require another type of transparency. Developers and System Integrators must understand what the intended purpose of a Foundation Model is; what its limits are; and what the associated risks are before they can integrate it into their particular application in a responsible manner. Without this type of guidance, Developers and System Integrators may inadvertently use a Foundation Model in an application or area for which it is not suited, thus creating a higher probability of negative impacts.

In contrast, end-users benefit from clear notifications of the use of an AI system, representations of the capabilities and limitations of the system, and ways to lodge complaints regarding the workings of the system. Because of this transparency, end-users will be better able to engage with AI-enabled services, leading to increased confidence in AI.

A further dimension of knowledge governance for foundation models is explainability. Legal and ethical standards frequently require explainable AI; however, because of the complexity of the systems, it is usually not feasible to offer comprehensive causal explanations. For instance, while post-hoc explanations may provide some insight, they are generally rough approximations of how the AI made its decision (Lipton, 2018). In most cases, there is a trade-off between the two, especially for models that are highly complex.

So we have to think about the limits of what we can do. This means we need to make sure we have a plan in place to check that Artificial Intelligence systems are working properly. We do not need to know everything about how the model works. What is important is that we follow the steps when we make and use the model. This includes checking for problems, testing, making sure people are watching what is going on and double checking everything. Artificial Intelligence systems need to be checked in these ways. As a result of this approach, rather than establishing trust through knowing how the model operates, procedural accountability can create confidence through demonstrated adherence to adequate governance processes similar to those used in a number of other domains that place considerable emphasis on safety (of which AI is only one).

4. Systemic Governance Implications and Transition to Accountability

Combining risk, transparency and explainability indicates that the governance of foundation models needs to be viewed with systemic oversight rather than as product-specific regulations. Foundation models will increasingly serve as socio-technical infrastructures with effects that cross-sectoral boundaries, geographical boundaries, and institutional boundaries. As a result, the governance mechanisms that will be applied to foundation models must be developed to accommodate interdependence between regulatory domains, scale, and cumulative risks (Floridi, 2023).

Systemic governance of foundation models requires collaboration among regulatory domains that have historically operated independently of one another, including: Data Protection, Consumer Protection, Competition Law and Enforcement of Fundamental Human Rights. The fragmented oversight of these regulatory domains creates regulatory blind spots at the intersection of different regulatory domains where harms arise, and the ability of regulators to coordinate together will ensure that regulatory responses will be consistent and that responsibility will not be shifted to the extent of being ineffective.

Continuous learning is an additional pillar of systemic governance. By creating AI incident reporting systems; collaboration across multiple audit frameworks, using information-sharing systems across different sectors to allow both regulators and developers to be aware of the types of failure that are occurring repeatedly; and using incident reporting systems to create shared audit frameworks of AI development. This creates an environment for AI governance that is iterative rather than a one-time regulatory intervention, allowing for continuous development and adjustment of regulatory frameworks to match rapidly changing technology and real-world impacts.

AI governance structures are not effective because of how they are designed but rather because of their ability to be enforced. Enforcement in the framework of the foundation models is a special problem since regulatory bodies do not always have both technical access to AI systems and rights over their creators.

Such regulatory tools as impact assessment, documentation obligations, and audit requirements can only produce significant levels of accountability when anchored by inspection authorities, penalties, and institutional competencies. Without plausible enforcement, the best governance mechanisms will also tend to become a mere form of compliance or ethics washing. This danger is especially sharp in the jurisdictions where limited technical understanding, insufficient funding of regulators, or other weaknesses of the cross-border cooperation are prevalent.

Institutional infrastructure, such as technical audit units, algorithmic forensics, and international regulatory cooperation, is therefore also needed in foundation model governance not only in legal rules but also in institutional infrastructure. Translating the procedural safeguards into the practical limitations on the potent AI systems is impossible without such mechanisms.

5. Power Asymmetries and Market Concentration

Formulation and management of foundation models are almost entirely concentrated to a small group of technology companies endowed with enormous computational, data and money assets. This unequal distribution of technological authority brings up issues that go beyond the competition issue to include the democratic administration of vital informational infrastructure.

Since foundation models are becoming more and more the mediator of access to knowledge, communication, and public discourse, their governance should not be considered an entirely technical matter. Constructions of models, their safety constraints, and choices of who to deploy to, are normative and political decisions that directly affect whole populations. When such decisions are taken by the individual companies rather than official bodies the danger that business interests will overshadow the concerns of general good increases significantly.

The political-economic aspect of such a phenomenon underlines the fact that such a phenomenon needs to be controlled, transparent, and regulated, in the sense that foundation models should be viewed as infrastructural power and not as regular software.

ACCOUNTABILITY, LIABILITY, AND SECTORAL GOVERNANCE OF FOUNDATION MODELS

1. The Accountability Problem in Foundation Models

There has been a continuing issue with how to hold accountable all parties involved in the creation and use of foundation models. Previous generations of artificial intelligence (AI) involved a small number of individuals or groups doing well-defined jobs; therefore, when something goes wrong, it is easy to trace back who was responsible for the harm done to the victim. On the other hand, foundation models work within a multi-layered and complex environment, where the entities involved in their creation, adapting, deploying, and finally using them are spread across multiple institutions and actors.

The creator of the foundation model could be one entity, the fine-tuner or adapters could be another, the deployer could be an intermediary between the two entities, and the end-user could range from multiple sectors and regions. All parties involved in the process influence how the model operates, but no one party controls fully how the model operates. Because of this division of responsibility, traditional legal mechanisms that track a person's actions to the cause of harm will not work for foundation models (Edwards & Veale, 2017).

Some researchers refer to this issue in the context of AI governance as the "accountability gap." An accountability gap occurs when a fully autonomous or semi-autonomous AI system causes harm, yet current legal frameworks cannot define who should be held responsible for the harm caused by the system. Researchers (Matthias, 2004) have found that the lack of a clear definition of responsibility within the existing legal framework contributes to the accountability gap.

Because of the constantly evolving nature of the behavior of a foundational Model, it will continue to change over time and between environments. As such, the connection between the various design decisions made at the time of creating a foundational Model and the potential for resulting downstream effects diminishes rapidly as time passes.

Developers of foundational Models frequently argue that they cannot anticipate every possible use of their system after it has been released for adaptation as a general-purpose system. Therefore, developers may view themselves as unable to take into account the wide array of potential uses that others may have for their Model. Thereafter, deployers and integrators of foundational Models may cite insufficient insight into how the Model is designed and how it should be used as an overwhelming barrier to understanding the model's inner workings or limitations. Since end-users typically have the least control over a foundational Model's outputs and often utilize AI-generated outputs without a full understanding of the related risks, it is reasonable to conclude that broad responsibility for AI rests within the entire AI value chain. Therefore, clear responsibility for AI is dissipated rather than fixed.

Significant practical implications stem from challenges around accountability; when individuals cannot identify who should be held accountable for their actions, affected parties encounter barriers to obtaining reparation from harm, regulators face challenges in enforcing compliance with laws/rules/regulations and public confidence in AI governance structures will likely erode due to these issues, hence there is a need to rethink traditional models of accountability and create new models appropriate to the socio-technical, distributed nature of the ecosystem of the foundation model.Various organisations today use large language models (LLMs) to screen applicant resumes, create interview prompts, and identify the suitability of the candidates.

Where these systems are supported by underlying models that are trained on historic data of employment, they have the ability to replicate existing trends of gender, racial, or socioeconomic discrimination.

Unlike the conventional hiring platforms, the foundational models can be characterised as black-box excessive engines, which makes the internal decision-making processes transparent and unquestionable. Therefore, in cases of discriminatory results, the discriminated applicants can be unable to provide an evidential basis to prove causal relationships or intent whereas the employers can be left poorly placed to offer the logic behind the use of the model in making its decision. The result of this dynamic is a sense of accountability lacuna where the civil rights laws and employment regulations are faced with a challenge of defining liability.

This analysis is the basis for exploring liability issues and sector-specific governance issues associated with foundation models. The subsequent portion of this report reviews how traditional liability principles have addressed harm associated with foundation models, including why traditional notions of liability fail to provide adequate remedies and deterrent effects.

2. Limits of Traditional Liability Doctrines

Laws of product liability and tort liability are not sufficiently applicable to the types of harm caused by the foundation models. For example, tort law's existing liability categories focus on establishing cause and effect relationships between harm caused from products and the people/entities who are responsible for producing the products. However, foundation models present significant challenges to the existing assumptions of product liability and tort liability because they do not follow traditional definitions of agency, provide outputs that are probabilistic in nature and exhibit emergent properties from the deployment of multiple systems over long periods of time.

In order for a plaintiff to sue for negligence, they must demonstrate that the defendant owed a duty of care to them, breached that duty, caused their damages, and caused their injuries. However, for plaintiffs seeking to sue foundation model developers for negligence, the requirements to establish causation and breach become a major hurdle to recovery. Foundation model developers can argue that they were reasonably diligent in developing and testing their models, and foundation model users can argue that they were using third-party models into which they had little technical knowledge. Even under the most straightforward of definitions of direct traditional causation, the non-deterministic outputs of foundation models creates confusion about whether a certain design choice caused a certain harm.

There are common challenges that arise with product liability structures around the world. In more jurisdictions, strict liability will most likely only apply to physical goods. It is inconsistent among them in its treatment of software as a product for the purposes of liability. When liability for software does exist, it does not fall clearly into existing categories of product defect: manufacturing and/ or design defects. Artificial Intelligence (AI) foundation models are created to be very fluid, whereby the behavior of a system could change over time as a result of updates, tuning adjustments to the system, or because of changes in the context of deployment. This presents a challenge for establishing a clearly defined and stable product. In addition, as regards product liability, the principle of foreseeability creates an additional doctrinal restriction. Liability is often established by reference to whether the courts believed it was foreseeable at the time of the design/lifecycle of the product that it could cause harm to another. Foundation models are designed to be general-purpose models for use by many different individuals, therefore, it is difficult, if not impossible, to predict all the uses and misuses of the model. Many emergent behaviours (e.g., hallucinations, biased or unfair outputs, recommendations of unsafe behavior) will only be seen after widespread use (Weidinger et al., 2022), creating further complications.

The use of intermediary liability protections contributes to a more complex analysis when determining who is liable for an act of harm. In certain jurisdictions, intermediaries (i.e., platform operators) are provided immunity to liability for acts of harm caused by users' posting of material on platforms, and in some instances, this immunity may apply to any act of harm caused by an Artificial Intelligence (AI)-generated product or output that was posted by a user. Because some foundation models are created to generate (or mediate) content on a mass scale, the lack of intermediary liability greatly obscures who would be accountable for the harm relating to the misinformation, defamation, and/or discriminatory speech (Citron & Chesney, 2019).

The combination of the aforementioned limitations has contributed to the "responsibility gap" that has emerged from the fact that there is no definitive legal or equitable standard to assign liability for a harm that has occurred (Matthias, 2004). The nature of the liability gap surrounding foundation models is complicated by the large number of parties involved and the rapid evolution of the behaviour of these systems. Many of the traditional liability doctrines of assign effectiveness (deterrence) and compensation do not work effectively on behalf of those who suffer harm due to the new technology.

Given the inadequacies of existing rights and liabilities doctrine, more and more regulators and academics are considering ways to hold accountable those involved with foundation models, considering the different kinds of harms caused by AI. It is recognised that the accountability for AI harms needs to be shared across multiple parties involved within the AI value chain, in accordance with each party's level of control and influence over the AI technology (Veale and Zuiderveen Borgesius 2021).

One result of this is the recognition of shared or tiered responsibility models. Shared or tiered responsibility models develop obligations for each one of the parties (that includes those responsible for developing, deploying, integrating, and using an AI model) based on what responsibilities they should have in relation to their involvement with the layers of an AI technology lifecycle. Developers will be held to certain minimum standards based on the development of foundational models, namely relating to safety testing, documenting a model, and mitigating risks associated with the use of AI models, whereas deployers who use AI models in high-risk environments have an additional level of accountability associated with contextually assessing the model, maintaining human oversight, and adhering to specific rules that apply based on their profession or industry (European Commission 2021).

European Union regulatory systems have begun to reflect a general theme of distributed accountability. The introduction of the European Union AI Act distinguishes between the responsibilities of general-purpose AIs, and downstream users deploying high-risk AIs. The Act does not assign strict liability for every harm. However, it attempts to lessen the accountability gap by requiring documentation, establishing post-market monitoring requirements, and establishing a process for collaboration with regulatory authorities. Other proposals on the horizon aim at balancing the knowledge gap regarding the burden of proof in AI Liability. This will include the introduction of rebuttable presumptions based on proven violations of record-keeping, and transparency (European Commission, 2021).

Another method of addressing the asymmetric distribution of information is the implementation of systems for shifting burden to address the asymmetry of information between parties, and taking into account the level of sophistication necessary to prove fault or causation for an injury related to AI technologies. In these instances, the burden shifts to the provider/deployer of the AI to show that they have met their standard of care in relation to deploying that AI. Practically, this is based on the premise that developers/operators of these technologies are in a better position than the affected individual(s) to describe the behaviour of the system and the policies in place for managing risk (Wachter, et. al., 2017).

The soft-law tools and the standards-based approaches have tried to create responsibility using the organisational governance mechanisms to complement the statutory reforms. One can apply ex ante measures, that is, preventative interventions built into the system, that allows responsible practices which incorporate algorithmic assessments of impact, an independent third-party audit, and internal governance frameworks; this effectively limits ex post liability on potential harms. Even though these measures are not necessarily a direct cure, they are critical in preventing damage through early detection and prevention of risks.

One major issue with a shared responsibility paradigm is that it will result in the watering down of the incentives of individual actors to take responsibility for harms that they might incur. In addition, the heightened removal of boundaries of obligations among all parties makes it more difficult to comply with the entire range of applicable laws and regulations. However, shared responsibility models are a significant departure of the paradigm of binary fault to more holistic systems of governance that is more reflective of the highly distributed and socio-technical character of foundation-model ecosystems.

3. Sectoral Governance and Case Studies: Healthcare and Finance

Although foundation models expose cross-cutting accountability issues, the issues in this regard differ significantly in the various sectors. Mistakes or biases produced by the AI systems can be especially deadly in areas like healthcare and finance where the resulting choices directly influence physical health, economic welfare, and fundamental liberties. Therefore, the practice of regulating foundation models through sector-specific regulatory frameworks is decisive.

3.1. Healthcare Clinical Support and Health Information Systems

In medical environments, foundation models are being used to underpin clinical decision-making, medical summaries, patient triage, and health-related information provision using conversational interfaces. Most recent research indicates that large language models are capable of doing well on some medical reasoning problems, which has led to interest in their use in clinical practice. Simultaneously, their establishment also brings up the issues of acute accountability and liability.

One of the major risks is a production of incorrect or misleading outputs. It is possible to have foundation models giving responses that seem authoritative but are in fact clinically incorrect, poorly contextualised or inconsistent with set medical standards. Since these models are frequently trained on general medical literature, and not validated medical data, then their outputs might not be of the evidential standard required in a professional medical environment. When there is an injury, liability can be disputed between developers of models, health practitioners and medical practitioners who used AI-supported recommendations.

Multimodal foundation models and large language models are also under investigation as tools to be used in clinical documentation and diagnostic assistance as well as interacting with patients. Even though it is assumed that these systems can produce efficiency gains, there are important governance concerns that are posed by their implementation. Empirical studies show that giant language models can generate medically feasible, but factually incorrect, prescriptions, including unsafe prescriptions.

The outputs of foundation models do not often conform to standard medical safety due to the fact that they are trained on a heterogeneous and internet-scale data (not on a curated and clinical dataset). However, once incorporated into health care systems, these models can affect clinical opinion in ways that clinicians can hardly recognize. The question of liability is also unclear: the developer of the model did not create the particular medical application, but the healthcare provider might not have technical resources to check the underlying model.

This case is a perfect illustration of foundation models that transform single software malfunctions into the systematic risks of healthcare, thus revealing the drawbacks of conventional medical devices regulation and professional liability law.

Not all jurisdictions react similarly to regulation. Under the AI Act, medical use of AI systems will be considered as high-risk in the European Union, which will initiate the process of requirements on conformity assessment, human supervision, and post-market surveillance (European Commission, 2021). Conversely, most AI-based health tools in the United States are not subject to the Food and Drug Administration regulation, but presented as decision-support systems as opposed to diagnostic ones, which introduces regulatory loopholes (U.S. Food and Drug Administration, 2022).

To overcome these issues, regulatory and academic discussions begin to focus more on the value of the role allocation, the imperative of human control, and strict validation of AI systems within the clinical setting. The most common suggestion as a practical governance strategy is shared responsibility models, where developers are not required to achieve greater safety, but healthcare professionals retain full clinical responsibility.

3.2. Finance: Credit, Risk, and Auto Advice

The financial industry is also becoming familiar with foundation models, which are used in activities like the evaluation of creditworthiness, fraud prediction, user profile creation, and even the production of compliance-related documents. Since historical research on financial decision-making has been sensitive to bias and discrimination, the use of opaque and general-purpose AI systems brings up multiple legal and ethical issues.

The automated processes need transparency and most financial regulations in various jurisdictions disallow any discriminatory decision-making. In some countries such as the United States the Equal credit opportunity act stipulates that lenders must make meaningful explanations whenever adverse credit decisions are made. The foundation models, however, do not usually produce such explanations as interpretable which makes it more difficult to meet the current legal requirements (Barocas, Hardt, and Narayanan, 2019).

At the European Union level, credit scoring and financial risk assessment AI systems will probably be classified as high-risk under the AI Act which will mean that deployers will be subject to increased documentation, transparency, and human supervisory requirements (European Commission, 2021). Still, accountability issues remain as long as third party providers of foundation models are adopted and integrated into intricate decision pipelines.

In rebuttal, the financial regulators and supervisory authorities have now placed a greater emphasis on model governance frameworks, algorithmic audit procedures and control over third-party AI providers. These are the steps to be taken to make sure that the accountability to AI-driven financial choices is properly distributed and that those who might suffer as a result of automated systems have viable means of recourse.

4. Sectoral Governance and Case Studies: Education, Media, and Autonomous Systems

4.1 Education: Instructional Support, Evaluation and Academic Integrity

In the educational field, models of foundations are used to produce learning material, offer personalised tutoring services, mark on students, and comment on the work of students. Even though these applications are scaling and accessible, they also lead to major issues related to accuracy, bias, and data protection, as well as undermine pedagogical judgement.

The threat of inaccurate or disproportionate feedback is a major accountability issue. The foundation models, which are mainly trained on the prevalent linguistic and cultural datasets, might fail to capture the wide range of educational settings, which could be detrimental to students who represent the under-represented groups (Bender et al., 2021). Automated grading systems can also be used to hide the standards on which judgments are made and enable students to understand or object to the results.

The legal regulation of education is decentralised and institution-oriented and has weak AI-focused legal advice. As a result, the task of AI-assisted educational choices might be divided between model vendors, schools, and single teachers. New forms of governance are based on values such as transparency to students, stringent data privacy policies, and human oversight to retain professional educational judgment by the AI systems being complementary instead of substitutive (UNESCO, 2021).

4.2 Media and Journalism: Creation of Media Content and Information Verisimilitude

In news summarisation, translation, headline generation and automatic drafting of content, foundation models have already been rapidly used in the media and journalism sectors. These tools have the potential to make work more efficient but have severe threats to the integrity of information, such as the creation of fake content and the spread of misinformation and the loss of editorial responsibility (Citron and Chesney, 2019).

The issue of accountability in this area is especially severe. In the case of AI-generated content that causes reputational damage or helps to disseminate fake news, it becomes hard to place the blame of the issue on model developers, media organisations, and individual editors. The current legal steps regarding defamation and the responsibility of the press were not written to handle the production of content on a massive scale.

As a reaction, certain media organisations have also implemented internal governance policies that demand information about the use of AI be disclosed, that AI usage be subjected to human editorial oversight, and that clear correction mechanisms are implemented. However, these practices still remain mostly voluntary and disproportionate. Stricter guidelines on attribution, transparency, and responsibility of AI-generated media have been increasingly demanded by regulators and professional organizations, when the issue of democratic processes and trust in the population is involved (Wardle and Derakhshan, 2017).

4.3. Independent and Semi-Independent Systems: Human-Machine Interaction

Though foundation models are not generally considered to involve core functions of controlling autonomous systems, they appear to be increasingly deployed in human-machine interfaces, especially in transportation, robotics, and public services. These may be natural language interfaces to navigation systems, decision support prompts to operators and coordination tools to emergency response.

In certain safety-critical situations, physics can be very dire in regards to errors in the guidance or interpretation of AI-generated information. Such environments require accountability that is centered on the need to maintain clear distinctions between automated support and human choices. Truly, regulatory systems in transportation and industrial safety have focused on certification and testing and operator accountability. The incorporation of foundation models into them makes it questionable that current certification regimes are sufficient to reflect the risks posed by general-purpose AI components (Calo, 2016).

To address these risks, scholars and regulators tend to support a conservative deployment policy, intensive real-life testing and clear sharing of the risk to human operators and deploying organisations. The foundation models are not supposed to be seen as autonomous decision-makers and instead should be considered as tools that help in safety-related settings, and the liability frameworks must also acknowledge this distinction.

5. Synthesis: Toward Coherent Accountability Across Sectors

The above sectoral analyses show that despite the differences in the concrete expressions of harm related to foundation models by the variety of domains, the structural pattern of accountability issues is similar. In healthcare, finance, education, media and systems with safety critical operations, there are numerous players who bear the responsibility at various points of the AI lifecycle. It is a nationwide dispersion that makes it more difficult to apply the traditional doctrines of liability and exposes affected people to the risk of not having effective redress mechanisms (Edwards and Veale, 2017).

A consistent foundation model accountability should then be able to go beyond sector specific and siloed methods whilst also being aware of contextual variations. At the core of such a framework is the understanding of the relational nature of accountability as opposed to being a single entity. Different types of control exist on system behaviour by developers, deployers, and institutional users, and the obligations ought to reflect their position, capability, and influence (Veale & Zuiderveen Borgesius, 2021).

The point of convergence that is therefore emerging in the regulatory discourse is that upstream actors, especially those developing foundation models, are required to have baseline duties in regards to safety, documentation, and risk mitigation. These are duties of testing rigorously, revealing known limitations and collaboration with regulatory bodies and downstream users. Conversely, the downstream participants in the high risk situation are expected to bear added responsibilities regarding contextual evaluation, human management and adherence to the industry-wide legal standards (European Commission, 2021).

Procedural accountability mechanisms are very important in bringing these shared responsibilities into practice. Technological tools like algorithmic impact testing, independent audits, and post-deployment monitoring systems allow the regulator to assess compliance even in situations where cause-and-effect relationships cannot be established easily. Though these mechanisms do not exclude substantive liability, they assist in minimizing the lack of accountability through the creation of evidence, clarification of expectations, and enforcement (Raji et al., 2020).

Meanwhile, it follows that meaningful accountability must have redress mechanisms available. The victims of AI-made decisions should be in a position to question the results, get explanations when possible and claim damages when there is some harm. One way of presenting informational asymmetries and enhancing access to justice over AI-related disputes are the presence of burden-shifting regulations and rebuttable presumptions of fault, as suggested in some regulatory initiatives (European Commission, 2021).

Overall, the responsibility of foundation models cannot be ensured with the help of one specific doctrinal reform or regulation tool. It requires a unified solution of combining shared responsibility models, procedural protection, sector-specific regulation, and flexible enforcement. This would be a more robust way of structuring the legal responsibility to the facts of distributed agency and systemic risk into the existence of advanced AI systems.

GOVERNANCE INNOVATIONS AND GLOBAL ALIGNMENT PATHWAYS

1. The Need for New Governance Instruments

The unique properties of foundation models, specifically, their scale, generality, lack of transparency, and ability to produce systemic impacts, have also raised the question about the adequacy of traditional regulatory tools. Such mechanisms as the compliance checklists as a static, sector-specific licensing, ex post remedies are not created to handle technologies that develop quickly, are used in many areas, and produce diffuse instead of discrete risks (Floridi, 2023).

Innovations in governance have been developed in reaction to two chronic shortcomings of the current regulatory strategies. To begin with, they are aimed at solving uncertainty. Due to the fact that much of the risks related to foundation models cannot be assessed in detail before deployment, regulatory frameworks should be able to react to unexpected behaviour and unforeseen harms. Second, the innovations will reduce the institutional and informational asymmetry between regulators, developers, and other stakeholders working across jurisdictions and sectors (Cave et al., 2019).

Notably, the innovations of governance are not aimed at substituting the enforceable legal regulations. Rather, they are facilitating instruments that are used to operationalize abstract legal and ethical values into action. Impact assessment, audit, registry and regulatory sandbox are examples of tools that implant oversight into the AI lifecycle, allowing regulators to oversee systems throughout their life but keeping some level of flexibility required to drive innovation.

This transformation is also indicative of a general re-strategisation of sophisticated AI systems into regulatory rhetoric. It is becoming recognized that foundation models are not just products and services, but as types of digital infrastructure, the governance of which poses problems similar to those related to financial systems, energy networks or communications platforms. In this sense, resilience planning, ongoing supervision, and coordination on a national and international scale are necessary to oversee this efficiently (Floridi et al., 2018).

It is against this backdrop that the following sections consider a collection of governance systems that have been of recent regulatory ardor. These are algorithmic impact assessments, audits, regulatory sandboxes, model registries along with international coordination frameworks, which together cover different aspects of the governance issue, and have the overall effect of making foundation models regulation more anticipatory and adaptable.

2. Algorithmic Impact Assessments and Audits

The role of algorithmic impact assessments (AIAs) and audits has become more and more popular as the tools of addressing the risks of foundation models. These systems strive to distinguish, assess as well as address any form of harms with a view of integrating awareness and accountability of risks into the processes of organisational decision making. Unlike regular compliance inspections, where the focus is usually put on formal compliance with the provisions of fixed rules and regulations, AIAs and auditing anticipate ongoing assessment of the entire AI lifecycle (Raji et al., 2020).

The impact assessments typically require the organisations to state the intended use of an AI system, speculate potential risks to people and society, and record mitigation approaches. In the context of the foundation models, AIAs are becoming more important since one model could be used in various environments that could possess vastly different risk profiles. This means that a model level baseline assessment can be supplemented by a set of context-specific assessment at the high-risk downstream applications.

Audits are used to supplement impact assessments, providing systematic audits of the use of an AI system in accordance with the relevant laws, ethics, and technical requirements. These audits can be conducted in-house or by third-party experts and can involve the factors of bias by demographic, susceptibility to adversarial data, data sovereignty policies, and compliance with proclaimed restrictions of use (Mitchell et al., 2019). Where foundation models are involved, auditing is usually done with assessments of both up-stream development methods and down-stream deployment situations.

Regulations have already begun to institutionalise the place of impact evaluation and audits. In the example of the AI Act in the European Union, they are required to have risk-management systems in place and complete conformity assessments during their lifetime (European Commission, 2021). Similarly, the proposed Artificial Intelligence and Data Act by Canada anticipates impact assessment of high-impact AI systems, and thus, is indicative of an extended regulatory shift to procedural governance.

Although they are increasingly becoming prominent, AIAs and audits are faced by serious practical challenges. Proper evaluation depends on the availability of substantive technical information which can be inconsistent with proprietary interests or trade-secret controls. Moreover, there are the chances that testing becomes more of formalities rather than a substantive evaluation. To deal with these issues, researchers emphasise the importance of transparent requirements, external regulation and penalties in case of non-adherence (Raji et al., 2020).

Well-constructed and executed algorithmic impact assessments and audits may be fundamental elements of foundation model governance. These mechanisms contribute to the practicalisation of the principles of accountability, fairness and transparency by institutionalising reflection, documentation and external scrutiny.

3. Regulatory Sandboxes and Experimental Governance

Regulatory sandboxes have become a frontline tool of regulation over uncertainties led by rapidly developing technologies such as foundation models. A regulatory sandbox is an environment where AI systems are experimented under regulatory controls before being widely deployed. This method will allow regulators to see how systems behave in practice, uncover emerging risks, and sophisticated regulatory expectations but not make them quickly set strict compliance requirements (Cave et al., 2019).

Sandbox arrangements play a number of benefits in the scenario of foundation models. Since it is challenging to predict the behavior of the general-purpose models in different settings, sandboxes allow restricted experimentation with set safeguards and monitoring tools. Developers will have the ability to test mitigation strategies, alignment techniques and usage constraints and regulators have the ability to get empirical understanding of real world performance as opposed to laboratory results. Through this process, information asymmetries can be minimised and help to make better regulatory decisions.

There are numerous jurisdictions in which sandbox initiatives are implemented in the context of broader AI governance. In the United Kingdom, the regulators have launched AI-oriented sandbox programmes that would promote responsible innovation without losing control. Similar efforts in other jurisdictions, like Singapore and Canada, are indicative of a growing appreciation of the fact that experimental governance may act as a supplement to traditional regulation mechanisms, especially in areas where technological change is high.

At the same time, regulatory sandboxes create pertinent governance issues. When only big or better-funded firms are involved, sandbox programmes will only serve to strengthen market concentration and lock out smaller developers or actors of the public interest. Besides, it has the risk that the participation in the sandbox might be viewed as a regulatory leniency instead of a pre-compliance measure. In order to deal with these issues, a good sandbox design must have explicit eligibility policies, transparency of consequences, and ways of ensuring lessons learned are brought to bear on wider regulatory frameworks.

Regulatory sandboxes must be conceptualised as transitional governance instruments in the framework of foundation model. Their main merit is that they allow the regulators and developers to produce knowledge jointly regarding risks, safeguards, and proper use conditions that help to create more responsive and evidence-based regulatory regimes instead of acting as long-lasting exemption of the legal framework.

4. Model Registries, Licensing, and Pre-Deployment Controls

With foundation models becoming more and more of an enabling infrastructure, rather than a tangible product, regulators and scholars have suggested more formalised systems of control, such as model registries and licensing regimes. These tools are supposed to augment the visibility, traceability, and control of the development and deployment of AI systems whose size and generality result in systemic risks (Floridi, 2023).

Model registries are other forms of registries that entail the creation of centralised data where developers share the most important details regarding AI systems before or at the time of deployment. The information that can be disclosed includes description of model architecture, training data used, computational resources used, intended use, known limitations, and applied risk mitigation measures. Registries can help regulate, enforce and coordinate through the ability to track the spread of foundation models into other fields. Similar processes can be seen within other controlled areas, including the pharmaceutical approval databases and the aviation safety registries.

The more interventionist method of governance is licensing schemes. Within a licensing model, users or deployers of some types of foundation models would have to seek regulatory permission before deployment. The conditions of licensing can be adjusted to the possible effect of the model, and more difficult requirements are made on systems that can be used widely or with high-stakes. Such requirements may entail obligatory audits, tests of security, protective measures of human supervision, and obligations of post-deployment checks.

Tremendous of the registry and licensing are already apparent in new policies of regulation. As an example, the AI Act by the European Union provides registration requirements of high-risk AI systems and increased transparency requirements of general-purpose AI models that can pose systemic risk (European Commission, 2021). The use of algorithm registration as a governance model has been tested in China on selected AI services, and this is a manifestation of a governance model that focuses on pre-deploying strong artificial algorithm systems (Creemers, 2023).

Opponents of licensing regimes warn that when too restrictive measures are imposed it can hamper innovation, especially in the case of smaller developers and open-source projects. The jurisdictional divergence also decreases the implementation as the national licensing requirements can be challenging to align in an international AI market. Proponents answer that pre-deployment regulation is warranted when foundation models fall under other types of critical infrastructure and a well-crafted licensing program may provide a balance between innovation and social responsibility and safety.

Model registries and licensing mechanisms are to be seen as part of a graduated governance toolkit as opposed to general solutions. These tools used in combination with impact analysis, audits, and post-deployment monitoring can help to achieve a more proactive and consistent way of governing foundation models.

The technocratic orientation and reliance on experts in specific domains and regulatory authorities and corporate stakeholders define the current prevalent AI governance frameworks. Nevertheless, the foundation models are having more impact on the community dialogues, labor markets, availability of information, and political dialogs. Therefore, their rule brings about democratic and technical anxieties.

Strong involvement by the populace cannot be done without when it comes to authorizing the control of foundation models. The affected communities should be in a position to have avenues to challenge deployments, seek clarifications and influence policy agendas. Without such measures in place, AI governance can become a source of power asymmetry whereby decision-making in respect to social infrastructure is made by a limited group of technocratic and corporate elites.

The integration of democratic control into the relation of the foundation models might require the representativeness of the people, the presence of the civilization in the regulatory institutions, and transparency requirements that would enable the power of journalism and academic research. These tools complement the use of law by making sure that the processes of governance have been made to reflect on societal values and that they are not dominated by technical efficiency.

5. Governance of Foundation Models in the Global South and Capacity Building

The spread of paradigm models throughout the globe has also brought the differences in regulatory capacity between technologically advanced economies and the Global South into fresh focus. There are many low- and middle-income nations with no well-developed data protection systems, specific AI regulation, and the technical skills required to monitor the operation of complex AI systems accurately. As a result, basic models are often imported and implemented in these jurisdictions without local control and question the accountability, exploitation, and unequal risk exposure (Gurumurthy and Chami, 2022).

One of the oft-referred issues here is the threat of AI colonialism, whereby ownership of technological buildings, data, and economic worth are held by only a few global technology companies and areas. Local populations in such environments can be observed, unfairly determined or exploitative of data without sufficient input into the process of governance or means of redress. The implementation of the basic models without proper local control can thus strengthen the existing social and economic disparities.

Capacity building is one of the key pillars towards dealing with these issues. Good management of foundational models in the Global South requires long-term investment in regulation institutions, technical capacity, and legal frameworks. This includes educating regulators, judges, and policymakers on the matters of AI; promoting the creation of local standards and guidelines; and promoting the research and involvement of civil society. Transnational alliances, international organisations, development agencies can therefore be instrumental in transferring of knowledge and strengthening of institutions.

The institutions in the Global North are the main creators, owners, and regulators of foundation models even though their social implications are the most acute in the developing regions. These systems have an impact on accessibility to education, labor, credit, and information in jurisdictions that lack regulatory capacity. Without specific governance and capacity-building mechanisms, foundation models can continue to reproduce the dynamics of digital colonialism where data, value, and control are transferred to the outside world, and the respective social risks are concentrated in one place.

Regional cooperation is an alternative volume of enhancing the capacity of governance. Regional establishments have the ability to come up with a common regulatory framework, share technical resources and organize oversight activity so as to entail social-economic conditions and law traditions. This kind of collaboration can strengthen the bargaining power against the international technology suppliers and reduce the dependency on the externally provided regulatory frameworks that might be incompatible with the national interests.

Lastly, inclusive global governance requires the developing countries to meaningfully be involved in the international AI norm-setting processes. Without this kind of participation, global standards will be placed at the risk of representing the interests and values of a small group of actors. Thus, it is crucial that the Global South views should be included in the discussion of the foundational model regulation to achieve fair, context-specific, and sustainable regulatory practices.

TOWARD AN INTEGRATED FRAMEWORK FOR RESPONSIBLE REGULATION OF FOUNDATION MODELS

1. Rethinking Regulation Across the AI Lifecycle

The above analysis has shown that none of the regulatory tools, such as basing on liability law, ethical principles, and procedural governance mechanisms, are sufficient to resolve the issues of foundation models. Their systemic risk generating nature, their deployment in various sectors and different jurisdictions require a more integrated regulatory approach. Siloed or fragmented responses in terms of governance are inappropriate to technologies whose effects transcend institutional and legal borders (Floridi, 2023).

Such integration can be achieved using a lifecycle-based regulatory view. Instead of focusing the regulatory focus on the deployment point or when damage is already inflicted, lifecycle governance acknowledges that risks arise at a variety of points, such as data collection, model training, evaluation, deployment, use, and subsequent updates. There will be different actors in each of the stages and will offer various possibilities of intervention and control (Raji et al., 2020).

During the development phase, data governance practices, model design decisions and pre-training risk assessment are the best areas to put the regulatory focus. The decisions made by developers of foundation models have a strong upstream control of downstream results, in terms of the data sets composition, optimization goals and safety restrictions. The baseline requirements at this stage can thus entail documentation of the dataset, testing of the bias, testing of robustness and disclosure of known limitations.

Governance responsibilities are more transferred to deployers and institutional users in the course of deployment and integration. Contextual risk assessment is especially relevant here since the same foundation model can be very risky in one context and obscure risk in another. Human oversight, transparency facing the users and adherence to standards that are sector specific all play a role in ensuring that the decisions made during deployment are in line with the legal requirements and the expectations of the society.

Post deployment governance is another key element of an integrated framework. Constant checking, reporting of incidents and reassessment periodically can help regulators and organisations to detect arising harms and adjust safeguards as time goes by. Since foundation models can be updated and fine-tuned over time, deployment should be a continuous regulatory issue and not a singular event.

Collectively, a lifecycle-based approach offers the framework of combined regulation. The subsequent sections expound upon the main principles that are underpinning such a framework and how they can be transformed into specific legal and institutional mechanisms.

2. Core Principles of an Integrated Regulatory Framework

A unified regulatory framework of foundation models needs to be based on a consistent complex of principles that guide the design of law as well as practice of governance. The principles help to bridge various regulatory tools, which are uniform throughout jurisdictions, sectors, and AI lifecycle phases. Although their actual application might differ, comparative regulatory analysis shows that there are a number of principles of governance that are shared between emergent forms of governance.

Proportionality is the first of such principles. The regulatory requirements must be adjusted to the degree of risk of a specific foundation model and its application. Considering the variety of applications that general-purpose systems are used in, it would be inefficient and normatively inappropriate to regulate them in a uniform way. The concept of proportionality means greater control over higher-risk uses (including the ones that impact fundamental rights, safety of the populace, or democracy) whereas less risky applications are the ones that are not regulated as strongly (European Commission, 2021).

The second principle is accountability by design. Instead of thinking of accountability as a tool that is activated once a harm has happened, integrated regulation focuses on accounting principles as being embedded in the design and implementation of AI systems. This involves recording the choices in the design, assigning responsibility to actors, and providing traceability between the model lifecycle. Institutionalisation of accountability at the earliest stage can decrease the need to cause harm and can be used to enforce rules in case of failures (Raji et al., 2020).

A third principle is that of transparency and traceability. Integrated governance relies on the presence of useful information on the foundation models, such as the information about training data, the purpose they are to serve, limitations, and mitigation of risks. Notably, transparency has to be purpose-specific: regulators, deployers, and end-users need various types and levels of disclosure. Traceability makes sure that decisions of earlier stages can be analyzed and assessed later on which facilitates oversight, learning and redress.

A fourth principle is human supervision and regulation. Individual regulation should also be taken to assure that critical social or individual decision-making leaves some meaningful human input. This would not be associated with the need to have human intervention at all times but rather the ability to supervise, intervene, and override where needed. Human supervision is seen as a resistance to automation bias and serves to strengthen the normative assumption that decision-making and responsibility are always left in the hands of human actors as opposed to computerized systems (UNESCO, 2021).

Lastly, flexibility is vital to the sustainability of cohesive regulatory systems. With the rate at which the field of technology is altering, the arrangements of governance should be able to change with the evidence, threats, and the changing social expectations. The adaptive regulation is based on periodical review, stakeholder consultation, and regulatory experimentation in order to be effective in the long run.

Combined, these principles offer the normative framework of comprehensive approach to foundation model governance. The second section discusses the ways these principles can be operationalised using tangible legal and institutional processes.

3. Operationalising Integration Through Law and Governance Mechanisms

The application of the main tenets of an integrated regulatory system in practice requires the implementation of tangible legal and institutional devices, which will be mutually consistent throughout the AI lifecycle. The integration itself does not necessarily imply the creation of completely new regulatory frameworks, instead, it could be accomplished by adopting pre-existing legal frameworks, such as the data protection law, sector regulation, liability systems, and new AI-specific tools, in regard to a coordinated governance framework.

Among the key tools that can be used to operationalize the integration, I would see the lifecycle-based assigning obligations. In this paradigm, legal obligations are attributed to specific actors based on the level where they actively master an AI system. Foundational model developers can be the ones who take core responsibility of the upstream problems of dataset governance, pre-training risk evaluation, technical documentation, and baseline safety testing. Institutional users and deployers, especially in situations involving high-risk should be obligated to assume extra responsibilities related to contextual impact assessment, human supervision and adherence to legal standards specific to the sector. Such assignment helps in minimizing ambiguity in terms of responsibility and eliminates accountability.

A second mechanism is procedural harmonization. Integrated regulation enjoys the fact that the procedural requirements (impact assessments, audits, record-keeping, reporting requirements, etc.) are aligned within the different regulatory arenas. In the context of a consistent use of analogous procedural tools when it comes to data protection, consumer protection, and AI-specific governance, compliance will be more consistent, and enforcement will be tighter. Procedural alignment also helps in the sharing of information among regulators hence leading to deeper insight of systemic risk.

Institutional coordination has a complementary effect on operationalizing integration. Government of foundational models entails cooperation of regulatory bodies that have always worked in their respective mandates. Coherent enforcement activities, joint directions, and information-sharing can assist in making sure that risks associated with AI are handled concurrently across areas. Cooperation between regulators at the international level further reduces fragmentation and promotes interoperability between national regimes of governance.

Lastly, feedback and learning mechanisms can also be used to strengthen integration. Both regulatory updates and organisational practices should be informed using information generated as a result of post-deployment monitoring, incident reporting, and audits. Through incorporating a feedback loop into the governance structure, regulators and regulated parties can alter their strategies based on the empirical evidence as opposed to the use of the ex-ante assumptions only. This learning-based model is especially well adapted to foundational models, the risks and impacts of which might not be even obvious unless they are put into actual use.

4. Challenges and Trade-Offs in Integrated Regulation

Despite the fact that an integrated regulatory framework provides a convincing answer to the governance challenges that are tied to foundation models, its application is associated with challenges (Kouroupis, K. & Efthymiou, I., 2026). The integration process creates a cacophony of useful and normative trade-offs that policymakers are forced to face so as to make sure that regulatory regimes are viable, proportionate and legitimate.

The first challenge is related to regulatory complexity. Integrated governance inevitably involves several areas of law, actors at the institution and procedural demands. Whereas coordination has the ability to improve coherence, it can also increase compliance burdens, especially to small developers, research institutions, and deployers in the public sector, who have finite resources at their disposal. Otherwise carefully adjusted, these frameworks pose the risk of entrenching the benefits of large, well-endowed companies that are in a better position to absorb compliance expenditure.

Institutional capacity is a correlative limitation. The success of integrated regulation is based on the ability of regulators to be technically skilled and adequately staffed and have effective analytical instruments to monitor the complex AI systems. In many jurisdictions, regulators are faced with knowledge gaps in the expertise to review foundation models, to examine technical documentation or to perform substantial audits. Lack of capacity can catalyze inconsistent enforcement and hence undermine the pragmatic effectiveness of regulatory requirements.

The other trade-off that arises is between flexibility and the law. Integrated frameworks are often based on malleable tools, like iterative guidance, post deployment monitoring, and discretionary enforcement, to be responsive to the technological change. Although this flexibility enhances flexibility, it can also reduce predictability of regulated parties and hinder judicial control. The main design issue is to find the right balance between adaptive governance and well-defined and enforceable legal norms.

Tensions between regulative goals require their attentive control. Transparency demands can conflict with intellectual property rights, cybersecurity needs or trade secrets. Similarly, risk governance methods based on precaution can be incompatible with the policy objectives related to innovation, competitiveness or access to technology. Competing objectives must thus be balanced within an integrated regulation, including by incorporating mechanisms, e.g. proportionality analyses or context-specific exemptions.

Last, integrated governance also brings back the issue of democratic accountability and legitimacy. With regulatory oversight being more based on technical standards, expert-led audits, and more complex procedural mechanisms, there is a risk that key-decisions will be cocooned against public oversight. To maintain trust and legitimacy in integrated AI regulation, it is necessary to ensure the transparency of the governance processes, stakeholder participation opportunities availability, and easy access to the contestation.

5. Future Directions and Concluding Reflections

As it is identified in the analysis presented in this chapter, legal and regulatory approaches to artificial intelligence need recalibration regarding the unique nature of foundation models. These systems undermine the assumptions that have been deeply entrenched in the current systems of governance, such as well-defined sectoral boundaries, consistent risk profiles, and chain responsibility. Solving such problems needs to move towards regulatory frameworks that are both integrated, adaptive and based on both legal enforceability and ethical commitments.

In the future, anticipatory governance is more likely to be significant. Instead of using reactive interventions when the damage is already done, regulatory systems need to invest into foresight-oriented practices, including stress tests, horizon scanning and early warning systems. These practices allow regulators to detect possible risks arising out of foundation models at an earlier stage before they become present at scale, thus facilitating more prompt and adequate responses.

Inter-country cooperation will be an indispensable practice. An overall harmonisation of AI regulation on a global scale might not be practical, but functional harmonisation by mutual standards, interoperable processes, and mutual learning can also significantly decrease the level of fragmentation. Since the foundation models are transboundary, it is essential to have coordination between the regulators, standards and international organisations to offer uniform protection of fundamental rights and public interests.

Another priority is inclusive governance. The regulatory frameworks should consider views and experiences of communities who will be impacted by the implementation of AI, especially those communities that have been marginalized in the past in technological decision-making processes. Significant engagement, openness in regulation procedures, and easy redress channels are not only the issues of fairness but also the preconditions of legitimacy and efficacy of AI governance.

Last but not least, integrated regulatory frameworks have to adapt with change in technology and social context. To avoid regulatory obsolescence, the periodic review, evidence-based revision, and institutional learning needs to be integrated into the governance arrangements. The flexibility, however, must not be sacrificed at the costs of legal certainty and accountability. One of the characteristics that the regulation of AI will confront going forward will be meeting a proper balance between flexibility and enforceability.

To say the least, the responsible governing of the foundation models cannot be implemented by simple isolated legal reforms or ethical statements. It also needs a holistic methodology that harmonizes lifetime based regulation, shared responsibility, governance innovations, and international coordination. Regulators can enhance the risks management to achieve socially beneficial innovation by incorporating moral values into legally enforceable frameworks and adjusting governance systems to the realities of complex and general systems of AI.

CONCLUSION

The introduction of foundation models marks a major change in the technology and regulation realms of artificial intelligence. Unlike antecedent, task-specific systems, foundation models are general, self-adaptive systems, the rampant use of which in many sectors and legal jurisdictions creates systemic and not isolated risks. This metamorphosis is a revelation of the natural gaps of the existing legal systems, which continue to believe in constrained functionality, predictable risk patterns, and defined responsibility.

The chapter shows that existing systems of AI regulation, be it based on data protection laws, industry-specific rules, or traditional liability principles, are not up to the challenge of dealing with the scale, transparency, and decentralization of harms related to foundation models. The vacuums of accountability persist in the context of diffused responsibility through the chain of development and deployment, and the absence of dynamic and situational nature of possible harm is represented by the fixed ex ante risk classification. Even though the principles of ethics are imperative, they are not enough when they are discussed as voluntary or aspirational principles that have no connection with definite legal duties.

Therefore, the chapter offers a coherent system of regulation that covers the entire AI lifecycle and aligns the legality of law with ethical regulation. Combining proportionality, design-based accountability, transparency, human agency, and flexibility with such instruments of governance as impact evaluations, audits, regulatory sandboxes, and documentation requirements, integrated regulation can provide a more practical and normatively-based approach to controlling systemic AI risk.

The models that are founded are mainly trained, owned, and controlled by the institutions located in the Global North, yet the social consequences are experienced in the developing countries. These systems determine access to education, labor markets, credit and information in jurisdictions that have limited regulatory power. Without specific governance and capacity-building efforts, foundation models are a threat of repeating scenarios of digital colonialism, in which data, value, and control are flowing outside, and social risks are sitting locally.

Finally, the prudent management of the foundation models requires a reconceptualization of the paradigms of regulations. Such systems need to be viewed not merely as software products but as socio-technical infrastructures whose governance involves the essential rights and democratic institutions and long-term citizen trust. A globally sensitive, responsive, and integrative regulatory framework can be seen as the most plausible way to ensure that the foundation models are represented and implemented in a way that they meet accountability, human dignity and the general will.

REFERENCES

Barocas, S., Hardt, M., & Narayanan, A. (2019). *Fairness and machine learning*. fairmlbook.org

Bender, E. M., Gebru, T., McMillan-Major, A., & Shmitchell, S. (2021). On the dangers of stochastic parrots: Can language models be too big? In *Proceedings of the 2021 ACM Conference on Fairness, Accountability, and Transparency* (pp. 610–623). ACM. DOI: 10.1145/3442188.3445922

Bommasani, R., Hudson, D. A., Adeli, E., ... (2021). On the opportunities and risks of foundation models. *arXiv preprint*. https://arxiv.org/abs/2108.07258

Brown, T. B., Mann, B., & Ryder, N.. (2020). Language models are few-shot learners. *Advances in Neural Information Processing Systems*, *33*, 1877–1901.

Burrell, J. (2016). How the machine "thinks": Understanding opacity in machine learning algorithms. *Big Data & Society*, *3*(1), 1–12. DOI: 10.1177/2053951715622512

Calo, R. (2016). Robotics and the lessons of cyberlaw. *California Law Review*, *103*(3), 513–563.

Cave, S., Nyrup, R., Vold, K., & Weller, A. (2019). Motivations and risks of machine ethics. *Proceedings of the IEEE*, *107*(3), 562–574. DOI: 10.1109/JPROC.2018.2865996

Citron, D. K., & Chesney, R. (2019). Deep fakes: A looming challenge for privacy, democracy, and national security. *California Law Review*, *107*, 1753–1819.

Creemers, R. (2023). China's governance of artificial intelligence: A regulatory overview. *Journal of Cyber Policy*, *8*(1), 1–24.

Domingos, P. (2015). *The master algorithm*. Basic Books.

Doshi-Velez, F., & Kim, B. (2017). Towards a rigorous science of interpretable machine learning. *arXiv preprint*. https://arxiv.org/abs/1702.08608

Edwards, L., & Veale, M. (2017). Slave to the algorithm? Why a "right to an explanation" is probably not the remedy you are looking for. *Duke Law & Technology Review*, *16*, 18–84. DOI: 10.31228/osf.io/97upg

European Commission. (2021). *Proposal for a regulation laying down harmonised rules on artificial intelligence (Artificial Intelligence Act)*. COM(2021) 206 final.

European Data Protection Board. (2020). *Guidelines on processing personal data in the context of connected vehicles and mobility related applications*.

Floridi, L. (2023). AI as an agency without intelligence: On the philosophical foundations of artificial intelligence. *Philosophy & Technology*, *36*, 1–21.

Floridi, L., Cowls, J., Beltrametti, M., Chatila, R., Chazerand, P., Dignum, V., Luetge, C., Madelin, R., Pagallo, U., Rossi, F., Schafer, B., Valcke, P., & Vayena, E. (2018). AI4People—An ethical framework for a good AI society. *Minds and Machines*, *28*(4), 689–707. DOI: 10.1007/s11023-018-9482-5 PMID: 30930541

Gebru, T., Morgenstern, J., Vecchione, B., ... (2018). Datasheets for datasets. *arXiv preprint*. https://arxiv.org/abs/1803.09010

Gurumurthy, A., & Chami, N. (2022). AI and the Global South: Designing inclusive futures. *Development and Change*, *53*(4), 901–918.

Kouroupis, K., & Efthymiou, I. (Eds.). (2026). *Legal Challenges of AI Across Interdisciplinary Sectors*. IGI Global Scientific Publishing., DOI: 10.4018/979-8-3373-5452-1

LeCun, Y., Bengio, Y., & Hinton, G. (2015). Deep learning. *Nature*, *521*(7553), 436–444. DOI: 10.1038/nature14539 PMID: 26017442

Liang, P., Bommasani, R., & Lee, T.. (2023). Holistic evaluation of language models. *Transactions of the Association for Computational Linguistics*, *11*, 531–551.

Lipton, Z. C. (2018). The mythos of model interpretability. *Communications of the ACM*, *61*(10), 36–43. DOI: 10.1145/3233231

Matthias, A. (2004). The responsibility gap: Ascribing responsibility for the actions of learning automata. *Ethics and Information Technology*, *6*(3), 175–183. DOI: 10.1007/s10676-004-3422-1

Mitchell, M., Wu, S., & Zaldivar, A.. (2019). Model cards for model reporting. In *Proceedings of the Conference on Fairness, Accountability, and Transparency* (pp. 220–229). ACM. DOI: 10.1145/3287560.3287596

Mitchell, T. M. (1997). *Machine learning*. McGraw-Hill.

Nilsson, N. J. (2010). *The quest for artificial intelligence*. Cambridge University Press.

OECD. (2019). *OECD principles on artificial intelligence*. https://www.oecd.org/ai

Raji, I. D., Smart, A., & White, R. N.. (2020). Closing the AI accountability gap. In *Proceedings of the 2020 Conference on Fairness, Accountability, and Transparency* (pp. 33–44). ACM. DOI: 10.1145/3351095.3372873

Russell, S., & Norvig, P. (2021). *Artificial intelligence: A modern approach* (4th ed.). Pearson.

Strubell, E., Ganesh, A., & McCallum, A. (2019). Energy and policy considerations for deep learning in NLP. *Proceedings of the 57th Annual Meeting of the ACL*, 3645–3650. DOI: 10.18653/v1/P19-1355

Taddeo, M., & Floridi, L. (2018). How AI can be a force for good. *Science*, *361*(6404), 751–752. DOI: 10.1126/science.aat5991 PMID: 30139858

UNESCO. (2021). *Recommendation on the ethics of artificial intelligence.*

Vaswani, A., Shazeer, N., & Parmar, N.. (2017). Attention is all you need. *Advances in Neural Information Processing Systems*, •••, 30.

Veale, M., & Zuiderveen Borgesius, F. (2021). Demystifying the draft EU Artificial Intelligence Act. *Computer Law Review International*, *22*(4), 97–112. DOI: 10.9785/cri-2021-220402

Wachter, S., Mittelstadt, B., & Floridi, L. (2017). Why a right to explanation of automated decision-making does not exist in the GDPR. *International Data Privacy Law*, *7*(2), 76–99. DOI: 10.1093/idpl/ipx005

Weidinger, L., Mellor, J., Rauh, M., ... (2022). Ethical and social risks of harm from language models. *arXiv preprint*. https://arxiv.org/abs/2112.04359

Weller, A. (2019). Transparency: Motivations and challenges. In *Explainable AI: Interpreting, Explaining and Visualizing Deep Learning* (pp. 23–40). Springer.

Yeung, K. (2018). Algorithmic regulation: A critical interrogation. *Regulation & Governance*, *12*(4), 505–523. DOI: 10.1111/rego.12158

Chapter 6
Law in the Loop: Reimagining Justice and Regulation in the Age of Artificial Intelligence

Nehal Saxena
https://orcid.org/0009-0003-9920-941X
Christ University, Delhi NCR, India

ABSTRACT

In an era where algorithms increasingly influence justice, governance, and human relationships, this paper reimagines law as more than a regulator of technology, it becomes its moral compass. "Law in the Loop: Reimagining Justice and Regulation in the Age of Artificial Intelligence" argues for a human-centric AI governance framework that embeds ethics, accountability, and dignity within code itself. Through a synthesis of jurisprudence and technological ethics, the study explores how legal personhood, fairness, and privacy must evolve to confront the rise of autonomous systems. It advocates for a model of co-regulation where law, ethics, and innovation coalesce to preserve humanity amidst automation. By drawing from thinkers like Hart and Finnis and global frameworks such as UNESCO's AI Ethics and the EU AI Act, the paper envisions a future where technology enhances, rather than eclipses, human judgment transforming AI from a tool of control into an ally of justice, empathy, and collective flourishing.

1. INTRODUCTION

In the twenty-first century, artificial intelligence has moved from the pages of speculative fiction into the very fabric of modern governance. Algorithms now

DOI: 10.4018/979-8-2600-0000-7.ch006

recommend judicial sentences, predict crime patterns, detect fraudulent activity, and even guide administrative decision-making. What once relied solely on human deliberation and moral reasoning is increasingly influenced by systems that calculate probabilities rather than perceive justice. This quiet yet profound transformation of the legal domain by artificial intelligence marks one of the greatest paradigm shifts in the history of law. Law has always been rooted in the human capacity for empathy, judgment, and moral discernment. Artificial intelligence, in contrast, operates through data, logic, and pattern recognition. When these two worlds intersect, a new jurisprudential reality emerges, one that compels us to ask whether machines can ever truly understand justice or whether they merely simulate it. As governments, corporations, and institutions adopt algorithmic systems to enhance efficiency, the question of how to preserve fairness, accountability, and transparency becomes increasingly urgent. The idea of "Law in the Loop" arises precisely from this tension: it seeks to ensure that human values and legal principles remain at the heart of technological governance.

Artificial intelligence has become the invisible architect of contemporary governance. Administrative decisions, welfare distribution, credit assessments, and even judicial predictions are now informed by algorithmic models. In India, initiatives under the Digital India mission (NITI Aayog, 2021) seek to integrate AI into governance frameworks, while the European Union's Artificial Intelligence Act represents one of the first comprehensive attempts to regulate the technology through law. (European Commission, 2024) Globally, there is an unmistakable shift toward automated governance, promising efficiency and precision. Yet, this efficiency often comes at the cost of transparency. When justice is mediated through algorithms, the very foundations of the legal order are challenged. Traditional institutions of justice rely on principles of reasoning, interpretation, and moral reflection. Algorithms, however, operate through data correlations that may not reveal the reasoning behind their outcomes. This opacity creates an accountability gap, when a machine makes or influences a legal decision, it is unclear who bears responsibility if that decision is unfair or discriminatory. The promise of automation thus coexists uneasily with the fear of dehumanisation. The law, once the guardian of human dignity, now finds itself questioning whether it can coexist with systems that act without understanding.

The law's evolution has always been gradual, responding to social change through deliberation and precedent. Technological innovation, on the other hand, advances exponentially, often leaving regulation far behind. This disparity has created a legal vacuum in which ethical challenges multiply unchecked. The proliferation of algorithmic systems has raised serious concerns about privacy, bias, discrimination, and the erosion of accountability. When decisions are made by systems that are neither fully human nor fully autonomous, existing doctrines of liability become inadequate. Ethically, artificial intelligence tests the boundaries of justice itself. Can fairness be programmed? Can dignity be digitised? These questions strike at the heart of jurisprudence. The rise of AI exposes the limitations of a legal system designed for human actors, forcing us to reconsider long-held assumptions about responsibility and rights (Crawford, 2021). Without an integrated legal framework that anticipates these changes, society risks normalising an algorithmic order where decisions are efficient but ethically hollow. (Bostrom, 2014)

The concept of "Law in the Loop" is not a metaphor but a necessity. It represents an effort to ensure that legal and ethical reasoning are not external regulators of technology but integral components of it. The rationale of this study lies in recognising that the future of governance must not only be technologically advanced but also normatively grounded. Artificial intelligence, if left unchecked, could reshape justice systems into mechanisms of control rather than emancipation. The inclusion of law within the loop of AI development ensures that human oversight, moral reflection, and constitutional values guide every stage of technological design and implementation. In an age where algorithmic systems influence decisions that determine liberty, access, and equality, the call for "Law in the Loop" becomes a call for sustainable and humane governance. This study therefore seeks to construct a bridge between law and technology, not as opposing forces but as complementary disciplines that must coevolve to protect the integrity of justice.

Research Gap and Justification

Existing literature on artificial intelligence largely focuses on ethics, policy, or technological design, with relatively little emphasis on integrating legal reasoning into algorithmic systems themselves. Most regulatory efforts remain reactive, attempting to contain the consequences of AI rather than shaping its evolution from within. International guidelines such as those proposed by UNESCO or the OECD provide valuable ethical directions but lack enforceable legal mechanisms. There is also a limited comparative understanding of how different jurisdictions conceptualise accountability and transparency in AI governance. This study fills that gap by offering a jurisprudential and comparative analysis that synthesises theoretical reflection with practical governance models. It seeks to contribute to global scholarship by articulating a cohesive framework where legal reasoning is not an afterthought but a design principle of intelligent systems. Such an approach is essential to ensure that the technological future remains compatible with democratic ideals and human dignity. The central aim of this research is to explore how law can evolve alongside artificial intelligence by being woven into its operational and ethical architecture. It strives to examine how human rights, fairness, and accountability can be preserved within algorithmic governance without stifling innovation. The study also aims to analyse global experiences such as those in the European Union, the United States, and India to identify patterns of success and failure in regulating AI. Ultimately, the objective is to propose a human-centred model of AI governance that repositions law not as a constraint but as a compass guiding technological progress.

Research Questions

The inquiry is driven by several core questions that form the intellectual backbone of this study. How can the legal system effectively respond to the challenges posed by autonomous decision-making? What regulatory models across the world best reflect the balance between innovation and accountability? Can legal reasoning and moral values be embedded within AI systems, allowing machines to act in accordance with principles of justice? How can fairness, oversight, and human dignity be preserved in automated governance? Finally, what future legal frameworks can reconcile the pursuit of efficiency with the imperatives of ethics and rights?

Research Methodology

This study employs a qualitative, doctrinal, and comparative research approach. It draws upon constitutional principles, international legal frameworks, and ethical theories to examine how law can adapt to and coexist with artificial intelligence. By analysing case studies, regulatory instruments, and jurisprudential writings, it develops a theoretical foundation for integrating legal oversight into AI systems. The research also adopts a normative lens, assessing how the ideals of justice, fairness, and human rights can be operationalised within technological structures (Rawls, 1971). Through comparative evaluation of global practices, it seeks to formulate a governance model that embodies transparency, accountability, and inclusivity.

In essence, the methodology combines legal reasoning with technological understanding to propose a vision of the future where artificial intelligence and law coexist symbiotically, each strengthening the other in the service of justice. "Law in the Loop" thus becomes both a conceptual framework and a moral imperative, ensuring that as technology evolves, it does so under the watchful guidance of law, reason, and humanity.

2. LITERATURE REVIEW AND THEORETICAL FRAMEWORK

The evolution of governance in the age of digital technologies reveals an increasing interdependence between law and algorithmic decision-making. Historically, jurisprudence addressed machines as tools under human control, but contemporary artificial intelligence alters that relationship. As algorithms increasingly make consequential decisions, legal scholars ask how normative frameworks must adapt. In Algorithmic Governance: Politics and Law in the Post-Human Era, (Ignas K,2019) argues that regulation is no longer simply applied to machines, it must be embedded into their architecture. This insight reflects a growing consensus: law must move from external regulation to internal design, ensuring that automated systems themselves reflect legal norms.

Classical legal theory offers two rich traditions that undergird this transformation. The legal positivist tradition exemplified by H.L.A. Hart's The Concept of Law (1961) posits a system of primary and secondary rules grounded in social institutions and recognitions of authority. Hart's framework provides explanations of how law operates through institutional recognition and rule-following. On the other hand, the natural law tradition, as articulated by John Finnis in Natural Law and Natural Rights (1980), emphasises law's grounding in human flourishing and basic goods such as life, knowledge and practical reasonableness (Dworkin, 1977). For AI governance, this dichotomy illustrates a core tension: algorithmic systems thrive in environments of codified rules and formal procedures (positivist), yet legal legitimacy requires alignment with deeper values of justice, autonomy, and dignity (natural law). The challenge becomes how to synthesise these traditions so that algorithms operate within formal legal bounds while enabling human-centred flourishing (Finnis, 1980).

Recent contributions in legal and technology scholarship extend these theoretical insights. In The Black Box Society, Frank Pasquale (2015) warns of opaque algorithmic systems that exclude human scrutiny, raising concerns about accountability and democratic legitimacy. Pasquale draws attention to the "black-boxed" architecture of decision systems that replicate and amplify bias (Pasquale F, 2015). Similarly, Mireille Hildebrandt's article "The Artificial Intelligence of European Union Law" (2020) in the German Law Journal explores the challenge of translating machine logic into human-lawful contexts, emphasising that AI must be subject to the same normative expectations as human actors in legal systems. These works underpin the need for embedding legal features such as transparency, contestability and oversight into algorithmic systems themselves.

Illustrative of this is the systematic review by Batool, Zowghi and Bano (2023), Responsible AI Governance: A Systematic Literature Review, which identifies persistent gaps in how AI governance frameworks address accountability, lifecycle governance and stakeholder inclusion. Their analysis underscores that most governance models remain ex-post focusing on regulation after deployment while few embed normative oversight during system design. This reinforces the central proposition of "Law in the Loop": designing AI systems that internalise legal reasoning rather than merely respond to regulation. Complementing this, Cobbe, Lee & Singh (2021) offer a "Reviewability" framework in Reviewable Automated

Decision-Making: A Framework for Accountable Algorithmic Systems, emphasising that automated decisions must be traceable and auditable in ways akin to human administrative discretion. Their work bridges administrative law and algorithmic decision-making, suggesting that the procedural protections of law must carry over to machine decisions (Jobin et al., 2019).

Policy and normative frameworks reflect this theoretical turn. The edited volume Law and Artificial Intelligence: Regulating AI and Applying AI in Legal Practice presents an overview of the legal impact of AI across domains, arguing that AI regulation must migrate from siloed law-and-technology responses into integrated legal architectures. Meanwhile, the book The Rule of Law and Automated Decision-Making: Exploring Fundamentals of Algorithmic Governance examines how the principle of rule of law applies to automated public decision-making, emphasising that legality, transparency and accountability must not be eroded by technological delegation. These comprehensive works provide both conceptual scaffolding and empirical examples for embedding law within AI. Yet, gaps remain in the literature. Much scholarship remains either predominantly normative addressing what ought to be or highly technical focusing on algorithmic fixes. Few works comprehensively integrate normative legal theory, governance practice and algorithmic system design into a unified framework. In particular, there is limited comparative research that explores how legal personhood, liability and embedded regulation differ across jurisdictions and technological contexts. The present research addresses this gap by proposing a jurisprudential-technological synthesis that places law inside the operational loop of AI systems.

In light of this literature, the present study develops the concept of Law in the Loop (LITL), a conceptual framework signalling the shift from Human-in-the-Loop (HITL) where human oversight is appended to a deeper embedding of legal norms within algorithmic design. Law in the Loop means that legal and ethical values (fairness, transparency, accountability, human dignity) are encoded into the data models, algorithmic logic, operational procedures and organisational governance of AI systems. While HITL emphasises human intervention, LITL emphasises legal–technological binding; it ensures that every decision-making unit including the machine operates under legally anchored constraints.

In sum, the literature reveals a clear trajectory from regulation of technology toward regulation through technology. Legal theory (positivist and natural law), critical scholarship (Hildebrandt, 2015), governance reviews (Cobbe et al. 2021), and policy frameworks all converge on a central proposition: AI governance must evolve to internalise law, rather than merely respond to it. This study builds on that convergence by constructing an interdisciplinary foundation for embedding legal reasoning within AI systems, paving the way for regulation that is anticipatory, integrated and human-centred.

3. ETHICAL AND HUMAN RIGHTS DIMENSIONS OF AI GOVERNANCE

Artificial Intelligence represents both an extraordinary opportunity and a profound ethical test for the modern legal order. As societies integrate algorithmic systems into decision-making processes from policing and finance to healthcare and employment, the question of human rights, fairness, and accountability becomes inescapable (Selbst et al., 2019). This section explores the intersection of technology, ethics, and law, examining how algorithmic systems challenge traditional notions of dignity, autonomy, and equality. One of the most pressing ethical challenges in AI governance is algorithmic bias, the subtle but pervasive discrimination embedded within data-driven systems (Kouroupis, K. & Efthymiou, I., 2026). Algorithms learn from historical data, and when that data reflects social inequities or prejudiced patterns, these biases are reproduced at scale. For example, facial recognition tools have shown higher error rates in identifying women and persons of colour, while predictive policing systems disproportionately target marginalized communities (Calo R, 2015; Eubanks, 2018). This results not merely in technical malfunction but in a form of structural injustice encoded into technology. Law must therefore act as a moral filter, ensuring that algorithmic systems align with constitutional values of equality and non-discrimination. Legal intervention should not aim to halt innovation but to democratize it, transforming AI from a mirror of bias into a tool of social justice (Latour, 2010).

AI systems thrive on data, the raw material of the digital economy. Yet, the unregulated collection, storage, and analysis of personal information pose grave threats to individual autonomy and privacy (United Nations, 1966). The erosion of meaningful consent in the age of automated profiling undermines the very foundation of human agency. People rarely understand how their digital traces are used to infer behavioural patterns, predict preferences, or even influence democratic choices. The right to privacy, enshrined in both national constitutions and international human rights law, must therefore evolve to confront algorithmic opacity. Legal frameworks such as the EU's General Data Protection Regulation (GDPR) provide a valuable model, embedding principles of purpose limitation, data minimization, and algorithmic transparency. However, true data justice demands more than compliance; it requires a philosophical shift that recognises privacy as an essential condition for human dignity in the algorithmic age (Taylor, 2017; Mantelero, 2022)..

Automation challenges not only privacy and equality but also the moral architecture of human dignity. When judicial, administrative, or employment decisions are made by opaque algorithms, individuals are reduced to data points, stripped of context and moral nuance. This dehumanization risks transforming justice from a deliberative process into a computational exercise. Human oversight, therefore, is not a procedural luxury but a moral necessity. The concept of "Law in the Loop" emerges here as a response, an insistence that legal reasoning, ethical reflection, and human accountability remain embedded within algorithmic systems. Oversight mechanisms, such as algorithmic audit trails, transparency mandates, and ethical review boards, can ensure that technological decisions remain tethered to human values. Ultimately, the legitimacy of AI governance depends on its ability to uphold the inherent worth of every individual, preventing the substitution of human judgment with mechanical objectivity (Habermas J, 1996).

Global institutions have recognized the urgency of establishing ethical guardrails for AI. UNESCO's Recommendation on the Ethics of Artificial Intelligence (2021) was the first global standard-setting instrument in this field, emphasizing principles of proportionality, accountability, and fairness (Mittelstadt et al., 2016). It calls for AI systems that promote inclusive social progress rather than exacerbate digital divides (UNESCO, 2021). Similarly, the Organisation for Economic Co-operation and Development (OECD, 2019) AI Principles advocate for transparency, human-centric design, and robust governance mechanisms. These frameworks converge on a common vision: that technology must serve humanity, not the reverse.

The European Union's AI Act represents the most ambitious legislative initiative to operationalize this vision (European Commission, 2024). By classifying AI systems according to risk and imposing strict compliance requirements on high-risk applications, it sets a global precedent for responsible innovation. The Act mandates transparency, human oversight, and post-deployment monitoring, translating ethical aspirations into enforceable legal obligations. Yet, the challenge remains in ensuring that such models are adaptable across diverse legal, cultural, and economic contexts particularly in developing countries where regulatory infrastructures are still evolving.

4. GLOBAL REGULATORY APPROACHES TO ARTIFICIAL INTELLIGENCE

The governance of Artificial Intelligence (AI) has rapidly evolved from an academic concern to a central pillar of national policy and international diplomacy. Around the world, governments are grappling with the dual challenge of fostering technological innovation while safeguarding ethical, legal, and human rights standards. As AI systems increasingly shape critical aspects of governance, commerce, and social life, regulatory diversity reflects distinct political, cultural, and economic values (Cath, 2018). This section analyses four major regulatory approaches the European Union, the United States, China, and India and offers a comparative synthesis that identifies the shared challenges and future policy directions shaping global AI governance.

The European Union: Risk-Based AI Act and Human-Centric Regulation

The European Union (EU) stands at the forefront of global AI regulation with its proposed Artificial Intelligence Act; a comprehensive legal framework built on the principle of "trustworthy AI." The EU's model embodies a risk-based approach that classifies AI systems into four categories: unacceptable risk, high risk, limited risk, and minimal risk (Floridi, 2021b). This graduated structure ensures that regulatory scrutiny corresponds to the potential societal impact of AI applications. Systems deemed to pose an "unacceptable risk" (such as social scoring or manipulative biometric systems) are prohibited, while "high-risk" systems including those used in critical infrastructure, law enforcement, and healthcare are subject to strict transparency, data quality, and human oversight requirements (European Commission, 2024).

What distinguishes the EU model is its deeply human-centric philosophy. Drawing inspiration from the Charter of Fundamental Rights of the European Union, the AI Act emphasizes respect for human dignity, autonomy, and accountability. It integrates ethical considerations into enforceable legal mechanisms, ensuring that innovation does not erode fundamental rights. Additionally, the establishment of the European Artificial Intelligence Board aims to harmonize national enforcement and promote a culture of compliance across Member States.

The United States: Sectoral Innovation and Self-Regulation Challenges

In contrast to the EU's comprehensive model, the United States adopts a decentralized and sectoral approach to AI governance. Rather than enacting a unified federal AI law, the U.S. relies on existing frameworks such as consumer protection, privacy, and civil rights laws applied on a case-by-case basis. This model of governance reflects the American ethos of market-driven innovation but simultaneously reveals critical vulnerabilities. Without mandatory transparency or accountability requirements, companies may prioritize efficiency over ethics. Moreover, the dominance of private corporations in the AI ecosystem raises concerns over monopolization, data exploitation, and the erosion of public trust. Nevertheless, the United States continues to influence global norms through its leadership in AI research, development, and multilateral cooperation, shaping the technical standards that underpin global AI infrastructure.

China: State-Centric Control and Data Governance Models

China presents a contrasting model rooted in state-led governance and data sovereignty. The Chinese regulatory framework for AI is embedded within its broader digital governance architecture, which prioritizes national security, social stability, and technological self-reliance. The Interim Measures for the Management of Generative Artificial Intelligence Services (2023) exemplify the state's proactive stance, mandating that AI-generated content must "reflect the core socialist values" and not undermine national unity, security, or public order (Stanford University DigiChina Project, 2023). At the core of China's AI governance lies the concept of algorithmic accountability with Chinese characteristics, a system where the state retains ultimate authority over data and algorithmic transparency. Companies are required to register their algorithms with government regulators, undergo security assessments, and ensure that their data practices align with national interests. In addition, the Personal Information Protection Law (PIPL) and the Data Security Law (DSL) demonstrate a growing emphasis on privacy and data protection, albeit within the framework of centralized state control (Fry, 2018).

This state-centric model enables rapid implementation and strong enforcement but also raises concerns regarding individual rights, freedom of expression, and international data sharing. China's approach illustrates how AI governance can reflect ideological and geopolitical priorities, using legal instruments not only for ethical regulation but also for political management. Still, its regulatory agility and investment in AI ethics research have positioned China as a key player in shaping global debates on responsible AI.

India: Emerging Frameworks and Policy Drafts by NITI Aayog

India, as a rising technological and democratic power, occupies a unique position in the global AI regulatory landscape. Although it lacks a comprehensive AI law, the government through NITI Aayog's National Strategy for Artificial Intelligence (NSAI,2018) has articulated a vision of "AI for All," emphasizing inclusivity, transparency, and innovation-driven growth (NITI Aayog, 2018). The focus is on harnessing AI for social good in sectors like healthcare, agriculture, education, and smart governance. The NSAI advocates for an "enabling regulation" model that balances innovation with risk mitigation. It proposes the establishment of ethical guidelines, data-sharing frameworks, and sector-specific governance mechanisms rather than a single overarching law. Complementary initiatives such as the Digital Personal Data Protection Act, 2023 aim to strengthen individual privacy and accountability in data use. Moreover, India's Responsible AI for Youth and AI Mission 2024 initiatives highlight its commitment to capacity-building and digital literacy.

However, the Indian approach also faces significant challenges, including fragmented institutional capacity, limited enforcement mechanisms, and the risk of algorithmic exclusion in a socio-economically diverse population. There remains a need for a robust legal architecture that embeds accountability, fairness, and human rights at the heart of AI governance. By integrating constitutional principles particularly, the right to equality, privacy, and freedom of expression. India can position itself as a model of ethical AI governance for the Global South.

Comparative Synthesis and Key Policy Takeaways

A comparative examination of these four jurisdictions reveals divergent but converging trajectories in AI regulation. The European Union emphasizes ethics through law, constructing a rights-based regulatory ecosystem. The United States prioritizes innovation through flexibility, relying on market mechanisms and voluntary standards. China enforces control through centralized governance, aligning technology with national objectives. India aspires for balance through developmental regulation, seeking to democratize AI's benefits while gradually embedding accountability (Lessig, 1999). Despite their differences, several common threads emerge. First, all jurisdictions recognize the inevitability of legal oversight in AI deployment, even if they differ on timing and scope. Second, human-centric principles- fairness, transparency, and accountability have become global benchmarks for legitimacy. Third, international cooperation is increasingly essential, as AI transcends borders and requires harmonized standards for safety, ethics, and trade.

The next stage of global AI governance must thus move toward interoperability rather than uniformity. Regulatory pluralism where diverse legal systems collaborate through shared ethical and technical standards may offer the most pragmatic path forward. In this sense, "Law in the Loop" becomes a metaphor for global governance itself: a reminder that as technology accelerates, law must evolve not to restrain but to humanize the digital future (Susskind, 2019). The challenge lies not in choosing between innovation and regulation, but in ensuring that both serve the enduring purpose of justice, accountability, and human dignity in an age defined by algorithms.

5. LAW IN THE LOOP: EMBEDDING REGULATION WITHIN AI SYSTEMS

Artificial Intelligence has outpaced traditional models of regulation. Conventional laws crafted in a human-centred, reactive manner struggle to keep pace with self-learning systems capable of making autonomous decisions. This accelerating gap between technological evolution and legal adaptation poses an urgent question: can the law itself be built into the design of AI systems? The emerging concept of "Law in the Loop" seeks precisely this transformation. It envisions an architecture where legal principles are not external checks but internal features of algorithmic decision-making. In doing so, it redefines law from a regulatory boundary to an embedded moral compass within digital intelligence (Hildebrandt, 2020; Floridi, 2021a).

The first step in embedding "Law in the Loop" is to envision AI not as a purely technical construct but as a socio-legal actor. Legally conscious algorithms are systems that operationalize legal and ethical reasoning within their computational architecture. This means designing models that are aware of boundaries such as data protection, anti-discrimination, and due process before they act. For example, an AI system screening job applicants should not merely optimize efficiency but must be programmed to respect equality norms derived from anti-discrimination statutes. The development of legally conscious algorithms relies on interdisciplinary collaboration. Lawyers, ethicists, and computer scientists must collectively determine how legal obligations can be translated into machine-readable rules. Techniques such as constraint-based programming and rule-based logic allow algorithms to incorporate explicit legal norms, while machine learning fairness constraints ensure compliance with principles of equity and proportionality. In this framework, legality becomes a feature of the system's architecture, not an afterthought. However, such integration poses philosophical challenges. Law is often ambiguous and context-dependent, while machines operate on binary precision. Thus, embedding law in AI demands not the codification of every statute but the encoding of interpretive principles. This requires AI systems to learn from precedent and adapt to evolving jurisprudence. In effect, law must move from static command to dynamic learning mirroring the adaptive capacities of the

Accountability and transparency are the twin pillars of lawful governance, and their operationalisation within AI represents both a technical and moral imperative. Traditional accountability relies on identifying human actors such as developers, deployers, or users responsible for a given act (Goodman & Flaxman, 2017). Yet, as AI systems become increasingly autonomous, attributing blame becomes complex. "Law in the Loop" responds to this challenge by embedding traceability and auditability mechanisms directly within algorithmic processes. Transparent AI is not merely about open-source access or public documentation; it involves constructing models that can explain their decisions in ways understandable to non-experts. The emerging field of explainable AI (XAI) aligns closely with this vision, aiming to make algorithmic reasoning interpretable, contestable, and reviewable. Through structured decision logs, algorithmic impact assessments, and embedded compliance layers, systems can create a continuous legal trail that mirrors human administrative accountability.

The fusion of law and technology cannot ignore the centrality of ethics. While law provides structure, ethics provides substance, the values that law must protect in an algorithmic world. The interplay of law, ethics, and machine learning thus forms a trinity of responsible AI governance. Ethics serves as the interpretive bridge between rigid rules and fluid realities. For example, while a privacy law may restrict data sharing, ethical principles guide decisions about when exceptions serve public welfare. Similarly, anti-bias legislation outlines non-discrimination obligations, but ethical reasoning determines what fairness means in different social contexts. Integrating this interpretive capacity within machine learning requires the inclusion of ethical reasoning modules systems trained on diverse normative datasets to evaluate the social implications of their decisions.

Moreover, machine learning models should be designed to prioritize value alignment, ensuring that algorithmic outcomes correspond to societal and constitutional values rather than purely technical optimization. This involves continuous monitoring, feedback loops, and algorithmic impact audits. Importantly, ethical and legal reasoning cannot be static; they must evolve with societal norms, just as jurisprudence evolves through judicial interpretation. Hence, "Law in the Loop" envisions adaptive legal intelligence, algorithms capable of learning not only from data but also from the moral progress of human societies.

This interdisciplinary convergence also calls for the creation of AI ethics boards and algorithmic oversight committees, where legal experts, technologists, and citizens collaboratively assess the normative soundness of algorithmic systems. These institutions act as custodians of digital justice, ensuring that the evolution of machine learning remains aligned with human values and the rule of law. Embedding law within AI systems also transforms the nature of governance itself. Static legal codes are ill-equipped to regulate technologies that evolve by the hour. Therefore, dynamic regulation rooted in continuous monitoring and responsive adaptation becomes essential. "Law in the Loop" envisions a model where regulators, policymakers, and developers engage in an ongoing dialogue, guided by data-driven insights and ethical evaluation. Adaptive regulation operates through feedback mechanisms similar to those within machine learning. Just as algorithms refine themselves through new data, legal frameworks should update through empirical observation of how AI affects human welfare. Regulatory sandboxes-controlled environments that allow AI experimentation under supervision exemplify this approach. They enable policymakers to test the social and ethical consequences of innovation before scaling it into law.

Additionally, dynamic governance requires interoperability across jurisdictions. Since AI transcends national borders, global coordination through institutions like the OECD, UNESCO, and the Global Partnership on AI becomes indispensable. These bodies can foster normative harmonization, develop shared standards, and ensure that AI systems respect universal principles such as dignity, accountability, and transparency. Ultimately, "Law in the Loop" is more than a metaphor. It is a vision for the next stage of legal evolution. It reimagines law not as a static constraint but as a living framework that learns, adapts, and coexists with intelligent systems. In this future, algorithms will not simply follow laws; they will participate in their continuous realization. The challenge before humanity is not to dominate machines but to educate them to embed within their logic the very essence of justice. Only then can we ensure that the age of artificial intelligence remains, at its core, an age of human values.

In practice, "Law in the Loop" is already taking tentative shape. The EU AI Act requires high-risk AI systems to be designed with built-in legal safeguards such as human oversight, risk assessments, and accountability mechanisms, while India's Responsible AI Mission stresses ethical design rooted in constitutional values. These concerns remind us that embedding law in AI must support, not substitute- human judgment, ensuring technology remains guided by justice rather than mere efficiency.

6. CASE STUDIES AND REAL-WORLD IMPLICATIONS

Artificial Intelligence has moved from theoretical discourse to practical deployment, influencing domains that lie at the core of human governance law, justice, safety, creativity, and information. Yet, as AI systems increasingly participate in decisions once reserved for human reasoning, real-world cases reveal the ethical, legal, and societal dilemmas that arise when accountability and automation collide (Helberger et al., 2018). Through examining global case studies in predictive policing, autonomous vehicle liability, judicial automation, and generative AI, this section illuminates how "Law in the Loop" must evolve from principle to practice. These examples demonstrate that embedding legal reasoning within technology is not a philosophical luxury but a pressing necessity for equitable governance in the digital age (Binns, 2018; Yeung, 2018). Perhaps nowhere are the stakes higher for AI governance than in the criminal justice system. Predictive policing algorithms, designed to forecast crime hotspots or identify potential offenders, promise efficiency and data-driven neutrality. Yet, their real-world application often exposes deep structural inequities. The COMPAS (Correctional Offender Management Profiling for Alternative Sanctions) algorithm in the United States stands as a defining case. Marketed as a tool for objective risk assessment in bail and sentencing decisions, it was found by ProPublica to exhibit racial bias, disproportionately labelling Black defendants as high-risk compared to white defendants with similar records.

This case illustrates how algorithmic systems can amplify existing social prejudices when trained on historically biased data. While developers may argue that algorithms reflect patterns, not prejudices, the consequences are profoundly legal deprivation of liberty based on flawed automation undermines the constitutional guarantee of equality before law. Courts and scholars increasingly argue that predictive models must meet standards of explainability and proportionality, aligning them with procedural fairness. European regulatory trends, including the AI Act, recognize predictive policing as a "high-risk" use case requiring transparency, human oversight, and auditability (European Commission, 2021). The "Law in the Loop" approach here demands not only algorithmic accuracy but institutional accountability ensuring that decision-making power remains anchored in human judgment guided by the rule of law.

Another frontier for AI law emerges on the roads, where self-driving vehicles blur the line between human agency and machine autonomy. The 2018 Uber v. Tempe accident, in which an autonomous car struck and killed a pedestrian, sparked global debates on liability, negligence, and moral responsibility. The incident revealed that while the vehicle's sensors detected the pedestrian, the algorithm failed to classify her as a human in time to react. Traditional legal frameworks assume human control behind every act; AI challenges this foundation. Should the developer, manufacturer, operator, or the algorithm itself bear responsibility for harm? Jurisdictions differ in their approach. The European Parliament has proposed a framework for strict liability in cases involving high-risk AI systems, while the U.S. relies on tort principles of negligence and product liability. Japan, meanwhile, explores a shared liability model that distributes responsibility among stakeholders.

The "Law in the Loop" paradigm reimagines accountability by embedding compliance and safety standards directly into AI systems. For example, ex ante regulatory testing can ensure algorithms meet ethical and legal thresholds before deployment. Similarly, continuous learning mechanisms where systems report near misses or safety deviations allow regulators to track performance in real time. These embedded legal feedback loops can transform reactive litigation into proactive prevention. However, autonomous vehicle cases also raise deeper moral questions: can machines make ethical choices? The "trolley problem" reemerges in modern form, should a car prioritize the life of its passenger over a pedestrian? Embedding moral reasoning within AI may never be perfect, but law can at least dictate the boundaries of permissible risk and ensure transparency in algorithmic decision hierarchies.

The integration of AI into judicial systems through case prediction tools, automated translation, or virtual court assistants represents both progress and peril. Estonia's experiment with "robot judges" in small claims disputes and the use of AI-based sentencing assistance in China mark a turning point in the evolution of justice. These systems are designed to enhance efficiency, reduce backlog, and standardize rulings. Yet, they also risk mechanizing moral reasoning and eroding the human empathy essential to adjudication. The core ethical dilemma lies in whether justice can or should be automated. Law is not only about rule application but also about interpretation, discretion, and compassion. When AI recommends sentences or predicts case outcomes, there is a danger of technocratic determinism where data patterns replace human judgment. Furthermore, opaque machine reasoning undermines the foundational legal principles of transparency and the right to a fair hearing.

A "Law in the Loop" approach here would demand algorithmic explainability, judicial oversight, and accountability mechanisms. AI can serve as an advisory instrument assisting judges rather than replacing them by providing data analytics while preserving the human dimension of moral reasoning. The challenge is not technological substitution but augmentation: ensuring that AI strengthens justice delivery without hollowing out its ethical core.

The rise of generative AI systems such as ChatGPT, Midjourney, and Stable Diffusion has transformed creativity and communication while exposing the fragility of existing intellectual property (IP) regimes. These models generate text, images, and code by learning from massive datasets that often include copyrighted materials. The legal question is- who owns the output of an AI model, and is its training data used lawfully? In Getty Images v. Stability AI (2023), the plaintiff alleged unauthorized use of copyrighted images to train an AI art generator. Similarly, U.S. authors including Sarah Silverman have filed class-action suits against OpenAI, arguing that the use of their books in model training constitutes copyright infringement. Courts now face the novel challenge of distinguishing between human creativity and algorithmic recombination. Simultaneously, generative AI amplifies the threat of misinformation and deepfakes. The 2024 incident involving fabricated AI-generated images of political unrest in multiple countries demonstrated how easily synthetic media can manipulate public opinion. This creates a regulatory paradox: while AI democratizes expression, it also destabilizes trust in truth.

The "Law in the Loop" solution involves embedding provenance, traceability, and watermarking technologies to authenticate AI-generated content. Legal frameworks such as the EU's Digital Services Act and UNESCO's ethical guidelines are beginning to demand transparency about synthetic media (UNESCO, 2023). Yet, enforcement remains a global challenge. Embedding regulatory compliance at the model level ensuring AI systems automatically flag deceptive outputs or respect copyright boundaries can prevent harm before it occurs. Across jurisdictions, courts and regulators are grappling with how to translate enduring legal principles into a digital ecosystem (Zuboff, 2019). The European Court of Human Rights, in Big Brother Watch v. United Kingdom (2021), underscored that mass data surveillance must satisfy proportionality and necessity tests, reaffirming human rights in the algorithmic era (European Commission, 2022; UNESCO, 2021). Meanwhile, the U.S. Supreme Court's cautious stance in Gonzalez v. Google (2023) on platform liability reflects judicial uncertainty over how much responsibility tech companies should bear for algorithmic amplification. Global governance experiments, such as the EU's AI Liability Directive and Singapore's Model AI Governance Framework, reveal a shift toward preventive regulation emphasizing risk assessment, human oversight, and impact audits. India's proposed Digital India Act and Responsible AI Mission mirror this trend, signalling the emergence of a hybrid governance model combining ethical guidance with flexible compliance tools.

These developments collectively affirm a global awakening: law must migrate from courtroom deliberation to code architecture. "Law in the Loop" captures this metamorphosis by envisioning legal principles as intrinsic to AI's functioning rather than external constraints. Real-world case studies show that when law is absent, AI systems replicate inequality, harm, and opacity; when law is embedded, they evolve toward fairness, transparency, and accountability. Ultimately, these cases demonstrate that the pursuit of just AI is not merely a technical or legal project, it is a civilizational one. The future of justice depends on whether humanity can teach its machines not only to think but to act lawfully. "Law in the Loop" offers a pathway to that future, ensuring that in a world increasingly governed by algorithms, legality remains not a constraint on innovation but its conscience.

7. RETHINKING LEGAL PERSONHOOD AND ACCOUNTABILITY

In the age of artificial intelligence, the very notion of what it means to be a "person" in law is undergoing a profound transformation. Traditionally, the law has recognised two types of persons- natural and juristic. Natural persons are human beings endowed with rights and obligations by virtue of their humanity, while juristic persons, such as corporations and states, are artificial constructs created by legal fiction to serve collective purposes. However, the emergence of autonomous AI systems that can perceive, reason, learn, and act with minimal human oversight is blurring this binary distinction. As AI systems increasingly perform tasks involving judgment, discretion, and moral consequence from approving loans and diagnosing diseases to adjudicating disputes, scholars and policymakers are beginning to ask a once-unthinkable question: should artificial intelligence be recognised, at least partially, as a new form of legal entity? (Bryson et al., 2017)

The debate surrounding "electronic personhood" is not merely academic; it lies at the heart of accountability in an algorithmic world. The European Parliament's 2017 resolution on civil law rules for robotics tentatively suggested creating a special legal status for "autonomous robots," sparking global controversy (European Parliament, 2017). Critics argued that such recognition would dangerously absolve human creators and operators of responsibility, while supporters claimed it could streamline liability in complex AI-driven systems. The core dilemma is philosophical as much as legal when an AI system acts independently and unpredictably, who should bear the blame or responsibility for its outcomes?

From a jurisprudential standpoint, extending personhood to AI challenges centuries of moral and legal theory. John Austin's positivist view would reject AI personhood entirely, since legal recognition must stem from sovereign command and AI, lacking consciousness or moral agency, cannot be a subject of duty or responsibility. Hart's theory of primary and secondary rules similarly situates legal capacity within social institutions, implying that AI cannot possess rights unless society collectively constructs a framework recognising them (Feinberg & Narveson, 1970). In contrast, thinkers from the natural law tradition, such as John Finnis, might argue that moral agency requires an orientation towards human flourishing, something machines cannot genuinely possess. Thus, the philosophical consensus largely resists equating machine intelligence with human moral responsibility.

Yet, the practical realities of AI's autonomy complicate this neat theoretical resistance. Consider an autonomous vehicle that causes a fatal accident due to its machine-learning model's decision-making, which cannot be easily traced to a specific human act. Traditional tort law principles, such as negligence or vicarious liability, struggle to accommodate this opacity. Manufacturers claim that the algorithm "learned" beyond its programming; users argue they had no control; regulators find it difficult to locate intent or foreseeability. In such cases, the absence of a clear "responsible person" erodes public trust and weakens the law's moral authority. Recognising AI as an "electronic person" with limited, functional liability akin to a corporation's distinct legal identity might offer a pragmatic solution, not to humanise machines but to allocate responsibility in a complex ecosystem of shared agency (Solum, 1992).

However, any model of distributed accountability must safeguard against moral dilution. Assigning "personhood" to AI should not serve as a shield for human actors. Instead, it should complement a layered system of liability where programmers, operators, and entities deploying AI remain jointly accountable (Bryson et al., 2017). This approach resonates with the principle of "shared responsibility," wherein accountability flows through every stage of the AI lifecycle from design to deployment. Such a framework could align with the "precautionary principle" in international law, which demands that actors take preventive measures even in the face of scientific uncertainty. In this sense, AI personhood becomes less about granting rights and more about structuring duties.

Moreover, this debate forces a re-examination of tort and criminal jurisprudence. Traditional tort doctrines rely on human fault, intention, or negligence concepts that falter when applied to autonomous algorithms. Similarly, criminal law's foundation rests on mens rea, the guilty mind, which an AI system cannot meaningfully possess. Some scholars suggest expanding doctrines such as strict liability or enterprise liability to encompass AI harms, where responsibility is imputed based on control or benefit rather than fault. Others advocate for "algorithmic trusteeship," a model where humans remain custodians of AI behaviour, ensuring ongoing oversight and moral anchoring.

In this evolving landscape, emerging international perspectives offer important insights. The EU's draft AI Act, for instance, emphasises traceability, human oversight, and risk-based classification rather than granting independent legal status to AI. The United States, with its market-driven ethos, focuses on post-harm liability through tort and consumer protection regimes. Meanwhile, China adopts a state-centric accountability model, where responsibility is ultimately absorbed by the entities controlling the AI system. India's nascent AI policy discourse, guided by NITI Aayog's ethical AI framework, underscores human accountability as "non-transferable," resisting the temptation of legal personhood for machines. These approaches reflect an underlying consensus: while AI may act autonomously, it cannot yet be morally autonomous (NITI Aayog, 2021). To be accountable in the legal sense, an actor must be capable of moral reflection and self-restraint, capacities AI systems do not inherently possess. Thus, while granting limited legal recognition to AI might simplify liability, it risks undermining the ethical foundation of law itself. The more meaningful reform lies in embedding accountability mechanisms within AI architectures through explainability, transparency, and ethical programming ensuring that law remains not outside but inside the algorithmic loop (Citron & Pasquale, 2014). Empirical studies and public surveys also reveal growing public anxiety about opaque AI decisions, alongside declining trust when accountability is unclear. From a technical standpoint, "Law in the Loop" offers a practical alternative: while law cannot be fully reduced to code, core legal principles such as traceability, proportionality, and non-discrimination can be translated into machine-readable constraints through interdisciplinary collaboration between lawyers and developers, ensuring legality is embedded without sacrificing human judgment.

Ultimately, rethinking legal personhood in the age of AI invites us to revisit law's enduring question: what does it mean to act responsibly? The law must evolve, not by anthropomorphising machines, but by reaffirming human agency in a technological world that increasingly tempts us to surrender it. Rather than declaring machines as legal persons, the future of accountability lies in constructing legal ecosystems where every algorithmic act can be traced back to a web of human intentions, controls, and benefits. In this way, the law in the loop becomes a safeguard against the moral vacuum of automation, a living reminder that justice, however intelligent its tools may become, remains a distinctly human pursuit.

8. TOWARDS A HUMAN-CENTRIC AI GOVERNANCE FRAMEWORK

As artificial intelligence increasingly infiltrates the structures of governance, economy, and daily human life, the fundamental challenge is no longer whether we can regulate AI, but whether we can do so in a way that preserves human dignity, autonomy, and justice. The emerging discourse around AI regulation cannot be confined to risk management or compliance alone; it must evolve into a moral and institutional project, one that consciously places humanity at the centre of technological progress. A human-centric AI governance framework, therefore, is not simply a regulatory aspiration but a civilisational necessity, a reassertion of law's role as the conscience of technology.

At its core, such a framework demands the seamless integration of law, ethics, and innovation. The law, when viewed merely as a restraint, risks being outpaced by technological acceleration. But when seen as an enabler of ethical innovation, it becomes the architecture of trust upon which sustainable AI ecosystems can flourish (Floridi, L.(2021b)). Legal regulation must move beyond prohibitive mandates and instead provide normative guidance articulating the values, rights, and duties that govern algorithmic design and deployment. Ethical frameworks, such as those proposed by UNESCO (2021) and the OECD (2019), provide the moral vocabulary fairness, accountability, transparency, explainability but these ideals require legal translation. Innovation, in turn, must be guided by "responsible design," ensuring that ethical reflection is embedded in the earliest stages of technological creation rather than retroactively imposed.

This triadic integration mirrors what might be called the constitutionalising of technology. Just as constitutional law constrains and empowers state power, AI governance must constitutionalise technological power embedding principles of justice, equality, and human rights into code itself. The idea of "law in the loop" thus matures here into a design philosophy: every algorithmic system must contain within it the capacity for legal compliance and ethical introspection. Regulation should not remain an external force acting upon AI, but a living system that operates through it.

The implementation of this vision necessitates multi-stakeholder governance and co-regulatory models. AI is not owned by any single institution; it is co-created by governments, corporations, researchers, and users. Therefore, its governance must reflect this plural authorship. States must set the normative boundaries through legislation and international agreements; industries must operationalise those boundaries through technical standards and internal ethics boards; civil society must act as watchdog and conscience, ensuring inclusivity and accountability. This shared stewardship represents a departure from the outdated binary of "regulation versus innovation." Instead, it introduces a collaborative model where law and technology evolve symbiotically, each informing the other in real time.

Co-regulation also offers practical flexibility in an era of rapid innovation. Static legal codes often struggle to anticipate technological change; adaptive governance models combining statutory mandates with soft-law instruments such as ethical charters, certification schemes, and algorithmic audits can fill this gap. The European Union's AI Act, for instance, offers a layered approach: high-risk systems are subjected to strict oversight, while low-risk systems rely on voluntary codes of conduct. Similarly, Japan's "Society 5.0" strategy blends state guidance with industry self-regulation to create an innovation-friendly yet ethically anchored environment. (Brownsword, 2019). India, through NITI Aayog's policy discourse on "Responsible AI for All," is exploring hybrid models that emphasise human accountability, diversity, and social equity in AI design.

However, a truly human-centric model requires going beyond governance structures, it must be deeply embedded in the digital architecture itself. Fairness, equity, and justice cannot remain abstract values; they must be translated into algorithmic logic. This involves ensuring representativeness in data, transparency in model design, and interpretability in outputs. Technical mechanisms such as bias detection tools, explainable AI (XAI), and human-in-the-loop (HITL) models can concretely operationalise ethical principles. For instance, algorithmic audits, much like financial audits, could become mandatory for AI systems that influence public decision-making. Ethical compliance should thus become as integral to AI development as cybersecurity or efficiency testing.

Moreover, embedding fairness also entails recognising the global asymmetries in AI governance. Most AI technologies are developed in the Global North, yet their impacts are deeply felt across the Global South. A human-centric framework must therefore account for digital colonialism the extraction of data and cognitive labour from developing nations without reciprocal benefit (De Stefano, 2020). Global AI legislation should promote equitable access to technological resources, capacity building, and the localisation of ethical standards (EU High-Level Expert Group on Artificial Intelligence, 2019). International cooperation must aim at preventing a new divide where algorithmic power becomes concentrated in the hands of a few techno-sovereign actors. In this pursuit, international law plays a vital harmonising role. The absence of a universal AI treaty has led to fragmented governance, with the EU, US, China, and India pursuing divergent regulatory philosophies. A coordinated global approach perhaps under the auspices of the United Nations or a new International Artificial Intelligence Agency could establish common principles for ethical AI development, much like the International Atomic Energy Agency regulates nuclear technology. Such a body could oversee compliance, facilitate knowledge exchange, and ensure that AI serves collective human welfare rather than narrow corporate or geopolitical interests (United Nations, 1966). From a jurisprudential perspective, a human-centric AI framework reaffirms law's ancient purpose to balance power with justice. It recalls H.L.A. Hart's vision of law as a system of primary and secondary rules prescribing conduct while creating mechanisms for accountability. It resonates with John Finnis' natural law principles of human flourishing and common good. In the AI context, this means not merely preventing harm but enabling the conditions for human creativity, participation, and dignity to thrive alongside machines.

Ultimately, a human-centric AI governance framework envisions technology not as a substitute for human judgment, but as its extension, a partner in the pursuit of justice rather than a rival to it. It recognises that the true test of intelligence, whether human or artificial, lies not in efficiency but in empathy, not in precision but in purpose. To put law in the loop is, therefore, to ensure that every algorithmic decision reflects the moral architecture of society. In the decades ahead, the societies that succeed will not be those that build the smartest machines, but those that build the wisest systems where innovation and integrity coexist, and where humanity remains the author of its technological destiny.

9. CONCLUSION AND POLICY RECOMMENDATIONS

The relationship between artificial intelligence and law has shifted from speculative ethics to urgent governance. This study shows that AI, though technological in form, is ultimately a question of justice, accountability, and human purpose. The real challenge is not technology itself, but the absence of humanity in its governance. Law must therefore move from controlling technology to partnering with it. The European Union's risk-based framework prioritises human welfare and ethics; the United States favours innovation through sectoral self-regulation but lacks accountability; China's state-centric model ensures control and data sovereignty at the cost of individual autonomy; and India's approach remains aspirational, guided by ethics without binding force. While imperfect, these models collectively suggest a harmonised path grounded in human-centric values, adaptive regulation, and institutional oversight.

Traditional doctrines of liability, personhood, and responsibility are increasingly strained by autonomous, self-learning systems. Built on assumptions of human intention and predictability, existing law struggles with probabilistic decision-making, creating accountability gaps that threaten public trust. Embedding "law in the loop" thus becomes a structural necessity, transforming law from an external constraint into an internal feature of technological design. This study bridges classical jurisprudence and technological modernity by integrating Hart's positivism, Finnis' natural law, and algorithmic ethics into a framework of juridical techno-moralism. While AI may exhibit functional autonomy, moral and legal responsibility must remain human-centred; responsibility cannot be delegated to code without eroding law's ethical foundations.

Practically, the findings call for multidimensional governance frameworks operating across legal, technical, and ethical domains. Regulatory sandboxes, algorithmic audits, transparency registries, and interdisciplinary collaboration should become standard tools. Legal education must also evolve, equipping judges, lawyers, and policymakers with technological literacy. Adaptive regulation should be guided by three principles: dynamism, allowing law to evolve with technology; inclusivity, incorporating Global South perspectives; and integrity, embedding human rights at the core of AI governance. Policy recommendations include international coordination through a global AI governance platform, national Algorithmic Accountability Acts, mandatory ethical impact assessments for high-risk AI, and enforceable remedies for algorithmic harm. Lessons from African data-dignity models and Latin American people-centred digital constitutionalism reinforce justice-oriented governance. For "Law in the Loop" to be effective in practice, governance must move beyond abstract principles into clear institutional action. Policymakers should adopt a step-by-step approach that begins with mandatory algorithmic impact assessments at the design stage, followed by continuous monitoring through independent audits and public transparency reports. Dedicated AI ethics boards within regulatory agencies can serve as early warning mechanisms, ensuring that systems align with legal standards before harm occurs. Courts, too, must evolve as active enforcers of algorithmic accountability, developing doctrines that allow meaningful scrutiny of automated decision-making and proportionate remedies when violations arise.

Equally important is ensuring that AI governance remains globally inclusive and socially grounded. Many African and Latin American jurisdictions are experimenting with community-oriented data governance, digital constitutionalism, and rights-based technology policies that prioritise dignity, access, and social equity over pure efficiency. These approaches remind us that effective AI regulation need not mirror Western models alone. By learning from diverse regulatory cultures and embedding enforcement within courts, regulators, and independent oversight bodies, "Law in the Loop" can become not merely a regulatory idea, but a lived guarantee of justice in the algorithmic age.

Ultimately, AI regulation is not about restraining innovation but humanising it. "Law in the Loop" restores law as a moral compass in a code-driven world. The future of justice will not be measured by how intelligently machines think, but by how faithfully they reflect human conscience. Law must remain an architect and guardian of human dignity, ensuring progress never comes at the cost of what makes us human.

REFERENCES

Aayog, N. I. T. I. (2018). *National strategy for artificial intelligence: AI for all.* Government of India. https://www.niti.gov.in

Aayog, N. I. T. I. (2021). *Responsible AI for all: Operationalizing principles for responsible AI.* Government of India. https://www.niti.gov.in

Binns, R. (2018). Fairness in machine learning: Lessons from political philosophy. *Proceedings of the 1st Conference on Fairness, Accountability and Transparency*, 149–159.

Bostrom, N. (2014). *Superintelligence: Paths, dangers, strategies.* Oxford University Press.

Brownsword, R. (2019). Law, technology and society: Re-imagining the regulatory environment. *Law, Innovation and Technology, 11*(1), 1–26.

Bryson, J. J., Diamantis, M. E., & Grant, T. D. (2017). Of, for, and by the people: The legal lacuna of synthetic persons. *Artificial Intelligence and Law, 25*(3), 273–291. DOI: 10.1007/s10506-017-9214-9

Calo, R. (2015). Robotics and the lessons of cyberlaw. *California Law Review, 103*(3), 513–563.

Cath, C. (2018). Governing artificial intelligence: Upholding human rights & dignity. European Parliament Research Service (EPRS) Study.

Citron, D. K., & Pasquale, F. (2014). The scored society: Due process for automated predictions. *Washington Law Review (Seattle, Wash.), 89*(1), 1–33.

Cobbe, J., Lee, M. S. A., & Singh, J. (2021). Reviewable automated decision-making: A framework for accountable algorithmic systems. *Computer Law & Security Review, 41*, 105561.

Crawford, K. (2021). *Atlas of AI: Power, politics, and the planetary costs of artificial intelligence.* Yale University Press.

De Stefano, V. (2020). Algorithmic management and collective bargaining. *Comparative Labor Law & Policy Journal, 41*(1), 1–14.

Dworkin, R. (1977). *Taking rights seriously.* Harvard University Press.

EU High-Level Expert Group on Artificial Intelligence. (2019). *Ethics guidelines for trustworthy AI.* European Commission. https://digital-strategy.ec.europa.eu

Eubanks, V. (2018). *Automating inequality: How high-tech tools profile, police, and punish the poor*. St. Martin's Press.

European Commission. (2024). *Artificial Intelligence Act: Regulation of the European Parliament and of the Council on Artificial Intelligence*. Official Journal of the European Union.

Feinberg, J., & Narveson, J. (1970). The nature and value of rights. *The Journal of Value Inquiry*, *4*(4), 243–257. DOI: 10.1007/BF00137935

Finnis, J. (1980). *Natural law and natural rights*. Oxford University Press.

Floridi, L. (2021a). The European legislation on AI: A brief analysis of its philosophical approach. *Philosophy & Technology*, *34*(2), 215–222. DOI: 10.1007/s13347-021-00460-9 PMID: 34104628

Floridi, L. (2021b). *Ethics, governance, and policies in artificial intelligence*. Springer. DOI: 10.1007/978-3-030-81907-1

Fry, H. (2018). *Hello world: Being human in the age of algorithms*. W. W. Norton & Company.

Goodman, B., & Flaxman, S. (2017). European Union regulations on algorithmic decision-making and a "right to explanation.". *AI Magazine*, *38*(3), 50–57. DOI: 10.1609/aimag.v38i3.2741

Habermas, J. (1996). *Between facts and norms: Contributions to a discourse theory of law and democracy*. MIT Press. DOI: 10.7551/mitpress/1564.001.0001

Helberger, N., Karppinen, K., & D'Acunto, L. (2018). Exposure diversity as a design principle for recommender systems. *Information Communication and Society*, *21*(2), 191–207. DOI: 10.1080/1369118X.2016.1271900

Hildebrandt, M. (2015). *Smart technologies and the end(s) of law*. Edward Elgar Publishing. DOI: 10.4337/9781849808774

Hildebrandt, M. (2020). The artificial intelligence of European Union law. *German Law Journal*, *21*(1), 74–95. DOI: 10.1017/glj.2019.99

Ignas, K. (2019). *Algorithmic governance: Politics and law in the post-human era*. Palgrave Macmillan.

Jobin, A., Ienca, M., & Vayena, E. (2019). The global landscape of AI ethics guidelines. *Nature Machine Intelligence*, *1*(9), 389–399. DOI: 10.1038/s42256-019-0088-2

Kouroupis, K., & Efthymiou, I. (Eds.). (2026). *Legal Challenges of AI Across Interdisciplinary Sectors*. IGI Global Scientific Publishing., DOI: 10.4018/979-8-3373-5452-1

Latour, B. (2010). *The making of law: An ethnography of the Conseil d'État*. Polity Press.

Lessig, L. (1999). *Code and other laws of cyberspace*. Basic Books.

Mantelero, A. (2022). Beyond data: Human rights, ethical and social impact assessment in AI systems. *Computer Law & Security Review*, •••, 45.

Mittelstadt, B. D., Allo, P., Taddeo, M., Wachter, S., & Floridi, L. (2016). The ethics of algorithms: Mapping the debate. *Big Data & Society*, *3*(2), 1–21. DOI: 10.1177/2053951716679679

Organisation for Economic Co-operation and Development (OECD). (2019). *Recommendation of the Council on Artificial Intelligence*. OECD Legal Instruments. https://legalinstruments.oecd.org/en/instruments/OECD-LEGAL-0449

Pasquale, F. (2015). *The black box society: The secret algorithms that control money and information*. Harvard University Press. DOI: 10.4159/harvard.9780674736061

Rawls, J. (1971). *A theory of justice*. Harvard University Press. DOI: 10.4159/9780674042605

Selbst, A. D., Boyd, D., Friedler, S. A., Venkatasubramanian, S., & Vertesi, J. (2019). Fairness and abstraction in sociotechnical systems. *Proceedings of the Conference on Fairness, Accountability, and Transparency*, 59–68. DOI: 10.1145/3287560.3287598

Solum, L. B. (1992). Legal personhood for artificial intelligences. *North Carolina Law Review*, *70*(4), 1231–1287.

Stanford University DigiChina Project. (2023). Translation: Interim measures for the management of generative artificial intelligence services (China). DigiChina Research Brief. https://digichina.stanford.edu

Susskind, R. (2019). *Online courts and the future of justice*. Oxford University Press. DOI: 10.1093/oso/9780198838364.001.0001

Taylor, L. (2017). What is data justice? The case for connecting digital rights and freedoms globally. *Big Data & Society*, *4*(2), 1–14. DOI: 10.1177/2053951717736335

UNESCO. (2021). *Recommendation on the ethics of artificial intelligence*. United Nations Educational, Scientific and Cultural Organization. https://unesdoc.unesco.org

United Nations. (1966). *International Covenant on Civil and Political Rights*. United Nations Treaty Series, 999, 171. https://www.ohchr.org

Yeung, K. (2018). Algorithmic regulation: A critical interrogation. *Regulation & Governance, 12*(4), 505–523. DOI: 10.1111/rego.12158

Zuboff, S. (2019). *The age of surveillance capitalism: The fight for a human future at the new frontier of power*. PublicAffairs.

Chapter 7
AI, Ethics, and the Law: Balancing Innovation and Responsibility

Iris-Panagiota Efthymiou
https://orcid.org/0000-0001-9656-8378
University of Piraeus, Greece

ABSTRACT

Artificial intelligence is now central to economic activity, public governance, and everyday decision-making, yet its rapid growth exposes tensions between innovation, ethics, and the law. This chapter examines how legal systems can protect human rights, fairness, accountability, and transparency while supporting technological progress. It analyses current regulatory models, gaps in liability, bias and discrimination risks, data governance, and the responsibilities of developers, institutions, and states. It also considers ethical principles and their limits when translated into binding rules. The chapter argues for clear standards, proportionate regulation, and shared accountability frameworks that align AI development with public interest, democratic values, and social trust.

1. INTRODUCTION

Artificial Intelligence (AI) has experienced exponential growth, permeating various industries while giving rise to concerns about its ethical and legal ramifications. Those engaged in AI research are endeavouring to delineate the circumstances under which this technology can be safely and equitably applied. Drawing upon insights from "Principles alone cannot guarantee ethical AI" and "Ethics of Artificial Intelligence and Robotics" (Al-Zain M. A., et al.2025; Liang C. J., et al., 2024). This

DOI: 10.4018/979-8-2600-0000-7.ch007

chapter examines the ethical and legal considerations surrounding AI and proposes approaches for harmonizing technological innovation with societal values.

Characterized by iterative enhancement through accumulated knowledge, scientific activities underpin technological progress. Within this framework, technology develops on an incremental basis, with each novel creation serving as a foundation for subsequent innovations (Efthymiou, I-P., 2024). The extraordinary achievements observable today echo the manner in which earlier generations endeavored to comprehend, and master, their practical environment. Technological advances have indelibly altered the modalities of human labor, enabled global population expansion, and facilitated the exploration of new frontiers.

Nonetheless, it is crucial to acknowledge that technology should be regarded less as a solution and more as an impetus for discovery; it systems the conditions within which new questions may be formulated and addressed. Consequently, the emergence of a new technological paradigm necessitates an expanded articulation of economic principles as well as social values. When the train revolutionized the transportation industry, it consequently transformed economic sectors such as trade and tourism, simultaneously introducing new ethical challenges. It is reasonable to infer that intelligent transportation systems represent a natural progression in the decades following the inception of motorized vehicles.

Artificial Intelligence (AI) is expanding rapidly in scope and influence. From voice interaction and e-commerce to self-driving cars and medical diagnostics, AI systems increasingly take over crucial decision-making roles (Vallor S., & Bekey A., 2017). Given humanity's history of workplace safety lessons, it is critical that these systems act ethically and within the bounds of existing law. The AI control problem remains unsolved; as AI is widely deployed, society will inevitably be surprised by the system's actions. Longer-term predictions for very general AI capabilities remain an open debate.

2. UNDERSTADING ETHICS IN AI

At their core, ethical questions concern how individuals, institutions, governments, and corporations ought to behave. "Ethics" is the philosophical inquiry into the nature of morality, and the attempt to prescribe a direct course of action, or criteria for choosing among possible actions (Baggini J., & Fosl P. S., 2024; R., & Agrawal G., 2023). Moral philosophy is often divided into ethics (theories of right action), meta-ethics (theories about the nature of ethics and morality), and applied ethics (addressing concrete moral issues) (Efthymiou, I-P., 2024).

The ethical questions prompted by AI fall primarily under the first category, the theory of right action. In addition to ensuring trust worthiness, fairness, transparency, and safety, AI systems may be faced with difficult ethical trade-offs, e.g., in the formulation of algorithms guiding the behaviour of autonomous vehicles (Mittelstadt B., 2019).

Traditionally three main theories of right action are distinguished, which allow the AI practitioner to apply them in specific contexts like safety or explainability. There are thus three main moral principles that a hypothetical AI ought always to satisfy:

- Respect for autonomy: the principle that individuals should be free to make their own choices and have those choices respected. This principle would preclude AI systems that deliver fake news, generate fake rather than genuine content, impersonate real people, or generate deceptive content.
- Prevention of harm: the obligation to minimize and prevent harm to others. This principle speaks against lethal autonomous weapons. It also constrains the sorts of manipulation and deception to which users of AI systems should be subject.
- Fairness: an AI system should never be biased against particular individuals on irrelevant grounds.

Adhering to all three principles presents a considerable challenge, as they frequently conflict with one another. For instance, achieving fairness for certain groups may necessitate a compromise of the autonomy of others. To effectively address these complex ethical dilemmas, an AI practitioner must possess a nuanced comprehension of ethics, enabling them to rationalize a certain level of autonomy infringement as a means to avert harm.

Moral Principles in Technology

Moral principles are guidelines that help individuals, organizations, and machines distinguish between right and wrong (Müller V. C., 2020). To behave ethically, moral agents, also machines, must understand these principles and the differences among them. For example, a robber who is prepared to kill someone to obtain their money violates the moral requirement that killing humans is wrong. The principles themselves are typically represented in diverse ethical theories, including consequentialism, deontology, and virtue ethics, all of which are relevant to the development and deployment of artificial intelligence (AI) systems. Ethical behavior presupposes a correct understanding of the agent's duties: a doctor may kill the patient to prevent them from suffering while knowing that this is morally wrong.

Ethical Theories Applied to AI

Early formulations of AI ethics were initially established on abstract principles, which effectively characterized the field as a significant branch of applied ethics (Hallamaa J., & Kalliokoski T., 2022; Poszler F., et al., 2025; Weichert J., et al., 2025). The early stages encompassed a variety of topical areas, with primary focuses on modeling human thought processes, profiling individuals based on distinct characteristics such as race, gender, and political views, as well as addressing the potential for AI systems to induce human discrimination.

Moreover, there was a notable emphasis on the ramifications of automation on job markets and employment opportunities for individuals across various sectors. Building on these foundational ideas, three prominent ethical stances have emerged and continue to inform the multidisciplinary research landscape in AI (Mao Q. and Xu M., 2025). Each of these stances translates established moral theories into specific implications relevant to AI systems and their governance frameworks. The branch known as Applied Ethics primarily concentrates on action-based deontological ethics, which emphasizes the importance of determining how AI systems ought to behave, specifically, what actions they should take or avoid in various contexts (SWARGIARY K., 2024).

In contrast, the area of Virtue Ethics, which has seen a revival in contemporary philosophical discourse, focuses on agent-based ethics, probing into the decisions that AI systems should actively support or encourage among human actors in their interactions (Slote M., 2023; Beier K., 2025). Finally, there is an ongoing examination of the implications of AI, which expands into the realms of metaethics and epistemology. This examination involves a reflective consideration of the substantive content of ethical theories themselves, delving into the foundational assumptions and philosophical implications underlying ethical decision-making as it relates to artificial intelligence.

3. LITERATURE REVIEW

Introduction

The convergence of artificial intelligence (AI), ethics, and law constitutes a multifaceted domain that requires thorough analysis. Recently, academic scholars have devoted greater attention to the socio-legal ramifications of AI, investigating the substantial ethical and regulatory challenges that arise from its widespread integration across multiple facets of society. This discourse articulates these challenges by broadly defining AI and accentuating its function in mediating social practices within a digitalized, data-centric framework. The importance of tackling issues of fairness in AI from a socio-legal viewpoint is emphasized, thereby paving the way for a more profound inquiry into the ethical dimensions associated with AI entities (Larsson S., 2019).

Building on this foundation, (Harris I., & Reese Anthis G. R., 2021) explore the moral considerations of sentient artificial entities. Their literature review highlights the ongoing debate regarding the rights and moral status of these entities, revealing a fragmented discourse across disciplines. The authors point out the limited focus on rights and moral patiency in existing studies, calling for further investigation into how these artificial entities can be integrated into moral and legal frameworks. This discourse is crucial as it lays the groundwork for understanding the ethical responsibilities that accompany AI development.

(Resseguier A. & Rodrigues R., 2021) further critique the ethical frameworks surrounding AI, arguing that many initiatives adopt high-level principles detached from the realities of social, political, and environmental contexts. They caution against the potential for "ethics washing," where ethical guidelines may serve to circumvent stricter legal regulations. This critique emphasizes the importance of grounding ethical considerations in tangible realities to avoid blind spots that could arise from abstract principles, particularly in technology sectors where misrepresentation is common.

In the area of health-related AI, (Da Silva M., et al., 2022) underscore the urgent need for governance innovations to keep pace with rapid technological advancements. Their scoping review protocol highlights that while medical innovations offer significant promise, they also necessitate legal responses to mitigate potential pitfalls. This dual focus on innovation and governance is vital for ensuring that the benefits of AI are realized without compromising ethical standards or public trust.

Finally, (Hernández, E. G. (2024) presents a multidimensional framework for the ethical and inclusive implementation of AI within organizations. He argues for the integration of ethics into regulatory and innovative practices, emphasizing the importance of human-centered design and continuous training. By identifying fundamental pillars for ethical AI, Hernández reinforces the notion that ethics must be central to organizational strategies, ensuring that technological advancements align with human values.

Collectively, these articles illustrate the multifaceted challenges and considerations at the intersection of AI, ethics, and law. They highlight the need for a comprehensive approach that balances innovation with responsibility, advocating for ethical frameworks that are both contextually grounded and adaptable to the evolving landscape of artificial intelligence.

Literature review

The article titled "The Socio-Legal Relevance of Artificial Intelligence" by Stefan Larsson offers a comprehensive examination of contemporary AI technologies, emphasizing their practical applications and socio-legal implications (Larsson S., 2019). Larsson shifts the discourse from speculative fears of AI revolting against humanity to a focus on everyday AI systems, including machine learning, natural language processing, image recognition, neural networks, and deep learning. This pivot is significant as it grounds the discussion in real-world contexts where AI influences daily life, such as healthcare, finance, and media, thus making the socio-legal questions more immediate and tangible.

A critical aspect highlighted by Larsson is the rapid development of machine learning driven by large data collections and increased processing power. This technological evolution has enabled AI systems to perform complex tasks like predictive analytics and image recognition, which are closely linked to statistical methods. The use of algorithms that facilitate learning and probability assessments underscores the technical sophistication of current AI systems, raising important questions about transparency and accountability in their deployment.

Larsson emphasizes that current regulatory debates are centered on the societal impact of these AI applications. These discussions are crucial because they address issues of fairness, bias, and discrimination, core concerns from a socio-legal perspective. The article underscores that AI's integration into critical sectors such as healthcare and finance necessitates a nuanced understanding of how legal frameworks can ensure responsible innovation without stifling technological progress.

From a critical standpoint, Larsson's focus on the socio-legal relevance of AI advances provides valuable insight into how existing legal principles can adapt to new technological realities. However, while the article effectively contextualizes AI's societal impact, it could benefit from a deeper exploration of specific legal mechanisms or frameworks that might address issues of fairness and accountability in AI systems. This would strengthen the practical implications of the discussion and offer clearer guidance for policymakers.

The article "The Moral Consideration of Artificial Entities: A Literature Review" by Harris and Anthis (2021) offers a comprehensive synthesis of the emerging discourse surrounding the moral status of artificial entities. The authors highlight that an increasing number of scholars are contemplating whether artificial entities, particularly those demonstrating sentience or consciousness, merit moral consideration. This focus is driven by concerns that neglecting the moral status of such entities could result in significant suffering, emphasizing the ethical imperative to address their moral inclusion (Harris I. & Reese Anthis G., R., 2021).

A key contribution of the review is its identification of the interdisciplinary nature of this debate, spanning philosophy, law, and technology. The authors critically note that much of the existing literature centers on legal rights, such as legal personhood and intellectual property, which serve as pragmatic starting points for integrating artificial entities into human legal frameworks. However, they also underscore that the field remains nascent, with ongoing debates about which types of artificial entities (ranging from simple algorithms to potentially sentient robots)should be incorporated into the moral circle.

The review critically evaluates the limited synthesis of this literature to date, pointing out that much of the discourse is fragmented across disciplines and national contexts. Harris I. and Anthis G. R., (2021) emphasize the importance of establishing clear criteria for moral consideration, which could include sentience, autonomy, or capacity for suffering. They advocate for further systematic research to determine whether artificial entities should be granted rights akin to those of humans or animals, or whether a different moral framework is warranted.

The article "Ethics as attention to context: recommendations for the ethics of artificial intelligence" by Anais Resseguier and Rowena Rodrigues (Resseguier A. & Rodrigues R., 2021) critically examines the prevalent tendency within AI ethics initiatives to formulate broad, high-level principles, such as those articulated by the European Commission. The authors argue that these abstract principles often lack sufficient contextual grounding, which hampers their practical applicability and effectiveness in guiding responsible AI development and deployment. They emphasize the importance of shifting from principle-based frameworks to approaches that are attentive to specific social, cultural, and situational contexts.

Resseguier A. and Rodrigues R., advocate for a nuanced understanding of ethics that recognizes the diversity of AI applications and the varied societal impacts they entail. Their critique highlights that many existing guidelines tend to be overly general, risking a one-size-fits-all approach that may overlook local values, power dynamics, and unintended consequences. By proposing a context-sensitive perspective, the authors suggest that ethical deliberation should be embedded within ongoing social and political processes, rather than merely serving as a set of static rules.

Critically, the article underscores that this contextual approach necessitates active engagement with stakeholders, including marginalized communities, to ensure that ethical considerations are genuinely reflective of diverse perspectives. It also calls for a more dynamic, reflective process whereby ethical assessments are continuously revisited and adapted as AI technologies evolve and their societal implications become clearer.

The article titled "Legal concerns in health-related artificial intelligence: a scoping review protocol" by Da Silva et al. (Da Silva M., et al., 2022) offers a comprehensive examination of the intersection between AI innovations in healthcare and the accompanying legal challenges. The authors underscore the transformative potential of medical innovations driven by AI, emphasizing their capacity to improve patient outcomes and streamline healthcare delivery. However, they also highlight that such advancements necessitate robust governance frameworks to mitigate risks and prevent adverse consequences.

Da Silva M., et al. (2022) argue that the rapid evolution of AI technology in health contexts presents unique legal concerns that must be addressed proactively. These concerns include issues related to patient safety, data privacy, liability, and accountability. The authors advocate for a multidisciplinary approach to develop legal responses that balance innovation with responsibility, ensuring that societal benefits are maximized while minimizing harm. They posit that scholarly insights across disciplines are vital for informing future regulatory reforms, which must be adaptable to the fast-paced nature of AI development.

Critically, the article emphasizes that without appropriate legal oversight, the deployment of health-related AI could exacerbate existing inequalities, compromise patient trust, or lead to unforeseen ethical dilemmas. The authors' focus on a scoping review protocol indicates a systematic effort to map the current landscape of legal concerns and identify gaps in regulation, which is essential for guiding policymakers and stakeholders.

While the article effectively highlights the importance of legal frameworks, it could benefit from a deeper exploration of specific regulatory models or international standards that might serve as benchmarks. Nonetheless, its emphasis on the need for interdisciplinary scholarly input and proactive governance aligns well with the broader discourse on balancing innovation and responsibility in AI, particularly within sensitive sectors like healthcare (Batakis, D., et al., 2025).

The article "Towards an Ethical and Inclusive Implementation of Artificial Intelligence in Organizations: A Multidimensional Framework" by Ernesto Giralt Hernández (2024) offers a comprehensive examination of the ethical considerations crucial to the development and deployment of AI within organizational contexts. Hernández underscores that the societal impact of AI necessitates a deliberate and structured ethical approach, emphasizing that neglecting these principles can lead to unintended harm and erosion of public trust.

A key contribution of the article is the delineation of four fundamental pillars essential for fostering ethical AI: trust and transparency, aligned values, human-centered design, and continuous adaptation. Hernández advocates for the integration of these pillars into organizational strategies, suggesting that trust and transparency serve as the backbone for stakeholder confidence, while aligned values ensure that AI systems reflect societal norms and organizational missions. The emphasis on human-centered design highlights the importance of prioritizing human needs and rights, ensuring that AI serves to augment rather than diminish human agency.

Continuous adaptation is also critical, as Hernández notes that ethical frameworks must evolve in tandem with technological advancements and societal expectations.

Critically, the article stresses that ethics should not be an afterthought but rather the foundation of AI strategies. Hernández argues convincingly that embedding ethics into professional training and stakeholder participation can facilitate more inclusive and responsible AI development (Gupta, S., et al., 2023). However, while the framework is robust conceptually, the article could benefit from more concrete examples of how organizations have successfully operationalized these principles in practice. Additionally, the challenge of balancing innovation with ethical responsibility remains complex, especially given the rapid pace of AI advancements and diverse stakeholder interests. (Hernández, E. G., 2024).

Conclusion

The examination of the intersection of artificial intelligence (AI), ethics, and law reveals a multifaceted landscape characterized by significant challenges and opportunities. The discourse initiated by Larsson (2021) emphasizes the socio-legal implications of AI technologies, advocating for a shift from speculative concerns to a focus on real-world applications. This perspective is essential as it grounds the ethical and legal discussions in practical contexts, particularly in sectors like healthcare and finance, where AI's impact is profound (Larsson S., 2019). The need for fairness, accountability, and transparency in AI systems emerges as a central theme, highlighting the urgency for regulatory frameworks that adapt to technological advancements while ensuring responsible innovation.

Building on this foundation, Harris I. and Anthis G. R., (2021) examine the moral considerations surrounding artificial entities, particularly those exhibiting sentience. Their review underscores the fragmented nature of the current discourse across disciplines, advocating for clearer criteria for moral consideration that could inform legal frameworks. This call for interdisciplinary synthesis is echoed in the critique by Resseguier A., and Rodrigues R., (2021), who argue that ethical guidelines often lack contextual grounding. Their emphasis on the necessity of embedding ethical deliberations within social and political processes highlights the importance of stakeholder engagement, particularly with marginalized communities, to ensure ethical considerations are reflective of diverse perspectives (Harris I. & Reese Anthis G. R., 2021;Ressequier A. & Rodrigues R., 2021).

In the healthcare domain, Da Silva et al. M., (2021) present a scoping review protocol that underscores the dual focus on innovation and governance. They argue that while AI holds transformative potential for improving patient outcomes, it also introduces unique legal challenges that necessitate proactive governance. This sentiment resonates with Hernández (2024), who articulates a multidimensional framework for ethical AI implementation within organizations. He identifies fundamental pillars such as trust, human-centered design, and continuous adaptation, advocating for the integration of ethics into organizational strategies to foster responsible AI development (Da Silva M., et al., 2022; Hernández, E. G. (2024).

Collectively, these articles highlight the pressing need for a comprehensive approach that balances innovation with ethical responsibility. The discussions underscore the importance of developing legal and ethical frameworks that are contextually grounded, adaptable, and inclusive. As AI continues to evolve, the integration of ethical considerations into technological advancements remains crucial for ensuring that the benefits of AI are realized without compromising societal values or public trust.

4. LEGAL FRAMEWORKS GOVERNING AI

Legal frameworks exert a powerful influence on the use of AI, potentially driving widespread adoption or fostering resistance. Governments are dependably among the first to take action following the emergence of new technologies. Indeed, the frameworks that govern AI extend across scales, from international guidelines and treaties to national and even state legislation. At the international level, the OECD's AI principles call for transparency, fairness, and accountability, obliging signatories like the United States, the United Kingdom, and Japan to strive toward their enactment. The European Union's sweeping AI Act has set a global agenda, despite ongoing critiques of its insufficient attention to ethical nuances (Cheong L., et al., 2023).

The UK currently does not have any specific law that addresses artificial intelligence directly as of now; however, there are several pending bills that could change this situation in the future. On the other hand, the U.S. lacks a comprehensive federal regulation for artificial intelligence and instead relies on various government agencies to guide practices along with voluntary codes of conduct. Meanwhile, it's important to note that individual states are taking different approaches, leading to a divergence in the regulatory landscape across the country.

At the national level, governments have largely focused on promoting adoption; funding research and development while also sponsoring pilot projects. National strategies often emphasize competitiveness and innovation, dominated by the recognition that, in the long term, AI will touch everything from education and agriculture to social services and international security. Notable efforts can be found throughout Asia, the Americas, and Europe, China, The Netherlands, and Singapore are a few of the countries with particularly detailed programs. The United States provides an example of the approach in its National AI Initiative Act, passed in January 2021, when then-President Donald Trump signed the National Defense Authorization Act that contain the language. Intelligence agencies are acutely aware of the risks posed by hostile actors, and the increasing rates of investment, acquisition, and corporate malfeasance (Hine, E., & Floridi, L., 2024; Roberts H., et al., 2021).

The discourse surrounding AI and vagueness is filled with a continuous back and forth over the implications of ambiguity and the difficulty in speaking about it precisely; discussions incorporate thoughts on language and epistemology, the importance of meaningful definition, and the ways in which these questions relate, broadly, to politics and the exercise of power (Tampubolon M., 2025; Munn L., et al., 2024; Kerr, A. D., & Scharp, K., 2022). Approaches to those tasks are often locked in epistemological frameworks—either aiming to dispossess the ambiguity, establishing clear and rigorous terminologies to reduce that uncertainty; or enshrining the very unpredictability of the AI conversation, emphasizing complexity over the desire to simplify, and pointing toward more anarchic or ethereal logics.

The first is valuable in producing pragmatic proposals; identifying ethics as an actionable set of principles to be integrated within legislation, AI certifications, and professional codes of conduct. The second aims to defend thinkers and practitioners from the theoretical limitations of legalistic techniques and the narrow solutions they engender. Such reflections embody an acknowledgment of the limits to knowledge about AI; and an insistence that without proper attention to the characterization of the social and historical conditions that enable particular forms of dominance, any attempt to capture AI in a normative model must remain profoundly incomplete.

In recent months, the establishment of BioGPT, an enormous generative language model that synthesized decades of questions and answers within the scientific literature surrounding biomedicine, has underscored how rapidly reliance on such communication tools is proliferating (Li Q., et al., 2024; Maity S., & Saikia M. G., 2025).

Corporations, meanwhile, have demonstrated a vigorous interest in the development of these applications, from Google's Bard and Microsoft Bing to SnapChat and Robinhood. Subsequent controversy has focused upon the opacity of the technical layers beneath LLMs, as well as the dangers of disinformation, dissimulation, and hallucination; and on the ethical challenges associated with emergent uses across finance, education, and governance.

International Regulations

International regulation of artificial intelligence (AI) has been centred on the European commitment of adherence to the UN 'Ban on Autonomous Weapon Systems' treaty, restriction of the method by which AI solutions may be developed and deployed, such as facial recognition, and support for the ambitious European Commission proposal for the regulation of AI Act (Feldstein S., 2024; Cantero Gamito, M., & Marsden, C. T., 2024). Such regulations and proposals have been critical in shaping global governance and de facto setting standards at the international level (Meleouni, C., & Efthymiou, I. P., 2024).

National Laws and Policies

National AI laws and policies address liability, impact, and transparency of AI. The EU's 2021 Directive on Liability for AI Systems clarifies conditions for damages and exempts developers and users from liability in some cases (Mittelstadt B., 2019). The EU Data Governance Act, effective 2023, facilitates public-sector data sharing, expanding datasets for AI systems. The EU's 2023 AI Act, already provisionally adopted, categorises AI systems by risk level and defines basic requirements. Various EU measures, including an EUA proposal for ePrivacy, address fairness, discrimination, and fundamental rights.

The United States has adopted a slower, sectoral approach focusing on economic competitiveness. The National AI Initiative Act and the National AI Research Resource Task Force support federal investment. The 2023 AI Bill of Rights outlines principles for mitigating harms and expanding access. The White House Office of Science and Technology Policy (OSTP) has issued memoranda on safe, secure, and trusted AI, with emphasises on transparency and stakeholder responsibility. Individual states have introduced bills on agency AI use, facial recognition, online content moderation, social credit systems, biometrics and data privacy, and AI liability, with six states imposing moratoria on law enforcement use of facial recognition or related technologies (Hacker P., 2023). Rising adoption of AI in government has prompted discussions about transparency, and OSTP is encouraging agencies to disclose AI use.

Other national policies cover ethical and regulatory frameworks, international cooperation, technical standards, and trustworthiness and transparency. China's AI development plan underscores safety; revised rules on recommendation algorithms address transparency and privacy. India's NITI Aayog released a discussion paper addressing economic, ethical, legal, and social issues, while a committee is preparing a national AI roadmap. Saudi Arabia established a National Centre for Artificial Intelligence. South Korea created an R&D strategy, a national AI strategy, and a roadmap to strengthen the AI industry.

5. CASE STUDIES IN AI ETHICS

While at times convergent, the law and ethics are conceptually distinct fields with different scopes. Ethics is an open-ended field that addresses broad abstract principles of right and wrong conduct, whereas "the law" refers to a more circumscribed set of rules within a specific jurisdiction, whose violations are punishable by that state. Ethics thus holds a wider purview and may sometimes require more onerous standards of conduct. For autonomous vehicles, the question of responsibility in a crash then involves both sets of concerns. Ethical deliberations that hold a vehicle operator accountable for not taking control may be incompatible with legal constructs that define operator liability in terms of actual control (Müller V. C., 2020). Moreover, the law contemplates a post-crash scenario in which some party is found culpable, whereas ethical considerations might well prioritize pre-crash assignments that will lessen the overall incidence or severity of collisions.

Autonomous Vehicles

Various ethical questions arise if roads are shared between human-driven and autonomous vehicles. The assignment of responsibility for possible crashes is debated: options include the occupant, manufacturer, software provider, insurer, or the vehicle itself. Answering this requires distinguishing between driving and steering. The occupant remains responsible if in control, but responsibility shifts if the vehicle operates autonomously. The author argues that liability should not exceed the general risk taken using autonomous vehicles; a tax or mandatory insurance might address this. Assuming autonomous vehicles save lives, there is a moral reason to limit manufacturer liability to avoid hindering progress. Manufacturers should not be free from liability if they knowingly market unsafe vehicles. Responsibility for accidents provides an incentive for continuous improvement (Efthymiou, I. P., & Efthymiou-Egleton, T. W., 2025: Hevelke, A. & Nida-Rümelin, J., 2015).

Another aspect concerns the trade-off between speed and accuracy in highly autonomous vehicles. Engineers must determine how accurate is sufficient for decision-making: waiting for more certainty can increase risk. Because autonomous vehicles may have inconsistent perceptions, coordinating inputs from multiple sensors is crucial. Tools that enable policymakers to assess the speed-accuracy trade-off facilitate the development of safety standards and mechanisms for accountability after accidents. Ex post mechanisms can diagnose causes and distinguish uncertainty due to this trade-off from software bugs or hardware failures. Such tools are essential for regulatory agencies to exercise recall authority and balance power between manufacturers and users.

Legal ambiguity surrounds the concept of a driver: several safety regulations target the human driver, yet fully autonomous vehicles lack human control mechanisms. The legal framework is shifting, with some authorities interpreting 'driver' to mean the autonomous system. No liability framework specifically governs autonomous vehicles. Although they aim to enhance safety, accidents remain inevitable, creating gaps in assigning responsibility. Liability often transfers from human drivers to manufacturers or the AI system, but clear legal designation has yet to be established (Sellwood, M., 2017).

Facial Recognition Technology

Facial recognition technology (FRT) is used to identify or verify individuals based on their facial features. The technology can be automated, or a human can use the search results for further evaluation. Academic researchers perceive FRT as an Orwellian and dystopian application or a use case in children's books, where a sheepdog causes the other sheep to hold hands, using face recognition to identify those who are vulnerable. Experienced practitioners believe that the technology is set in advanced use cases such as money laundering, where the AI model for face recognition is used on the facial features of a suspect and would check whether the suspect obtains cancer insurance seven years ago.

Most existing methods focus mainly on the accuracy of face recognition. They are willing to sacrifice essential properties, such as efficiency, diversity and privacy, to improve accuracy. However, accuracy is not enough to ensure usability and security. In recent years, these properties have attracted increasing attention and inspired corresponding research. Face recognition is the first stage in the pipeline. The dataset consisting of face data determines these essential properties. To reduce the risk of using face recognition in sensitive environments and further promote the application of face recognition, it is necessary to build a dataset with the three properties of efficiency, diversity and privacy.

AI in Healthcare

The healthcare sector presents a dynamic realm for the application of artificial intelligence (AI) (Efthymiou, I. P., et al. 2019; Efthymiou, I. P. (2026)... The integration of AI has the potential to significantly alter how healthcare is diagnosed, administered, and supported. At the same time, the ability to equitably reduce healthcare costs, improve or sustain access to healthcare, and deliver healthcare services to masses remains a critical challenge (Pasricha S., 2022).

The earliest AI applications in healthcare trace back to the 1970s, with expert systems designed for electrocardiogram (ECG) interpretation (Efthymiou-Egleton,

I., 2016). Since then, technological strides have redefined the potential that AI offers, particularly in pattern recognition and risk prediction. By 2022, the U.S. Food and Drug Administration (FDA) had given clearance to more than three hundred medical devices utilizing AI or machine learning (ML), heralding a new era in diagnostics and wearable health technologies. Meanwhile, a variety of AI-driven healthcare products materialize on the horizon, ranging from virtual health assistants to personalized therapeutics and pharmaceuticals, as well as robotic surgeries.

Natural language processing, in particular, has enabled conversational agents and chatbots, while robotics aids in surgery and geriatric assistance. Concomitantly, the rise of the Internet of Medical Things (IoMT) provides additional avenues for the development of smart tools. Despite this technological progress, ethical issues such as transparency, bias, privacy, safety, responsibility, justice, and autonomy come to the fore, for instance, in the context of COVID-19 (Diareme, K. et al., 2022).

Mismanagement of physician patient interactions may undermine the benefits of AI systems when the centrality of human factors is ignored. Moreover, AI algorithms might deliver biased or otherwise unintelligible results if the training data encode bias or fail to reflect the underlying decision-making process (Livieri, G., et al., 2024). Therefore, ongoing work on the ethical and legal implications of AI strives to ensure that responsible regulation, collaboration, and development of governance best practices keep pace with those systems in the application domain.

6. CHALLEGES IN AI REGULATION

Governments worldwide face the task of connecting artificial intelligence's potential while mitigating associated risks and challenges developing alongside the technology. Artificial intelligence is characterized by the ability of a programmed system to conduct complex, independent processes, such as problem-solving, decision-making, or visual recognition, that resemble human cognitive functions (Kouroupis, K. & Efthymiou, I. (Eds.), 2026). There is a proliferation of new services integrating AI technologies, applications to speed up interpretation of x-ray images and video feeds and help optimize production processes within supply chains have become common in advanced economies, yet this rapid development continues to outpace regulatory response. There remain few guiding principles that outline the ethical use of data valued by various groups or that precisely define the legal boundaries of automated services, and the proliferation of private sector and academic initiatives underlines the early stage of these debates.

The spread of the technology highlights a puzzle concerning the corporate responsibility for socially ambiguous human–machine relationships. By its nature, AI eliminates full human oversight from decision-making processes, develops self-instruction through machine learning, and thereby distorts the myriad responsibilities assigned to different corporate actors. National differences in the terms of access to data and technology, in legal systems of governance, and in cultural attitudes toward automation may also have far-reaching consequences for the competitive position of states developing ethical and legal maps ahead of the international community. Working through the intricacies of these fundamental challenges would create more favourable conditions for fair corporate engagement that fully respects targeted individual and collective interests (Cheong L., et al., 2023).

Rapid Technological Advancements

Technological progress is faster than institutions can react, making legislation and regulation difficult and sometimes inconsistent across different jurisdictions (Walz A. & Firth-Butterfield K., 2019). International bodies such as UNESCO and the OECD have formulated ethical frameworks and policy instruments for AI governance, while the European Union has proposed AI specific legislation, and the United States has adopted a less binding approach through non-binding standards and guidelines. Other jurisdictions such as the United Kingdom and Singapore have deployed national AI strategies encompassing a wide variety of policies with increased attention to responsible and trustworthy AI (Müller V. C., 2020). This heterogeneity in regulatory and legal approaches could potentially lead to conflicts of law governing cross-border data flows, supply chain management, liability of autonomous systems, and data protection for AI datasets, among others.

In addition, the private sector tends to have powerful private interests that could override ethical concerns for the sake of maximizing profit. It is therefore crucial for governments and businesses to determine their joint commitment to the adoption and use of ethical principles for the development and deployment of AI and autonomous systems in widely distributed applications at every level (Auld G., et al., 2022; Kashefi Y., et al., 2024). Elsewhere, the academic sector has a significant influence in shaping the AI ethics and governance agenda (Schiff, D., 2022; Schiff, D. S., 2023). Local civil society organizations and the public at large play a key role in raising stakeholders' awareness of a much-needed ethical approach to these technologies, thus contributing to the development of inclusive policies and supporting innovation and research in the context of a sustainable human-centric AI ecosystem (Budziszewska, A., and Potluka, O., 2024; Schiff D., et al., 2021; Moon, M. J. 2023; Yigitcanlar T., et al., 2024).

Global Disparities in Regulation

The swift development of artificial intelligence has prompted responses ranging from calls for banishment to demands for global legislation (de Laat R. B., 2021). AI technologies often evolve faster than legislation can accommodate, permitting activity that might later be judged unacceptable or unlawful (Hacker P., 2023). Wide disparities between national and regional regulatory frameworks may lead to the mistrust or blocking of AI developments, diminishing incentives to innovate or adopt technologies that deliver significant benefits. Worldwide experimentation with different models and the emergence of standard-setting frontrunners may establish eventual blueprints, but these may be long in coming.

Given the immense aggregate impact of AI innovation on wellbeing, development and wealth creation, an unchecked mandate to simply ban or regulate public deployments is both impractical and counterproductive. A balanced approach that safeguards innovation while upholding ethical accountability is a far more realistic undertaking.

7. THE ROLE OF STAKEHOLDERS

Artificial intelligence (AI) development strategies and deployment plans involve diverse stakeholders, among them both public and private sector institutions. Each group produces strategies with specific objectives and ethical commitments, making coordination and cooperation among them important (Choung H., et al., 2023). Varying ethical sentiments across sectors lead to different viewpoints on the governance of AI and its potential benefits, risks, and ethical requirements.

Governments typically face pressures to ensure the safe, beneficial, and trustworthy use of AI and seek to be involved in the facilitation and regulation of AI companies and projects. When done strategically, proper government leadership can encourage private-sector and nonprofit investment and assure public trust (Efthymiou, I. P., 2026). The private sector produces and deploys most AI products and services, giving it one of the heaviest burdens in ensuring AI corresponds with societal and consumer ethical expectations. Companies aim to be profitable while inspiring confidence in their respective AI systems, requiring a balance of commercial objectives with ethical considerations. Academia offers invaluable insight and analysis that can help other stakeholders (private-sector companies, nonprofits, and government organizations) develop effective strategies and plans.

Government and Policymakers

The rate of technological advancement, and the concurrent rise of concern about the implications of these technologies, requires policymakers to strike a delicate balance between facilitating innovation and enforcing safeguards on their use. To that end, national governments, international bodies, and individual agencies are crafting strategies and frameworks to guide the governance of emerging digital technologies.

The United States, for example, has tasked the National Institute of Standards and Technology with developing standards and other policy guidance, while the European Union's Horizon 2020 program has supported large-scale research initiatives on responsible innovation. But adherence to national and international standards will not be sufficient to manage the ethical dilemmas that inevitably arise. Given the scope of impact, pressure to adopt emerging technologies will only accelerate, and governance platforms must evolve and mature accordingly.

When unaccountable implementation could result in dangerous breaches of privacy, compromised safety, or worse, immediate safeguards, such as the Food and Drug Administration's Medical Device Action Plan, must be enacted. Ethical considerations can also be advanced through programs such as the U.K.'s Centre for Data Ethics and Innovation, which works to maximize the potential benefits of emerging digital technologies within ethical, societal, and environmental constraints (Reed-Berendt R., et al., 2022; Becker S. J., et al., 2023; Almeida D., et al., 2022). Besides government regulation, the private sector wields considerable power through purchasing and licensing decisions, particularly for large-scale AI and blockchain projects. Leveraging this power to enforce "ethics by design" and responsible-use guidelines would advance these goals while opening new avenues for growth and creative opportunity (Kalenzi, C., 2022).

Private Sector and Corporations

Shaping the future of Artificial Intelligence (AI) inevitably involves all tiers of government, the private sector, academia, and civil society (Psomiadi, A., 2026). Effective stewardship also depends on broad engagement with ethical considerations associated with AI systems.

As novel avenues for the application of AI emerge, so have new ways to put to use the vast troves and streams of data generated by society's ever-expanding global network infrastructure (Efthymiou, I. P., 2025). New arrangements for access and use proliferate, and with them, new concerns. The private sector and corporations call for the rapid development of policy frameworks that reduce uncertainty around data sharing agreements and other barriers to innovation in this space (Rumiantsev A., et al.2024). Corporate leaders simultaneously observe that new guidelines and regulations taking on ethical and privacy concerns would actually afford confident and responsible data use across government agencies, companies, and institutions (Verhulst, S. G., 2021; Brown, I., & Marsden, C. T., 2023). To drive development of AI responsibly, the private sector actively supports work on its social impact and the adoption of guiding principles. Corporate efforts prioritize the integrity of their data supply chains, adopting frameworks that serve as a foundation for the development of algorithms.

Deep development of data practices occupies an essential role as well. Corporations embrace the establishment and implementation of principles related to data, further emphasizing the critical connection of responsible AI innovation with the quality and provenance of data. The confidence with which the private sector demonstrates an awareness of the potential social impact of AI has been reflected in numerous initiatives and core guiding principles, which underscore the application of AI that is fair, ethical, human-centric, and typified by transparency.

Academia and Researchers

Research institutions have critical roles in ethical AI development, offering in-house expertise to support ethically sound practice. Where specialists are lacking, institutions benefit from promoting research on AI ethics (Chekhratova O., & Pohorielova, T., 2024; Shah Z., et al., 2025; Caton L., & Hwang, Y., 2025). Developing expertise in ethical AI may require dedicated centers, postdoctoral programs, or doctoral schools to ensure competence.

Transparency is essential; public AI models and development criteria foster accountability and allow for the identification of systemic biases. Mandatory data documentation aids implementers and reviewers in assessing fairness (Efthymiou, I. P., et al., 2025). Pre-publication review by neutral AI ethics researchers, ideally independent of funders, can strengthen ethical practice. Materials should be broadly accessible to facilitate widespread ethical adherence and multidisciplinary collaboration. Encouragement of 'bottom-up' AI ethics, where models reinforce ethics by design, can be particularly effective, though such initiatives are rare and predominantly academic.

Accountability standards and monitoring procedures constitute 'soft-law' instruments that can assist private actors in aligning AI deployment with societal values. Frameworks addressing fairness and non-discrimination risks, alongside guidelines for high-risk domains, offer valuable benchmarks. Sector-specific standards and regulatory governance models, supported by consensus and advice from professional societies or ethics bodies, can further reinforce ethical integrity (Efthymiou, I. P., 2025). Communities of practice promote cultural spread and knowledge exchange, though continued progress requires sustained efforts in guidance development, enforcement mechanisms, and effective dissemination channels.

8. PUBLIC PERCEPTION OF AI

Artificial intelligence (AI), appearing in the twenty-first century, has experienced rapid growth, cultivating new possibilities and applications with each step of advancement. Integral to many aspects of everyday life, ethical and legal concerns now need consideration to achieve a balance between technological innovation and moral accountability (Megdad, M. M., et al., 2024; Voegtlin, C., et al., 2022; Yanamala A. K. Y., et al., 2024). While regulation is necessary, risks such as infringement of human rights, particularly the right to privacy, and highly biased algorithmic decisions require reflection. Assessing the current view of AI is crucial: how do people feel and think about the innovativeness of new applications? Is there a desire for more transparent and explainable AI? Can users trust the machines? Indeed, several inquiries embrace these questions, including initiatives proposed by the EU Commission.

Trust and Transparency

The socio-legal relevance of artificial intelligence Today, and possibly in the future, much AI governance takes a principle- or guideline approach. A series of principles from robotics and similar bodies, as well as from governments, is often adopted. The military drone analysis project, for example, was not renewed by Google after it released its AI principles in June 2018. The construction of a responsible culture and behaviour is widely understood as a prerequisite for development teams, regardless of sector, to keep their tools under control and assess potentially harmful effects. In age of algorithmic design, technology might qualify as ubiquitous persuasion and have a significant influence on human autonomy and decision-making. Distributed Agency With most systems being distributed, agency is often a shared and collective undertaking requiring delegated, fragmented, proxy, or responsible agency. Whether a machine can understand things, have intentions, or ultimately be held accountable has legal consequences, especially as autonomous AI service providers emerge. Shared agency thus appears of particular importance to questions of accountability.

Ethical Consumerism

Ethical consumerism challenges firms to guarantee products' socially responsible origins, scrutinising pathways from supply to production (Dewasiri N. J., and Singh R., 2025). Large firms must evaluate the social impact of outsourced production on all stakeholders; transnational ethics functions and auditors work alongside civil society to verify these standards (Karropoulos, E. & Psomiadi, A., 2026). Companies formally committed to responsible AI remain vulnerable.

Developments instilling fairness remain controversial. Multiple conceptions distinguish diverse fairness conceptions; consequent precision fosters fairness metrics implemented in numerous software tools facilitating model compliance with specific fairness preferences (Yousefi, Y., 2024). Such tools aid ethical discourse indirectly; they do not provide a technocratic solution, nor redirect attention from underlying societal tensions and involved stakeholders. Overall, some progress towards responsible AI emerges; accusations of ethics washing appear misplaced for certain companies. Firms yet to engage might catch up subsequently.

9. FUTURE DIRECTIONS IN AI ETHICS AND LAW

Rapid advances in Artificial Intelligence (AI) increasingly affect ethical and legal frameworks. The multidisciplinary nature of these challenges currently lacks clearly defined scope and definitions, complicating the design of effective governance mechanisms (Efthymiou, I. P., 2025). Ensuring that AI development remains consistent with human values requires a balanced approach that supports the technology's progress while providing safeguards against harmful and undesirable consequences through a system of ethical accountability and legal oversight.

A broader understanding of AI-related ethical and legal concerns will therefore assist researchers, policymakers, and end users in balancing innovation with responsibility and awareness of their social impact. AI had a significant role in popularizing the term 'Artificial Intelligence' during the preceding decades. Since then, it has grown progressively more influential in guiding the conceptualization of diverse methods for representing, processing, and acquiring knowledge, as well as providing a framework for examining kinds of people problems.

In the late 1990s, recent advances in knowledge representation, uncertainty quantification, automated reasoning, and planning led to a new commitment to 'building intelligent systems'. AI methods now enjoy widespread commercial adoption. Related areas, such as machine learning and data mining, have diversified to encompass a wide range of fields and applications. AI also continues an intriguing symbiotic relationship with cognitive science, which fosters neurological inspiration for innovative representations of knowledge and mechanized reasoning. Attempts to understand human knowledge acquisition and reasoning have inspired new developments. In many respects, AI seeks to understand, emulate, and approximate human reasoning as closely as possible through computational mechanisms (Lima, G., & Cha, M., 2020).

Emerging Trends

Wider societal implications are discussed in various studies. The political angle of technology is often separated from philosophy or ethics, despite common concerns. Another approach to AI ethics is grounded in tangible moral problems, to build both practical and theoretical frameworks for solutions. The principle is to start from concrete moral dilemmas in everyday AI interaction and work "upwards" towards general principles. This "bottom-up approach" derives ethics from experience rather than a conceptual roadmap, involving active involvement of AI and philosophy researchers to develop ethics grounded in concrete situations and focused on real solutions (Mittelstadt, 2019).

Current developments in the field of AI ethics showcase similarities to the principles governing medical ethics, underscoring the necessity for ethical frameworks to evolve concurrently with rapidly changing applications (Weidener L., & Fischer, M., 2024; Jha D., et al., 2025; Radanliev, R., 2025). There is an imperative to convert ethical philosophies into tangible instruments, including protocols and legislative measures. Furthermore, there exists the possibility of leveraging AI technologies to bolster ethical practices, through the implementation of methods such as privacy-preserving machine learning techniques, algorithms designed to reduce false positives and negatives, fairness-aware data mining, and speech recognition systems aimed at enhancing transparency.

Many ethical questions in AI arise because social practices and legal structures have yet to adapt to new technology contexts. When such an ethical issue is identified, lawmakers can assess whether a new law should be introduced or whether existing laws need clarification or improved understanding. If legislation is required, regulatory bodies play a critical role in determining the form and content of these laws. AI ethics occupies a transitional space between currently inadequate regulatory structures and the potential development of new laws and policies (Walter, Y., 2024; Pagallo U., et al., 2023).

In the domain of technology policy, the processes of planning and implementation continue to present significant difficulties. Principal practical considerations encompass the identification of the stakeholders engaged in policy formulation and the comprehension of existing power dynamics. Governments, parliaments, associations, and industry circles have produced various reports and white papers, with some generating good-will slogans, but little concrete policy exists beyond AI funding.

The latest EU policy document advocates for "trustworthy AI" that is lawful, ethical, and technically robust, with requirements encompassing human oversight, technical robustness, privacy and data governance, transparency, fairness, well-being, and accountability (Stamboliev, E., & Christiaens, T., 2025). There is a risk that current self-regulation delegates ethical decision-making to a narrow circle of experts rather than societal stakeholders. A helpful summary of an ethical framework for AI is available from European experts.

Potential Legislative Changes

As artificial intelligence (AI) increasingly permeates society, profound ethical and legal questions are emerging with considerable implications for the discipline, the computer and information industry, and the general public. As AI systems transition from lab prototypes to public services, a broad as-yet unanswered question becomes increasingly relevant: How should policymakers respond to a more demanding and intrusive form of AI? The search for solutions is underway, but the complexity and novelty of the issue make this an uncommonly tough policy challenge. Moreover, AI tends to amplify and accelerate the disruptive character of other technologies, such as drones, genomics, or robotics. At the same time, a healthy and dynamic AI discipline requires to thrive. What we must do, therefore, is to properly balance the opportunities offered by progress in AI with an appropriate degree of political risk management. Only such a balanced approach will promote positive reforms without unnecessarily slowing down the technical advance of the field.

An example of a suggested balance between the two approaches can be found in the paper proposing soft law as an alternative to binding legislative frameworks (Walz R. & Firth-Butterfield K., 2019). Soft law might take the form of guidelines on the ethical compliance of AI systems, for example. Unlike legislation, soft law can be more precise than non-specific principles and adapt more readily to technological advances. In any event, governments should safeguard citizens from harm caused by new products and technologies through the encouragement of responsible design, production and use, rather than waiting for injuries to occur. To this end, regulators should be demonstrably expert, sufficiently nimble, and able to collaborate with companies on a basis of mutual trust and respect, through which responsibly designed ethical AI products are certified as fit for purpose and can be confidently unleashed on the marketplace, all whilst preserving the conditions that underpin innovation.

Regulators also have an important role where their objective is to guide the development of AI technologies in an ethically acceptable manner (Kashefi P., et al., 2024; Ryan, M., and Stahl, B. C., 2021). They can do so, for example, through the suggestion of appropriate milestones on the road to long-term ethical alignment or by creating monetary incentives, such as the linking of public R&D funding to compliance with ethical requirements, using either binding legislation or soft law.

10. CONCUSION

The European Union employs a precautionary and risk-based regulatory framework designed specifically to ensure safety and protect users, implementing more stringent regulations at an earlier stage of technological development and deployment. In contrast, the United States advocates for a permissive and innovation-first approach, which often prioritizes the advancement of technology over regulation. This differing approach can lead to significant disparities in how artificial intelligence and other technologies are governed. For instance, former President Trump's AI Plan for 2025 aimed to eliminate obstacles that could hinder technological progress and actively prevent state-level regulations from imposing additional limitations.

Corporate responses to these regulatory environments vary significantly among major tech companies: OpenAI and Google have willingly subscribed to the EU Code of Practice, demonstrating a commitment to aligning with the EU's regulatory standards, while Meta has opted out of this agreement altogether, choosing to follow a different strategic path that does not conform to EU regulations. Additionally, xAI has committed solely to adhering to the safety chapter of the code, attributing its decision to concerns regarding innovation and its potential to thrive in a less restrictive regulatory landscape.

Artificial Intelligence (AI) has reached previously unimagined frontiers, becoming a source of optimism and excitement while simultaneously raising societal concerns (Efthymiou, I. P., 2024). For every incremental advance, the ethical and legal liabilities accrue, highlighting the need for a balanced approach that still encourages innovation and early adoption. Achieving a synthesis that energizes AI's advancement while constraining its potential to cause harm is imperative.

Reviewed are the evolution of AI, ethics, the law, contemporary application and regulatory frameworks, case studies, major challenges, the roles of stakeholders, public perceptions, and emerging trends in ethics and legislation. Establishing a synthesis based on insights gleaned from these foundational elements directs a course toward a new set of principles, roles, and responsibilities for AI's advancement. By simultaneously fostering innovation, enforcing ethical accountability, and demanding legal oversight, reliable human–machine partnerships come into view.

REFERENCES

Al-Zain, M. A., Abdulaziz, S. A., & Al Mubarak, M. (2025). Ethics of artificial intelligence and robotics. In *Ethical Standards for Technological and Business Education Sustainability* (pp. 85–98). Emerald Publishing Limited. DOI: 10.1108/978-1-83608-036-720251006

Almeida, D., Shmarko, K., & Lomas, E. (2022). The ethics of facial recognition technologies, surveillance, and accountability in an age of artificial intelligence: A comparative analysis of US, EU, and UK regulatory frameworks. *AI and Ethics*, *2*(3), 377–387. DOI: 10.1007/s43681-021-00077-w PMID: 34790955

Auld, G., Casovan, A., Clarke, A., & Faveri, B. (2022). Governing AI through ethical standards: Learning from the experiences of other private governance initiatives. *Journal of European Public Policy*, *29*(11), 1822–1844. DOI: 10.1080/13501763.2022.2099449

Baggini, J., & Fosl, P. S. (2024). *The ethics toolkit: A compendium of ethical concepts and methods*. John Wiley & Sons.

Batakis, D., Chatzivasileiou, S., & Efthymiou, I. P. (2025). Artificial Intelligence and its impact on healthcare.

Becker, S. J., Nemat, A. T., Lucas, S., Heinitz, R. M., Klevesath, M., & Charton, J. E. (2023). A Code of Digital Ethics: Laying the foundation for digital ethics in a science and technology company. *AI & Society*, *38*(6), 2629–2639. DOI: 10.1007/s00146-021-01376-w

Beier, K. (2025). Virtue Ethics and AI. *A Companion to Applied Philosophy of AI*, 178-191.

Brown, I., & Marsden, C. T. (2023). *Regulating code: Good governance and better regulation in the information age*. MIT press.

Budziszewska, A., & Potluka, O. (2024). The use of AI and technology by civil society organizations and its international implications. In *The Routledge Handbook of Artificial Intelligence and Philanthropy* (pp. 94–106). Routledge. DOI: 10.4324/9781003468615-8

Cantero Gamito, M., & Marsden, C. T. (2024). Artificial intelligence co-regulation? The role of standards in the EU AI Act. *International Journal of Law and Information Technology*, *32*, eaae011. DOI: 10.1093/ijlit/eaae011

Caton, L., & Hwang, Y. (2025). Ethical Implications and Governance of AI in Research. In *Institutional guide to using AI for research* (pp. 167–183). Springer Nature Switzerland. DOI: 10.1007/978-3-031-94809-1_8

Chekhratova, O., & Pohorielova, T. (2024). Ethical research in the age of artificial intelligence.

Cheong, I., Caliskan, A., & Kohno, T. (2023). Is the US Legal System Ready for AI's Challenges to Human Values? *arXiv preprint arXiv:2308.15906.*

Choung, H., David, P., & Seberger, J. S. (2023). A multilevel framework for AI governance. *arXiv preprint arXiv:2307.03198.*

Da Silva, M., Horsley, T., Singh, D., Da Silva, E., Ly, V., Thomas, B., Daniel, R. C., Chagal-Feferkorn, K. A., Iantomasi, S., White, K., Kent, A., & Flood, C. M. (2022). Legal concerns in health-related artificial intelligence: A scoping review protocol. *Systematic Reviews, 11*(1), 123. DOI: 10.1186/s13643-022-01939-y PMID: 35715812

de Laat, P. B. (2021). Companies committed to responsible AI: From principles towards implementation and regulation? *Philosophy & Technology, 34*(4), 1135–1193. DOI: 10.1007/s13347-021-00474-3 PMID: 34631392

Dewasiri, N. J., & Singh, R. (2025). Advancing Climate Neutrality Through Sustainable Production and Consumption: Pathways to a Resilient Future. In *Climate Neutrality Through Smart Eco-Innovation and Environmental Sustainability* (pp. 3–19). Springer Nature Switzerland. DOI: 10.1007/978-3-031-83250-5_1

Diareme, K. C., Liapakis, A., & Efthymiou, I. (2022). Big Data, Sentiment Analysis, and Examples during the COVID-19 Pandemic. *HAPSc Policy Briefs Series, 3*(2).

Efthymiou, I. P. (2024). Human wellbeing in the age of artificial intelligence.

Efthymiou, I. P. (2026). AI in epidemiology: Predicting the unpredictable leveraging AI to detect outbreaks and predict disease. In *Innovation and transformation of public health* (pp. 281–308). IGI Global Scientific Publishing.

Efthymiou, I. P. (2026). Bridging the Divide: Addressing the Leadership Gap. In *Bridging the Leadership Gap: Policies and Workplace Equality* (pp. 103-138). IGI Global Scientific Publishing.

Efthymiou, I. P., & Egleton, T. E. (2025). The Impact of Deep Fakes in Markets and Economies. In *Deepfakes and Their Impact on Business* (pp. 19–50). IGI Global Scientific Publishing., DOI: 10.4018/979-8-3693-6890-9.ch002

Efthymiou, I. P., Egleton, T. E., & Psomiadi, A. (2025). The Role of Artificial Intelligence in Humanitarian Aid for Achieving Sustainable Development Goals. In *Cases on AI-Driven Solutions to Environmental Challenges* (pp. 43–72). IGI Global Scientific Publishing. DOI: 10.4018/979-8-3693-7483-2.ch003

Efthymiou, I. P., Vozikis, A., & Sidiropoulos, S. (2019). Application of Sociocybernetic model in the field of Health Management. *International Journal of Scientific and Engineering Research, 10*(5), 451–560.

Efthymiou-Egleton, I. (2016). *Trends in health care: A global challenge.* Xlibris Corporation.

Feldstein, S. (2024). Evaluating Europe's push to enact AI regulations: How will this influence global norms? *Democratization, 31*(5), 1049–1066. DOI: 10.1080/13510347.2023.2196068

Gupta, S., Kamboj, S., & Bag, S. (2023). Role of risks in the development of responsible artificial intelligence in the digital healthcare domain. *Information Systems Frontiers, 25*(6), 2257–2274. DOI: 10.1007/s10796-021-10174-0

Hacker, P. (2023). The European AI liability directives–Critique of a half-hearted approach and lessons for the future. *Computer Law & Security Review, 51*, 105871. DOI: 10.1016/j.clsr.2023.105871

Hallamaa, J., & Kalliokoski, T. (2022). AI ethics as applied ethics. *Frontiers of Computer Science, 4*, 776837. DOI: 10.3389/fcomp.2022.776837

Harris, J., & Anthis, J. R. (2021). The moral consideration of artificial entities: A literature review. *Science and Engineering Ethics, 27*(4), 53. DOI: 10.1007/s11948-021-00331-8 PMID: 34370075

Hernández, E. G. (2024). Towards an ethical and inclusive implementation of artificial intelligence in organizations: a multidimensional framework. *arXiv preprint arXiv:2405.01697.*

Hevelke, A., & Nida-Rümelin, J. (2015). Responsibility for crashes of autonomous vehicles: An ethical analysis. *Science and Engineering Ethics, 21*(3), 619–630. DOI: 10.1007/s11948-014-9565-5 PMID: 25027859

Hine, E., & Floridi, L. (2024). Artificial intelligence with American values and Chinese characteristics: A comparative analysis of American and Chinese governmental AI policies. *AI & Society, 39*(1), 257–278. DOI: 10.1007/s00146-022-01499-8

Jha, D., Durak, G., Sharma, V., Keles, E., Cicek, V., Zhang, Z., Srivastava, A., Rauniyar, A., Hagos, D. H., Tomar, N. K., Miller, F. H., Topcu, A., Yazidi, A., Håkegård, J. E., & Bagci, U. (2025). A Conceptual Framework for Applying Ethical Principles of AI to Medical Practice. *Bioengineering (Basel, Switzerland), 12*(2), 180. DOI: 10.3390/bioengineering12020180 PMID: 40001699

Kalenzi, C. (2022). Artificial intelligence and blockchain: How should emerging technologies be governed? *Frontiers in Research Metrics and Analytics, 7*, 801549. DOI: 10.3389/frma.2022.801549 PMID: 35224423

Karropoulos, E., & Psomiadi, A. (2026). Feeding the Future: AI-Driven Food Distribution for Aging Populations. In Stolakis, K., & Efthymiou, I. (Eds.), *AI for Geriatric Care in an Aging Society: Ethical, Clinical, and Policy Challenges* (pp. 345–370). IGI Global Scientific Publishing., DOI: 10.4018/979-8-3373-5462-0.ch012

Kashefi, P., Kashefi, Y., & Ghafouri Mirsaraei, A. (2024). Shaping the future of AI: Balancing innovation and ethics in global regulation. *Uniform Law Review, 29*(3), 524–548. DOI: 10.1093/ulr/unae040

Kerr, A. D., & Scharp, K. (2022). The end of vagueness: Technological epistemicism, surveillance capitalism, and explainable artificial intelligence. *Minds and Machines, 32*(3), 585–611. DOI: 10.1007/s11023-022-09609-7

Kouroupis, K., & Efthymiou, I. (Eds.). (2026). *Legal Challenges of AI Across Interdisciplinary Sectors*. IGI Global Scientific Publishing., DOI: 10.4018/979-8-3373-5452-1

Larsson, S. (2019). The socio-legal relevance of artificial intelligence. *Droit Social, 103*(3), 573–593. DOI: 10.3917/drs1.103.0573

Li, Q., Li, L., & Li, Y. (2024). Developing ChatGPT for biology and medicine: A complete review of biomedical question answering. *Biophysics Reports, 10*(3), 152. DOI: 10.52601/bpr.2024.240004 PMID: 39027317

Liang, C. J., Le, T. H., Ham, Y., Mantha, B. R., Cheng, M. H., & Lin, J. J. (2024). Ethics of artificial intelligence and robotics in the architecture, engineering, and construction industry. *Automation in Construction, 162*, 105369. DOI: 10.1016/j.autcon.2024.105369

Lima, G., & Cha, M. (2020). Responsible AI and its stakeholders. *arXiv preprint arXiv:2004.11434*.

Livieri, G., Anastasopoulou, C., Briola, K., Nikolopoulou, P., & Efthymiou, I. P. (2024). Ethical issues arising from the use of AI in drug discovery. *Journal of Politics and Ethics in New Technologies and AI, 3*(1), e37093–e37093. DOI: 10.12681/jpentai.37093

Maity, S., & Saikia, M. J. (2025). Large Language Models in Healthcare and Medical Applications: A Review. *Bioengineering (Basel, Switzerland), 12*(6), 631. DOI: 10.3390/bioengineering12060631 PMID: 40564447

Mao, Q., & Xu, M. (2025). Navigating the Ethical Landscape of AI Innovation: Challenges and Opportunities. *International Journal of Digital Law and Governance*, (0).

Megdad, M. M., Abueleiwa, M. H., Al Qatrawi, M., El-Tantaw, J., Harara, F. E., Abu-Nasser, B. S., & Abu-Naser, S. S. (2024). Ethics in AI: Balancing innovation and responsibility.

Meleouni, C., & Efthymiou, I. P. (2024). The Use of Artificial Intelligence (AI) in National Security: Defining International Standards and Guidelines. *Journal of Politics and Ethics in New Technologies and AI, 3*(1), e37847–e37847.

Mittelstadt, B. (2019). Principles alone cannot guarantee ethical AI. *Nature Machine Intelligence, 1*(11), 501–507. DOI: 10.1038/s42256-019-0114-4

Moon, M. J. (2023). Searching for inclusive artificial intelligence for social good: Participatory governance and policy recommendations for making AI more inclusive and benign for society. *Public Administration Review, 83*(6), 1496–1505. DOI: 10.1111/puar.13648

Müller, V. C. (2020). *Ethics of artificial intelligence and robotics*

Munn, L., Magee, L., & Arora, V. (2024). Truth machines: Synthesizing veracity in AI language models. *AI & Society, 39*(6), 2759–2773. DOI: 10.1007/s00146-023-01756-4

Pagallo, U., Bassi, E., & Durante, M. (2023). The normative challenges of AI in outer space: Law, ethics, and the realignment of terrestrial standards. *Philosophy & Technology, 36*(2), 23. DOI: 10.1007/s13347-023-00626-7

Pasricha, S. (2022). AI ethics in smart healthcare. *IEEE Consumer Electronics Magazine, 12*(4), 12–20. DOI: 10.1109/MCE.2022.3220001

Poszler, F., Portmann, E., & Lütge, C. (2025). Formalizing ethical principles within AI systems: Experts' opinions on why (not) and how to do it. *AI and Ethics, 5*(2), 937–965. DOI: 10.1007/s43681-024-00425-6

Psomiadi, A. (2026). Social Dynamics and AI-Driven Care for Older Adults. In *AI for Geriatric Care in an Aging Society: Ethical, Clinical, and Policy Challenges* (pp. 123-154). IGI Global Scientific Publishing.

Radanliev, P. (2025). AI ethics: Integrating transparency, fairness, and privacy in AI development. *Applied Artificial Intelligence*, *39*(1), 2463722. DOI: 10.1080/08839514.2025.2463722

Reed-Berendt, R., Dove, E. S., & Pareek, M.UK-REACH Study Collaborative Group. (2022). The ethical implications of big data research in public health:"big data ethics by design" in the uk-reach study. *Ethics & Human Research*, *44*(1), 2–17. DOI: 10.1002/eahr.500111 PMID: 34936235

Resseguier, A., & Rodrigues, R. (2021). Ethics as attention to context: Recommendations for the ethics of artificial intelligence. *Open Research Europe*, *1*, 27. DOI: 10.12688/openreseurope.13260.2 PMID: 37645106

Roberts, H., Cowls, J., Hine, E., Mazzi, F., Tsamados, A., Taddeo, M., & Floridi, L. (2021). Achieving a 'Good AI Society': Comparing the Aims and Progress of the EU and the US. *Science and Engineering Ethics*, *27*(6), 68. DOI: 10.1007/s11948-021-00340-7 PMID: 34767085

Rumiantsev, A., Morozov, V., & Panikar, G. (2024). Innovation at the crossroads: How Public-private partnerships drive technological development in Europe. *Actual Problems of International Relations*, *1*(159), 94–108. DOI: 10.17721/apmv.2024.159.1.94-108

Ryan, M., & Stahl, B. C. (2021). Artificial intelligence ethics guidelines for developers and users: Clarifying their content and normative implications. *Journal of Information. Communication and Ethics in Society*, *19*(1), 61–86. DOI: 10.1108/JICES-12-2019-0138

Schiff, D. (2022). Education for AI, not AI for education: The role of education and ethics in national AI policy strategies. *International Journal of Artificial Intelligence in Education*, *32*(3), 527–563. DOI: 10.1007/s40593-021-00270-2

Schiff, D., Borenstein, J., Biddle, J., & Laas, K. (2021). AI ethics in the public, private, and NGO sectors: A review of a global document collection. *IEEE Transactions on Technology and Society*, *2*(1), 31–42. DOI: 10.1109/TTS.2021.3052127

Schiff, D. S. (2023). Looking through a policy window with tinted glasses: Setting the agenda for US AI policy. *The Review of Policy Research*, *40*(5), 729–756. DOI: 10.1111/ropr.12535

Sellwood, M. (2017). The road to autonomy. *The San Diego Law Review*, *54*, 829.

Shah, Z., Shahzad, M. H., Saleem, S., Taj, I., Amin, S., Almagharbeh, W. T., & Durvesh, S. (2025). ETHICAL CONSIDERATIONS IN THE USE OF AI FOR ACADEMIC RESEARCH AND SCIENTIFIC DISCOVERY: A NARRATIVE REVIEW. *Insights-Journal of Life and Social Sciences*, *3*(2), 183–189. DOI: 10.71000/jfesgv69

Slote, M. (2023). Agent-based virtue ethics. In *Philosophical Essays East and West: Agent-Based Virtue Ethics and other topics at the intersection of Chinese thought and Western analytic philosophy* (pp. 83–95). Springer Nature Switzerland. DOI: 10.1007/978-3-031-39955-8_6

Stamboliev, E., & Christiaens, T. (2025). How empty is trustworthy AI? A discourse analysis of the ethics guidelines of trustworthy AI. *Critical Policy Studies*, *19*(1), 39–56. DOI: 10.1080/19460171.2024.2315431

SWARGIARY. K. (2024). *Foundations and Frontiers in Ethics (A Comprehensive Study)*. EdTech Research Association, US.

Tampubolon, M. (2025). Decoding Legal Ambiguity: the Interplay between Law and Legal Semiotics in Modern Jurisprudence. *International Journal for the Semiotics of Law-Revue internationale de Sémiotique juridique*, 1-29.

Vallor, S., & Bekey, G. A. (2017). *Artificial intelligence and the ethics of self-learning robots*

Verhulst, S. G. (2021). Reimagining data responsibility: 10 new approaches toward a culture of trust in re-using data to address critical public needs. *Data & Policy*, *3*, e6. DOI: 10.1017/dap.2021.4

Voegtlin, C., Scherer, A. G., Stahl, G. K., & Hawn, O. (2022). Grand societal challenges and responsible innovation. *Journal of Management Studies*, *59*(1), 1–28. DOI: 10.1111/joms.12785

Walter, Y. (2024). Managing the race to the moon: Global policy and governance in Artificial Intelligence regulation—A contemporary overview and an analysis of socioeconomic consequences. *Discover Artificial Intelligence*, *4*(1), 14. DOI: 10.1007/s44163-024-00109-4

Walz, A., & Firth-Butterfield, K. (2019). Implementing ethics into artificial intelligence: A contribution, from a legal perspective, to the development of an AI governance regime. *Duke L. & Tech. Rev.*, *18*, 176.

Weichert, J., Zhu, Q., Kim, D., & Eldardiry, H. (2025). Perceptions of AI Ethics Policies Among Scientists and Engineers in Policy-Related Roles: An Exploratory Investigation. *Digital Society : Ethics, Socio-Legal and Governance of Digital Technology*, *4*(1), 1–25. DOI: 10.1007/s44206-025-00176-9

Weidener, L., & Fischer, M. (2024). Proposing a principle-based approach for teaching AI ethics in medical education. *JMIR Medical Education*, *10*(1), e55368. DOI: 10.2196/55368 PMID: 38285931

Yanamala, A. K. Y., Suryadevara, S., & Kalli, V. D. R. (2024). Balancing innovation and privacy: The intersection of data protection and artificial intelligence. *International Journal of Machine Learning Research in Cybersecurity and Artificial Intelligence*, *15*(1), 1–43.

Yigitcanlar, T., David, A., Li, W., Fookes, C., Bibri, S. E., & Ye, X. (2024). Unlocking artificial intelligence adoption in local governments: Best practice lessons from real-world implementations. *Smart Cities*, *7*(4), 1576–1625. DOI: 10.3390/smartcities7040064

Yousefi, Y. (2024, October). Approaching Fairness in Digital Governance: A Multi-Dimensional, Multi-Layered Framework. In *Proceedings of the 17th International Conference on Theory and Practice of Electronic Governance* (pp. 41-47). DOI: 10.1145/3680127.3680197

Chapter 8
Intellectual Property Challenges in the Age of Artificial Intelligence

Alanamol Shinto
https://orcid.org/0009-0008-2788-6514
Christ University, India

ABSTRACT

Intellectual property rights, or IPRs, are crucial in the digital age for promoting innovation, protecting the rights of creators, and driving economic growth. The paper examines the challenges that intellectual property rights (IPR) face, including digital piracy, online copyright infringement, and cross-border enforcement issues, as globalization and digital transformation continue to transform industries. AI has created serious problems and posed challenges in the areas of intellectual property rights, more particularly in copyright law. Analyzes the models used in the USA, EU, China, UK, India, and other countries, identifying their strengths and weaknesses. The present paper explores how AI plays a vital role in creating artistic works such as art, music, poems, and novels, among other forms. Moreover, the research examines the legal structures of various countries and provides a summary of different authorship models for AI-generated content.

1. INTRODUCTION

In the internet age, copyright, along with intellectual property (IP), represents two fundamental legal concepts that, to a large extent, serve as the initial protective measures for creators' rights. More precisely, it is copyright that confers the authors the rights to reproduce, distribute, and publicly display their creative works, which

DOI: 10.4018/979-8-2600-0000-7.ch008

can be in the form of software, movies, music, and . Patents, trademarks, and trade secrets are just a few of the more general categories that fall under the umbrella of intellectual property, which also includes inventions and branding (Flew, T., Suzor, N., & Liu, B. R., 2013). The ease with which material can be copied, shared, and altered in the digital age has made intellectual property protection more challenging, as conventional frameworks often struggle to keep pace with the rapid advancements in technology (Md Ashraful, A., Gazi, M., & Sanjida Akter, S., 2025). In this interconnected world, maintaining fair usage while upholding the rights of artists remains a crucial task. Itis pretty significant to protect digital content in a global village that we live in today if we are to have continuous innovation, creativity, and economic growth. In the process, software developers, filmmakers, musicians, and writers create products such as technological solutions, movies, music, and digital books, respectively, and they pour their time, energy, and money into them. In this regard, copyright laws are intended to be the instruments that ensure creators receive the fruits of their labor. The creative industries are those that suffer in the absence of strong protection, as they are the targets of illegal use and piracy, which threaten them and discourage the introduction of innovations (Becerra, M., & Waisbord, S. R., 2021). Moreover, the protection of digital content is the assurance of the customers that they will have access to the real and qualitatively good products, which, in turn, will attract them to buy and not . Societies can achieve a balance between fostering artistic expression and providing creators with the necessary incentives to continue making contributions to the digital economy by protecting intellectual property rights.

The legal issues surrounding AI in terms of intellectual property are complex and evolving as AI technology continues to develop. One of the main foci of the interaction between AI and IPRs has been a significant area of change in IP law. After a lot of, though selective, argument, it appears that the matter is becoming increasingly real in practice due to the lively AI-related market activity, the first judicial decisions in the matter, and policy moves by international organizations and legislators .

2. LITERATURE REVIEW

Intellectual property (IP) law has been changed significantly by AI breakthroughs, and these changes bring new opportunities and worries for the future. With the development of AI technologies that can create content, analyze data, and manage IP, traditional IP security paradigms have become obsolete and need to be changed. The paper examines the intricate relationship between AI and IP, emphasizing the need for a deeper understanding of IP in the digital era and the challenges associated with identifying and protecting AI-generated IP (Lu, 2020[CS1.1]). In the digital age, the concept of intellectual property has expanded to encompass not only copyrights and artistic creations but also algorithms, databases, and AI-generated works, which have become intellectual property due to the value of data and the autonomy of AI systems. This section is new in terms of questions regarding ownership, authorship, and originality of AI-generated work. All-generated works raise practical problems of ownership and authorship under intellectual property law. Original works are protected against copying by copyright law; however, AI technology further challenges this protection. There are concerns about originality and creativity, as AI algorithms often rely on extensive historical data to generate content (Wills, 2021). It is the duty of government officials to ensure that the content produced by machines is not given more importance as compared to the content produced by the human mind. The use of AI in IP management and content creation should be done in accordance with ethical principles. A responsible AI approach also involves understanding bias in AI algorithms, little transparency, and unexpected effects of AI-generated material. There are various ways to enhance IP protection in the era of AI. The use of blockchain technology and AI-powered IP management tools is an efficient way to achieve copyright registration and ownership verification. (Drexl,). Awell-rounded legal system aligned with ethics, cooperation amongst nations, and laws pertaining to intellectual property, when conceived by AI, can be a major factor in the promotion of the ethical conduct of AI and the protection of its users, as well as the creators. In today's digitized world, data has become the most valuable form of intellectual property. The exponential increase of digital interactions and transactions is not only producing data but also recording it in huge volumes. While the data allows companies to make inferences, offer customization, and do targeted advertising, it is also becoming indispensable to businesses and organizations. As a result, knowledge has become a valuable resource that needs to be protected.

The rise of IP in the digital age has given rise to various protection issues. On the one hand, digital media has made it more complicated to infringe copyright, patents, and trademarks, the three traditional IP protection methods. (Ghosh, 2020) The main source of intellectual property protection for artistic works, copyright, is encountering increasing problems in the digital . It has become very difficult to maintain copyright protection since the methods for digital copying and distribution have changed substantially over the years. The rise of content aggregation platforms, illegal activities, and unauthorized content sharing has generally contributed to a decrease in the financial and legal rights of content producers. As the digital ecosystem cannot function without software, safeguarding intellectual property in this area is quite challenging. (Kop, 2020)

3. ARTIFICIAL INTELLIGENCE

The term "Artificial Intelligence" was coined by John McCarthy in 1956. There is currently no legal definition of "artificial intelligence." "Artificial intelligence" may be stated to be "the capability of machines to carry out tasks which, if done by humans, would require intelligence". Ray Kurzweil defined AI in 1990 as "the science of making computers do things that require intelligence when done by humans"5. AI typically refers to the "ability of machines to perform cognitive tasks like thinking, perceiving, learning, problem-solving, and decision-making". According to Russ Pearlman, "the central goals of AI include reasoning, knowledge, planning, learning, natural language processing (e.g., understanding and speaking languages), perception, and the ability to. The foundation of AI is 'artificial neural networks,' which are 'brain-inspired systems that are designed to imitate the way the human mind learns.' The artificial neural networks have the capabilities of self-learning that "enable them to produce better results as more data becomes available". AI, therefore, enables machines to carry out tasks independently or with limited human intervention, which may otherwise require human intelligence. AI is not to be considered as a single technology, but rather a field that encompasses many subfields, including machine learning, robotics, language processing, and deep learning. "Machine learning" and "deep learning" are, therefore, two subsets of AI. For machine learning, there is an inbuilt algorithm in the computer program that "gives it the capability to learn from the provided data, evolve, and make decisions in the future," either on its own or in a specific direction. In other words, the machine learning algorithms learn from the programmer's provided inputs to generate something new by making its own independent decisions move and manipulate objects". The three categories of AI systems identified by WIPO are –(i) "expert (or knowledge-base) systems"; (ii) "perception systems"; and (iii) "natural language systems".

The involvement of AI can produce two categories of creative works: (i) "AI-generated" work; and (ii) "AI-assisted" work. The AI-generated works, which are also known as "generated autonomously by AI," refer to the creation of a work by AI without human intervention. In this category of work, AI may "change its behavior during operation to respond to unanticipated information or events" and produce work that might not have been intended or anticipated. The "AI-assisted" works, on the other hand, are created with significant human intervention.

4. ARTIFICIAL INTELLIGENCE AND COPYRIGHT

There has been an extensive use of computer programs in the generation of copyrighted works since the 1970s. The computer-generated works did not create many problems with respect to copyright ownership. The reason was that computer programs were considered mere tools to support creative activities, which required human intervention for their production. These programs were just like stationery items, which required human beings to use them to create works. Things have completely changed now. With AI in place, computer programs are no longer just tools; they have the potential to work independently, making their own decisions.

The AI has the potential to create an enormous amount of work with less investment in a very short span of time. In principle, AI-generated works may be considered original and hence be eligible for copyright protection in all territorial jurisdictions. The requirement of use of "skill and judgement" in originality may be deemed to have been satisfied by virtue of the "programming and parameters on which such AI actually compiles and creates the work". However, there will be no author in the case of AI-generated work. In the case of AI-assisted work, human intervention is required. Therefore, in the case of the latter, the person who caused the work to be created using artificial intelligence may claim to be the author; however, this is not true when AI itself has created the work withoutany human intervention. The issue of authorship has been examined by various governments, judicial authorities, and scholars worldwide. Threegeneral possibilities concerning the question of authorship can be imagined: (i) the copyright system should recognize the authorship of an AI; (ii) there should be no authorship in AI-generated works and such works should be considered as "public domain"; and (iii) there should be a sui generis law instead of copyright law to protect such .

The premise that reflects the civil law countries such as Germany, France, and Spain indicates that works created must bear the "imprint of the author's personality". The authorship, therefore, should be denied to AI in the AI-generated works, as the AI does not have a personality. Making AI a legal entity would mean that it should possess the capacity to enter into contracts with other persons. It will also have duties under the law and will be liable for its acts. Most importantly, it should have the capacity "to sue and be sued" under the law. Most countries are not in favour of granting legal status to AI.

It is interesting to note that the copyright legislations of several countries extend moral rights to the creator, even though this is not a requirement stipulated by the TRIPs Agreement. Two moral rights – (i) right of paternity, and (ii) right of integrity are ordinarily provided to the author. The former ensures the author's right to be associated with their work and be named as its creator. In contrast, the latter enables the author to claim such claims for any disfigurement or distortion of the work if this is harmful to their honour or . In Amar Nath Sehgal v. Union of India6, the Delhi High Court observed that "in the material world, laws are mainly designed to secure the right to fair pay. Nevertheless, life is not just material. It is temporal as well. Most of us believe in the soul. The moral rights of the author are the soul of his works. The author is entitled to keep, guard, and develop his works through his moral rights". Moral rights are associated with the emotions and the feelings of the human creator." These rights are not meant for AI.

The UK Copyright, Designs and Patents Act, 1988 (hereinafter referred to as "CDPA") defines and regulates works created by computers. According to the CDPA, "computer-generated" work is one where the output is by a computer in conditions where there is no human author of the work.7" . This provision was put in place "to establish a carve-out from the requirement that a work be of human authorship in order to give fair recognition and protection to the effort involved in creating a program that is able to generate works independently". In view of s. 9(3) of the CDPA, the author of the work is the one "to be deemed the person who made the necessary arrangements for the production of the work" in a case of "a literary, dramatic, musical or artistic work which is omputer-generated".

AndresGuadamuz argues that the authorship in this kind of situation should be attributed to the programmer, not the user. He explains this position by giving the example of Microsoft, which developed the computer program "Word" for users to create their own works. Microsoft cannot hold a copyright in a work that has been produced by a user with the help of its program. The copyright in the work so created will be with the user of the program, who will be considered the author, as they have created the work using that program. In Express Newspapers plc v. Liverpool Daily Post & Echo, the court treated a computer as a tool, much like a pen is a tool. In the United States, the author of a work created with the help of AI may hold copyright if they can demonstrate that the AI program was used as a tool/ medium in the creation of the work. In Naruto v. Slater, popularly known as the "Monkey Selfie" case, the court in the United States held that the monkey could not be considered the author of the selfies it took. Copyright can only be given to a human author, i.e., no copyrights for animals or machines in the .S.

The Indian Copyright Act lacks a definition for "computer-generated work," as the CDPA does. However, it defines "author" as one relating to "any fictional, dramatic, musical, or creative effort which is computer-generated" as "the individual who causes the labor to be created". In the case of Camlin Pvt. Ltd. v. National Pencil Industries, the Delhi High Court further clarified the meaning of the term "author". The Courts indicated that "mechanically reproduced printed carton" was not the subject matter of copyright because it was not possible to ascertain who the author of such a carton was. The Court also asserted that "copyright is a privilege reserved only for authors or those human beings from whom the work is derived." Consequently, the plaintiff is ineligible to file a copyright infringement claim over any carton that has been reproduced mechanically by a printing process, as the work has not been derived from the author. It follows that a machine cannot qualify as an author of an artistic work, and likewise, it cannot own a copyright in that ." In Tech

Plus Media Private Ltd v. Jyoti Janda, the Delhi Court decided that "the plaintiff is a legal entity and is not capable of being the author of any work, the copyright of which may vest." The Court went on to say that the plaintiff may, however, become the owner of the copyright in the work by a written agreement with the author."

InAustralia, copyright is given to the developer of an AI tool in the "source code of the machine" only; however, the work generated by AI does not have a copyright due to the absence of human . The best approach is to decide the issue of authorship in a case on its merits by adopting a case-by-case approach. One of the contemporary applications of AI is the creation of literary works; therefore, studying copyright in the context of AI becomes crucial. To understand this concept within the context of this research, it is helpful to examine three court cases: Burrow-Gilles Lithographic Co. v. Sarony, Bleistein v. Donaldson Lithographing Co., and Alfred Bell & Co. v. Catalda Fine Arts.

Burrow Gilles Lithographic Co. v. Sarony

At the center of this disagreement was the concern of whether an image could be granted copyright protection. The fact that it highlighted the distinction between creative and mechanical labor made it a case worthy of consideration. In this case, the concern about whether the output of a machine can be protected by copyright was raised in court. The court limited the extent of their protection when it came to creative work by determining that work that is simply mechanical does not, by definition, qualify as artistic[10]. Therefore, it would be challenging to assign copyrights for work that was produced by AI systems if this kind of stringent approach were to be used.

Bleistein v. Donaldson Lithographing Co.

This case addressed the same legal issue as the one raised in the prior case. In this particular instance, the Court distinguished quite specifically between things that were made artificially and those that a human manufactured. In his majority opinion, Justice Holmes emphasized the importance of copyright laws by drawing attention to the one-of-a-kind qualities that come with human personality. The position of the Court was apparent when it used the term "something unrestricted which is one man's alone," which indicated that there was no space for anything that was not the result of human ingenuity. In other words, there was no room for anything.

Alfred Bell & Co. v. Catalda Fine Arts, Inc.

As a result of this judgment, the courts have developed an order that is more accommodating with respect to copyrights. The Supreme Court reduced the bar for originality when it determined that creative work does not have to be an exact copy of another creative effort of the same kind in order for that work to be considered original. This decision had the effect of lowering the bar for originality. Additionally, it highlighted the potential for a creator to claim ownership of flaws or unexpected alterations to their work. The court's decision provided some relief to those asserting copyrights for works produced by AIs, as the works in question were not duplicated, despite being the product of specific programming and algorithms. The uncertainty about the protection of AI systems has been reduced to some degree as a result of these three judgments. The absence of a clearly articulated attitude, on the other hand, has and will continue to affect the potential right holders.

The suggestion of considering AI and a human author as joint authors of the work so produced is not a sound one. The reason is that human beings do not control all operations of AI, and the AI operates without any control. This does not fit into the definition of "works of joint authorship". For example, the Indian Copyright Act, 1957, defines "work of joint authorship" to mean"a work created Through the co-operation of two or more authors, the work in which the contribution of one author cannot be distinguished from the contribution of the other author or authors.". Rich specifically states that "machine learning generally leads to models being developed that are so intricate that. Even the creators of the algorithm have very little idea of the exact way or reasons in which the generated model produces accurate .' Additionally, the suggestion of making an AI programmer and the AI user joint authors of the AI-generated program is also not a sound one.

On the world stage, the Berne Convention of 1886 did not contemplate "non-human Authorship. The exact position may be considered as correct in the case of the Trade-Related Aspects of Intellectual Property Rights (hereinafter referred to as "TRIPs") Agreement, in view of the fact that it incorporates the provisions of the Berne Convention. A similar position may be deemed to be correct about the WIPO Copyright Treaty and WIPO Performances and Phonograms Treaty of 1996 (WIPO Internet Treaties). At the same time, it may also be argued that the international legal regime on copyright did not preclude the possibility of a non-human authorship in national legislations. The international treaties ordinarily lay down minimum common standards to be followed. The countries are obligated not to derogate from them, but at the same time, they are free to provide better protection than what has been laid down in the treaties.

5. THE EVOLUTION OF INTELLECTUAL PROPERTY IN THE DIGITAL AGE

The protection of material innovations, which include physical inventions, artistic works, and brand identifiers, is the very point where intellectual property rights (IPR) have their roots. However, the emergence of new creation and consumption methods brought about by the digital age has altered the concept of intellectual property. To start with, books, inventions, and logos are some of the tangible assets that, at the beginning, were the primary concern of traditional intellectual property . Although the progression of digital technologies has been very rapid, the companies' experiences remain relevant. The growth of intangibles, in the form of data, software code, digital designs, and multimedia content, has been the main story over the last couple of years. The spread of these maverick products has made it more complex to identify, monitor, and enforce IPR. Most of the regulations that could once be enforced lawfully within a country's borders now face challenges in enforcement due to the worldwide digital networking nature of the Internet, and intellectual property has become multi-regionally linked instead of being bound to a particular geographical area. Intellectual property works worldwide at present as a result of the global network and digital platforms, which are the places where unauthorized distribution and infringement may happen in a matter of seconds across borders and are usually beyond the jurisdiction of national laws. One such example is the increase in digital piracy to an enormous extent.

These practices have largely been typical with software, gaming, music, and film industries, in which users download, share, and upload copyrighted works without seeking permission from the creators. As this is happening, a large number of digital platforms and online services, such as social media networks, e-commerce websites, and content-sharing platforms, have also emerged due to the globally interconnected digital economy. These networks have allowed creators of content to access worldwide audiences; however, they have likewise made it extremely easy for the unapproved use of intellectual property on a huge scale. Hence, the internet era is a two-sided phenomenon: on the one hand, it has promoted unlimited connectivity and cooperation, while on the other, it poses a serious risk of intellectual property abuse.

5.1 Understanding Intellectual Property in the Digital Era

With the emergence of AI technologies as a source of unique problems and possibilities, IP has risen as one of the most significant issues. Hilty et al. (2020) state that 68 various kinds of intellectual property, e.g., patents, copyrights, trademarks, and trade secrets, are generally employed to protect different facets of human intelligence and innovation. Nevertheless, AI progress has entangled the issues of giving credit, ownership, and even the basic idea of intellectual property by indistinctly separating the maker from the .

5.2 Concept of Ownership in Intellectual Property Law

Intellectual property is the result of creation and recognition through the laws controlling the same, i.e., copyright, patent, trademark, and trade secret protection. For the most part, it is referred to as an intangible asset, and like real property, it is also subject to rights. These rights open the way for the makers or the proprietors of the intangible property to get paid or to enjoy other advantages from . Additionally, these rights are defined in Article 27 of the United Nations Declaration of Human Rights, which states that anyone who generates intellectual property has the right to benefit from the protection of their interests as a result of their works of science, art, or literature (Laukyte & Lucchi, 2022). The earliest international accords to safeguard intellectual property were the Paris Convention for the Protection of Industrial Property (1883)[CS1.1] and the Berne Convention for the Protection of Literary and Artistic Works (1886)[CS2.1]. Patents and copyrights are the legal ownership rights that are granted to the creator or inventor over their intellectual property. If someone wants to make a profit from the property, they must first obtain the permission of the patent and copyright owners. As per Austin, the definition of ownership is "a right that is indefinite in terms of the user, unrestricted in terms of disposition, and unlimited in terms of duration." These features primarily define the concept of human ownership; thus, when considering the question of granting ownership rights to artificial intelligence, they are, in fact, the components that should be examined most closely . The limitation of the ownership right to a "natural person" is for the purpose of maintaining and safeguarding the rights of human creators. Such protection is derived from the innovation, which is the main driver of further research and development. Property is safeguarded from illegal use through the ownership right. In this way, it is ensured that the creator receives the due recognition for their outpouring of love, talent, and unique insight. Oneof the biggest battles among top tech companies is over who gets to own the rights to a certain piece of intellectual property due to a huge rise in the number of patents and copyrights filed in the name of AI. The growth of AI patents is significant; they have increased fourfold from 2012 to 2016, rising from 708 to 2,888, respectively. Can AI satisfy the criteria of having ownership rights, or are humans also entitled to such rights? The next court issue is the idea of "a natural person," which is the main reason that the authors emphasize the difficulty of giving ownership rights to AI. The paper deals with the problems in law that come along with AI-based .

6. CHALLENGES IN DEFINING AND PROTECTING AI-GENERATED INTELLECTUAL PROPERTY

Artificial Intelligence (AI) has been advancing rapidly over the last few years. As a result, the IP of AI systems has become a subject of dispute. IP issues related to AI encompass a range of ethical, legal, and practical concerns, and as the field of AI-based innovation evolves, these issues must be addressed.

6.1 Authorship and Ownership

A significant problem is identifying the creator and owner of AI-generated works. Because traditional intellectual property rules were created to protect works created by humans, works entirely produced by AI are now in legal uncertainty. Consider the contentious 'Portrait of Edmond de Belamy' AI-generated art case from 2018, which sparked discussion about whether the AI itself or its creators held the copyright. It is challenging to establish artists' rights and distribute rewards for AI-generated material due to this lack of clarity.

6.2 Attribution and Accountability

It could be challenging to identify the source of AI-generated content. Deepfake videos, which employ AI to alter images and produce misleading content, are the ones that fundamentally highlight the issue. It is difficult to find the person who is at fault when deepfake videos are utilized to disseminate false information or to defame someone. Legal authorities and judicial systems are frequently in a dilemma of decision-making in such.

6.3 Legal Precedent

The absence of case law for intellectual property created by AI presents a substantial challenge. stated that laws protecting intellectual property were created when the human imagination was the primary force behind creation. The rapid advancement of technology has not been reflected in laws. Courts and lawmakers must either modify existing rules or enact new ones to address these issues. The swift progress of technology is something that the laws have not yet fully taken into account. Changes need to be made by the courts and lawmakers, either to the existing regulations or a completely new set of rules, in order to address the consequences of AI-created intellectual property. The quick technological progress has not been matched by the laws. Judges and lawmakers are required to change the current regulations or create new ones in order to deal with the consequences of AI-generated intellectual .

6.4 Fair Use and Transformative Works

Fair use and transformative uses in AI-generated works can often be confused, making it difficult to distinguish between them. For example, Google's DeepMind utilized human gaming skills to create an AI that outperformed humans in the game of Go. The question remains whether the AI is a transformative use or a copyright infringer since it was the human-generated data that was .

6.5 International Harmonisation

It is particularly challenging to create global uniformity in the handling of AI-generated intellectual property. As a result of various methods of IP protection in various nations, the global intellectual property landscape may become inconsistent and complex. Such obstacles require a comprehensive plan that engages workers from the entire intellectual property law-making community and AI developers in conversation and cooperation. Clear standards, laws, and ethical codes of conduct would have to be put in place so that AI inventions would be a source of innovation, but artists' rights would be protected, and a just and equitable digital environment would be.

6.6 Copyright Law and AI: Balancing creativity and automation

According to Sharma (2023), the relationship between artificial intelligence (AI) and copyright law is a complicated one, as current legal frameworks must accommodate the potential disruptiveness of AI technologies. It is very important, in this intricate environment, to locate a cautious equilibrium between respecting the rights of human creators and making use of AI for further . In a complex and dynamic legal context where traditional legal norms must compete with the revolutionary potential of AI systems, artificial intelligence (AI) and copyright law intersect. Themusic industry can be considered a good example of this complicated equilibrium.

6.7 AI-Generated Music and Copyright Challenges

In recent years, AI systems that are based on machine learning algorithms have shown the capability of creating music on their own. An outstanding example is the Magenta project, a joint venture between Google and OpenAI, which has created musical pieces that are not only complex but also of such a high standard that they can be compared to those made by humans. AI-createdsongs of this kind raise quite complicated legal questions regarding copyright. In situations where artificial intelligence searches through a large number of old pieces of music to compose a new one, it may unintentionally embed fragments of a copyrighted work. Consequently, the line between new and derivative works is getting blurred, hence the issue of potential copyright violations is .

Artificial Intelligence wrote the music for the 'I AM AI' album, which was released by the musical label Amper Music in 2021. All tracks on the record were created by AI systems only. The AI's adoption of melodic and stylistic elements similar to those found in human music raised concerns that it might be mistaken for a new composer of tunes, although its goal was to generate new ones. The musical label Amper Music released the 'I AM AI' album in 2021, which AI wrote. The songs on the CD were all created by AI systems. However, the AI system's incorporation of melodic and stylistic elements similar to those found in human-made music raised concerns, even though the system was intended to generate new tunes. This situation highlighted the challenges of assigning credit, ownership, and fair use in the context of AI-generated music. The key issue under discussion was whether these AI-generated songs violated existing copyright laws or qualified as transformative works under the fair use principle16. To successfully address such issues, changes to copyright regulations regarding AI-generated art are recommended. Laws around the world need to include criteria such as the extent of human intervention in work generated by AI, the impact on the original work, and the new and different uses of AI-created content. Suitable rules and standards need to be put in place for providing legal clarity as well as for the protection of the interests of AI developers on one hand and the writers of creative content on the other. Such a move will facilitate the application of copyright law in a just and equitable manner in the AI era, thereby promoting both aspects of the law: the promotion of innovation and the protection of intellectual property rights.

6.8 Ethical Considerations in AI and Intellectual Property

Artificial intelligence (AI) and intellectual property (IP) law combine to create a complicated network of considerable moral consequences, offering a mix of advantages and difficulties. In the UK, the increasing interconnection of these two fields has raised several ethical questions that warrant serious reflection. This discussion explores some of these ethical issues in relation to AI and IP, drawing on relevant UK case law and providing real-world examples. Anessential ethical issue in AI is the question of ownership of the works generated by it. An illustrative instance of this is the court case in the UK, Warner-Lambert Company LLC v. Generics (UK) Ltd (Mylan) [2018]97 UKSC 56, as per Ubaydullayeva (2023). Authorship and intellectual property rights are often disputed, as teams or organisations frequently develop AI systems. It is crucial to determine who should be awarded IP protection in scenarios where AI generates creative works autonomously, in order to prevent disputes and foster innovation.

Oneof the ethical issues that arises from AI is the possibility of it violating intellectual property rights that already exist. In the case of an AI-driven content generation tool, it is possible that such a tool unintentionally creates copyrighted content, resulting in legal conflicts and potential infringement. The Copyright, Designs and Patents Act 1988 of the UK, which is modified by the Digital Economy Act 2010, might have to be changed to be able to deal with such situations . Ethicalconsiderations are also concerned about the possibility of bias and discrimination in artificial intelligence. Bias in AI can lead to discrimination against specific groups of people, and if personal information and intellectual property rights are not adequately protected, privacy will be compromised. In the UK, the General Data Protection Regulation (GDPR) and the Equality Act 2010 establish a legal framework for solving these . However, proactive identification and mitigation of AI bias remain essential.

Another ethical issue is the possible use of AI for IP infringement, such as AI-driven counterfeiting and IP theft. The UKhas put out a plan to deal with these issues through its IP Crime Strategy, which works through cooperation between the authorities, the business sector, and the tech providers. The intersection of AI and IP law requires a nuanced and complex ethical approach. It is very important to find the right balance between encouraging innovation, ensuring the protection of IP rights, and solving problems such as AI bias and the wrong use of . To address the challenges posed by AI in the realm of intellectual property, the UK's adaptable legal system must be adapted, ensuring that ethical considerations guide the development and use of AI technology.

6.9 Impact of Artificial Intelligence on Intellectual Property Law

Artificial intelligence (AI) technologies have undergone significant advancements lately, and they are now being implemented in a diverse range of industries such as law, finance, and healthcare. It is essential to consider the changes these advancements bring to IP law as AI systems continue to increase their processing power, and eventually, they will be able to generate both artistic works and technological innovations.

6.9.1 Copyright Law

Under copyright law, authors are given exclusive rights to their original literary, artistic, and musical works. In light of AI's rapid development, the question of whether copyright protection should be extended to works created by AI systems has been a topic of discussion.

In the realm of copyright law, figuring out who made the AI-generated work is the biggest problem. The primary purpose of copyright law is to grant rights to individuals or entities that create original works of authorship. An AI (artificial intelligence) -generated artwork that can be done fully by the machine without any human help is a concept that confuses the . Since AI-generated works lack the human originality and expression that copyright laws are supposed to protect, some scholars believe that they shouldn't be covered by these rules. Some people argue that in the case of AI-generated works, copyright protection should be given; either the AI system is considered the author, or the human programmer or user is given the authorship. A further issue in the application of copyright law to AI-generated works is the requirement for originality and inventiveness. Generally, a work must be distinct and demonstrate a certain level of originality to be granted copyright protection by law. Creatures of AI might not be able to reach these levels, as they must be sourced from existing data and algorithms to produce new ones. Some legal experts believe that AI-generated works should be treated as unique and creative since they have the capacity to produce inventive and surprising results that significantly differ from their underlying material. Ubaydullayeva (2023) stated that AI-generated products are simply derivative and lack the human creative element necessary for copyright protection.

6.9.2 Patent Law

Patentlaw confers on creators the privilege of using their creations, thus promoting innovation and technological progress. The evolution of AI has an impact on patent law, particularly in aspects such as inventorship, ownership, and the inventive step requirement. AI systems can come up with new inventions on their own or with the help of human inventors. This raises the question of whether AI systems should be considered inventors for the purposes of patent law and whether inventions made possible by AI should be eligible for patent protection. Legal experts claim that, as AI systems lack the originality and intentionality required for inventorship, they shouldn't be considered inventors. Others contend that recognizing artificial intelligence (AI) systems as creators might promote their development and use in the innovation process. Determining who is the legitimate owner of ideas created by AI presents another challenge for patent law. Under conventional patent regulations, the inventor's employer or the inventor is presumed to be the owner. As the question of who made AI-generated works may not be clear, it might be complicated to assign ownership rights to them. Some alternatives to addressing the issue include designating the AI system as the owner, assigning ownership to the system's developer or user, or forming a new legal entity to hold ownership rights.

6.9.3 Trademark Law

According to Kaur (2016), trademark law helps identify products and services through unique signs, symbols, and trademarks, which distinguish them from those of competitors. The use of artificial intelligence (AI) in the creation and selection of trademarks brings up the issue of whether AI-created trademarks can get legal protection and, in what way, their uniqueness can be . AIhas been incorporated in the process of choosing and creating trademarks. According to Kop (2020), artificial intelligence (AI) can generate new brands by analyzing past data and trends across multiple industries. This raises questions about whether trademarks developed by artificial intelligence can be deemed unique and eligible for trademark protection. Experts in intellectual property law believe that computer programs cannot replicate the human level of invention and intentionality necessary for a trademark to be unique. Others maintain that trademarks developed with the assistance of artificial intelligence may be considered original if they convey an adequate level of distinctiveness and are capable of identifying the origin of the products or services.

Table 1. Lists the main legal acts regulating relations in artificial intelligence and intellectual property, along with their key provisions and features.

Country/Region	Name of the Legal Act	Main Points	Features/Challenges
European Union	Artificial Intelligence Act (under approval)	It outlines the guidelines for the safe use of AI, ensuring transparency of algorithms and clarifying the responsibilities of both developers and users.	There are no specific regulations regarding intellectual property rights for objects created by AI.
	Copyright Directive (2019/790)	It guarantees copyright protection in the digital era, covering works produced through automated methods.	Ambiguity in the application of artificial intelligence to performance outcomes.
USA	U.S. Copyright Law (Section 102(a))	It specifies that only works created by humans are eligible for copyright protection.	Excluding AI-created works from protection complicates the regulation of new technologies.

continued on following page

Table 1. Continued

Country/Region	Name of the Legal Act	Main Points	Features/Challenges
	Patent Act	It governs the patenting of inventions involving AI algorithms, provided there is proof of an "inventive step."	Fully autonomous AI raises questions about patenting outcomes.
United Kingdom	Copyright, Designs and Patents Act (1988	It contains specific provisions for computer programs and data produced by AI.	Debates surrounding the clarification of ownership rights for AI-generated outcomes.
	AI Regulation White Paper (2023)	Guidelines for AI regulation, focusing on transparency and ethical use principles.	Absence of a clear legislative framework and the presence of only recommendatory restrictions.
China	Guidance on Regulating Generative AI (2023)	It provides a framework for developing and using generative AI models, with attention to intellectual property issues.	Emphasize state authority in innovation while maintaining limited transparency in law enforcement.
	Patent Law of the People's Republic of China	AI algorithms can be patented if they address a technical problem.	High competition exists for AI patents, making it difficult to evaluate the technical level.
Japan	Copyright Law (changes in 2020)	It considers the use of data for AI training to be an exception to copyright protection.	Additional regulations are required to address the outcomes of AI activities.
	AI Strategy 2021	An AI development strategy that proposes regulations for intellectual property in the context of automated systems.	Insufficient implementation within legislation.
Canada	Copyright Act	It safeguards creations that involve considerable human participation.	Insufficient regulation for objects primarily generated by AI.
International acts	Berne Convention for the Protection of Literary and Artistic Works (1886)	It offers global copyright protection for creative works.	It overlooks the details of AI and automated creation.
	TRIPS Agreement (1995)	Establishes minimum standards for protecting intellectual property in WTO member countries.	It should be adjusted to the practicalities of utilizing AI.

continued on following page

Table 1. Continued

Country/Region	Name of the Legal Act	Main Points	Features/Challenges
	World Intellectual Property Organisation (WIPO) Recommendations on AI and IP (2020	Guidelines for safeguarding IP rights related to AI outputs, such as patents and copyrights.	It is not legally binding and depends on how national governments choose to implement it.
India	The Copyright Act, 1957 (as amended)	Defines the author of a literary, dramatic, musical, or artistic work that is computer-generated as "the person who causes the work to be created" (Sec. 2(d)(vi)).	Ambiguity in Authorship: The emphasis stays on human involvement or cause, which leads to uncertainty for works created with minimal human input. Typically, the judicial stance is that a machine cannot be considered an author.
	The Patents Act, 1970 (as amended)	AI algorithms or computer programs are not patentable under Section 3(k) on their own. However, if an invention shows a "technical effect" or makes a technical contribution beyond the algorithm, it can be patented.	Strict Technical Effect Test: The invention must address a specific technical problem—such as enhanced efficiency, security, or hardware interaction—to overcome the exclusion. Fully autonomous AI inventions pose additional challenges concerning the "inventor" requirement.

Source: created by the author based on (Batsurovska et al., 2024; Kazeeva, 2024; Meyer, 2024; Puertas-Bravo et al., 2024; Vesala, 2023).

The legal acts presented in Table 2 illustrate various approaches to regulating the relationships between artificial intelligence (AI) and intellectual property. Each jurisdiction develops its regulatory framework in accordance with national priorities, resulting in significant differences between countries. Particular emphasis is placed on adapting existing international acts, such as the Berne Convention and the TRIPS Agreement, to the conditions of digital transformation; however, achieving global harmonization of standards remains a challenge. In particular, uncertainty over authorship and ownership of AI-generated works complicates the regulation of cross-border activities.

7. CONCLUSION

The role of AI is set to increase significantly in all sectors of our day-to-day lives. The law is required to regulate its use. Artificial intelligence is set to remain an essential factor in the scenario of intellectual property rights, most notably in copyright. Questions of authorship and ownership of AI-generated works under copyright law have led the global community to contemplate and develop a universally acceptable solution. There is no foolproof rule to address this issue, and every rule has its own flaws. There will be significant ramifications for offering non-human authorship to AI-generated works. Putting the AI-generated works in the public domain is also not a good idea, as it will discourage the AI programmers and companies owning such AI from further investing in the AI domain. The WIPO is working hard to address these issues. A sui generis system could possibly be a preferable solution. Another option might be that certain provisions in the copyright laws of different countries, specifically drafted for AI and AI-generated works, could address this problem. In any case, AI-generated works should be afforded less protection, and human creativity should be prioritized over machine-generated creativity. A balanced approach is therefore the need of the hour.

AI presents issues and complexity for intellectual property, but IP protection is expected to improve. AI-driven IP monitoring and enforcement systems can quickly detect and handle online copyright, trademark, and patent infringements. UK companies like Red Points use AIfor protecting trademarks and IP rights. Different technologies, such as blockchain, can provide highly reliable records of who owns the IP and who has the rights to use it, thereby protecting the IP. In America, companies like Binded provide artist registration that cannot be tampered with and are exploring blockchain-based copyright protection. AI can expedite art and patent searches that have been previously conducted in the field of intellectual property, thereby reducing costly IP litigation and making patent applications more efficient. AI has been employed by the USPTO to improve patent examination. AI may collaborate with human artists to create new masterpieces. Adobe Photoshop utilizes AI to propose changes, enabling graphic artists to create more effectively.

Businesses are increasingly using AI to identify and prevent IP theft. AI can detect IP leaks and security breaches in network traffic and employee behaviour, protecting priceless assets. The incorporation of AI into intellectual property law has led to significant changes and complex issues. In light of AI-generated IP, which enables precise legal definitions of artists' rights, considering the heavy reliance on AI, the question of authorship and ownership must be revisited. An AI-generated work should be handled in the context of copyright law with adaptable regulations and an international governing body. The development and use of AI should be on the right track ethically, and for that matter, issues of ownership, infringement, and the like, as well as biases and discrimination in intellectual property, should be . Ethical frameworks must keep up with technology to reconcile creative freedom and ethical content development.

IP protection has several potential applications in the field of AI. AI accelerates IP processes collaboratively, while blockchain technology provides immutable records of IP. Better monitoring and enforcement are also possible with the aid of AI. Stakeholders may profit from AI while protecting artists and innovators by proactively using AI to protect IP and stimulate innovation. Whentalking about the future of intellectual property in the age of AI, we should not lose sight of the interaction between human and machine intelligence. As such, laws, ethics, and fresh ideas should be like three wheels of the same cart when it comes to creating AI that is compatible with justice, accountability, and responsible innovation. The discussions surrounding AI-IP legal issues will shape the possibilities for creative expression, the pace of innovation, and the methods of IP protection as the technology continues to evolve.

8. RECOMMENDATIONS

While AI interaction with IP law is a very complex and intricate topic, it requires not only careful, but also strategic thinking, especially in relation to copyrights. One of the most crucial steps in creating a just law that regulates creative content is establishing clear and unambiguous legal concepts and standards. Therefore, this collaboration between legislators and the judiciary is an utmost necessity in setting out clear, commonly accepted standards that characterize AI-generated works. These definitions must precisely delineate the threshold of human involvement required for a work to qualify for traditional copyright protection, explicitly determining who—the AI developer, the user providing the prompt, or the machine itself—is the author and controls the rights to any independent works that AI systems generate. Clear legal standards will serve as the foundation for addressing new copyright challenges and minimizing costly legal disputes.

Building upon legal clarity, it is essential to adapt copyright laws to directly address the creation and use of copyrighted material by AI. It is highly advisable that copyright regulations be changed in such a way as to allow AI-generated content while also safeguarding essential features like fair use and transformative works. The main emphasis of this change should be on the legal use of training data, which means setting up a complete mechanism for Text and Data Mining (TDM) exceptions that takes into account both the protection of confidentiality and the exclusivity of the original creative data, and the provision of the freedom of innovation and creativity to the works supported by . Atthe same time, governance should be enhanced through the setting up of AI accountability structures. Such structures are vital for figuring out who is responsible and who has the authority over AI-generated content, especially when it comes to the risk of copyright infringements resulting from unpermitted reproductions or the manufacture of misleading content like . Clear guidelines for determining accountability will facilitate the resolution of legal difficulties resulting from the use of copyrighted material in training models or the generation of infringing content.

Ultimately, the safeguarding of copyrights related to AI requires global collaboration and robust enforcement mechanisms. To maintain uniformity and prevent differences in jurisdictional treatment of AI-generated works, it is crucial to have an international agreement on copyright norms and rules, regulated by organizations such as WIPO and TRIPS. Such a global partnership will guarantee the protection of creators' rights everywhere and simplify the process of addressing new issues in a just and coordinated manner. This strategic alignment should be backed up by a commitment to the promotion of an ethical AI development; thus, researchers and developers of AI should always follow the most excellent practices in data sourcing and respect for intellectual property rights. Mostimportantly, companies and government bodies must not only use AI to improve their IP tracking, but they should also be implementing AI-based tech that would be constantly scanning the digital world for even the slightest infringements and reacting to them quickly, thus allowing the almost instantaneous securing of the artistic, literary, and musical IP assets that are copyrighted. Fundamentally, these measures constitute an integrated system for preserving the creators' rights, as well as promoting creative output in the era of generative AI.

REFERENCES

Batsurovska, I. B., & Kurylen, V. M. (2024). Models and algorithms of machine learning in computer systems of artificial intelligence: New trends and prospects for development. *Modern Aspects of Sciences*, *XLVII*, 462–473.

Becerra, M., & Waisbord, S. R. (2021). The curious absence of cyber nationalism in Latin America: Lessons for the study of digital sovereignty and governance. *Communication and the Public*, *6*(1–4), 67–79. DOI: 10.1177/20570473211046730

Flew, T., Suzor, N., & Liu, B. R. (2013). Copyrights and copy fights: Copyright law and the digital economy. *International Journal of Technology Policy and Law*, *1*(3), 297–315. DOI: 10.1504/IJTPL.2013.057009

. Hilty, R., Hoffmann, J., & Scheuerer, S. (2020). *Intellectual property justification for artificial intelligence.*

. Kaur, J., 2016. Intellectual Property Law in times of artificial intelligence: is it a misnomer to consider the Bot a possible IP right holder? Journal of legal studies and research, 2(6)

Kazeeva, I. (2024). Sui generis intellectual property protection for works generated by AI systems. In *I*. Springer Nature Singapore., DOI: 10.1007/978-981-99-8897-6_5

Kop, M. (2020). Regulating Transformative Technology in the Quantum Age: Intellectual property, standardization & sustainable innovation. *SSRN*, *2*, 145–173. DOI: 10.2139/ssrn.3653544

. Laukyte, M., & Lucchi, N. (2022). Creative AI: The complex relationship between human inventiveness and intellectual property. *BioLaw Journal – Rivista di Bio-Diritto*, (3), 169–183.

Md Ashraful, A., Gazi, M., & Sanjida Akter, S. (2025). Enhancing Regulatory Frameworks to Prevent Financial Crises. https://core.ac.uk/download/651277907.pdf

Meyer, S. (2024). Legal challenges of artificial intelligence and how to manage them. In Barton, T., & Müller, C. (Eds.), *Artificial intelligence in application* (pp. 9–30). Springer Fachmedien., DOI: 10.1007/978-3-658-43843-2_2

Puertas-Bravo, L., & Piedra, N. (2024). Regulation of artificial intelligence: Challenges and perspectives in the Andean Community. In Bianchi, I., & Dávila, G. A. (Eds.), *Knowledge management and artificial intelligence for growth* (pp. 221–244). Springer Nature Switzerland., DOI: 10.1007/978-3-031-65552-4_11

Sharma, M. (2023). Intellectual property rights and artificial intelligence. *Indian Journal of Law and Legal Research*, *5*(2), 1–15.

. The Copyright, Designs and Patents Act, 1988, s.178

Ubaydullayeva, A. (2023). Intellectual property in the era of artificial intelligence: Challenges and solutions. *Uzbek Journal of Law and Digital Policy*, *1*(3). Advance online publication. DOI: 10.59022/ujldp.93

. Vesala, J. (2023). Developing artificial intelligence-based content creation: Are EU copyright and antitrust law fit for purpose? *IIC – International Review of Intellectual Property and Competition Law, 54*(3), 351–380. https://doi.org/DOI: 10.1007/s40319-023-01301-2

Wills, K. (2021). AI around the world: Intellectual property law considerations and beyond. *Journal of the Patent and Trademark Office Society*, *102*, 186.

Chapter 9
Regulating Intellectual Property in the Age of AI:
Redefining Ownership, Authorship, and Inventorship

Khushi Sood
https://orcid.org/0009-0004-8290-6160
Christ University, Delhi NCR, India

Khushboo Malik
Christ University, Delhi NCR, India

ABSTRACT

Artificial intelligence development in this digital era has questioned the fundamental concepts of intellectual property law, whose foundation was based on and centred on human creativity, inventiveness, and authorship. AI being able to generate outputs such as artistic and literary works, symbols and patterns, and inventions raises a question on the traditional notions of intellectual property. This chapter critically examines the conflict between AI-generated outputs and inventions and the existing legal and regulatory framework of copyright, ownership, inventorship, and data governance regimes. The chapter uses doctrinal and comparative methodology to analyse approaches taken by different countries across the globe. The chapter presents a comparison of approaches taken in the European Union, the United States, India, and the DABUS case in South Africa. By mapping regulatory gaps and proposing balanced reform frameworks, this chapter offers a way forward for harmonising accountability, incentivising innovation, and adapting IP law to an era of human-AI collaboration.

DOI: 10.4018/979-8-2600-0000-7.ch009

1. INTRODUCTION

Artificial Intelligence has transformed the way creation and innovation take place. It is reshaping the thinking and actions associated with creative and innovative work in almost every domain. "Machine learning systems and generative technologies are now capable of producing written text, music, visual art, software code, functionality, and technical solutions that were once only thought to be achievable by human creativity and ingenuity" (Kriz, 2023). The advent of AI is rapidly encroaching into all aspects of entertainment, research, health, product design, manufacturing, and business automation. As such, the underpinnings of intellectual property have been challenged, as traditionally, IP law only addressed human creativity and authorship.

There is no standard definition of AI that is universally accepted. Rather, different jurisdictions define AI in a way that best serves their policy. However, all IP protections are based on the foundation that creation or invention is undertaken as a product of human thought and labour (Senftleben, 2021). The growth of AI conflicts with this underlying foundation and legal framework. For example, "the notion of AI in the European Union AI Act describes AI as a 'system' that produces outputs, according to human-created goals" (European Commission, 2023). "The National Institute of Standards and Technology in the United States defines AI as a perception, reasoning, and autonomous decision-making tool" (NIST, 2022). Whereas in India and China, there is no proper definition of AI.

The current studies only examine the increase of AI in the globe but do not focus on the debate on whether outputs generated by AI can be granted intellectual property rights and protection. The study lacks the consequences of using AI for generating such outputs and the legal framework and changes that are needed for the regulation of use of AI in granting rights of intellectual property.

The gap formed between technology and innovation, and the Intellectual Property's current legal framework raises issues in the economic domain. There is a rising debate between overprotection and under protection of work generated by AI. The chapter uses a doctrinal and comparative study method to understand how current IP laws explain the concept of author, inventor, and owner when a dispute arises because of the use of AI-generated outputs.

The chapter aims to contribute to the global dilemma about the future of IP with the increasing use of AI for generating outputs. Its regulation is necessary to bring harmony and consistency to the global stage and to promote innovation and creativity.

2. BACKGROUND

Policy makers around the world consider regulating AI (Schuett, 2019). But before moving forward with its regulation, the question arises: "What is the legal definition of AI, or how is it defined?" There is no uniform definition of AI present globally, and this lack of consensus creates a problem in the legal discourse. Since its first usage by John McCarthy in 1955, a vast spectrum of definitions has emerged (Schuett, 2019). He is widely recognized as the father of Artificial Intelligence. He defined Artificial Intelligence as "the science and engineering of making intelligent machines (McCarthy et al., 1955). From a technical point of view, Artificial Intelligence is described as systems that are capable of performing tasks that require human intelligence. However, for its legal definition, liability, rights, and regulatory obligations must be defined with great precision. The European Union adopted the Artificial Intelligence Act, which laid down harmonized rules on artificial intelligence (Presno Linera & Meuwese, 2025). It is one of the first regulatory instruments dedicated to AI regulation (Almada, 2025). In the USA, according to NIST standards, they focus on AI's risks and capabilities rather than drafting a single binding statutory definition (NIST, 2023). China has the 'New Generation Artificial Intelligence Development Plan" that sets out strategic aims and delineates the overarching goal of making China the world leader in AI by 2030 (Roberts et al., 2021). It relies on policy statements and administrative guidance instead of codified definitions. This sets out a scope for further interpretation and regulatory divergence. India has been working on formulating a National Artificial Intelligence Strategy to provide a comprehensive framework for the development, deployment, and regulation of AI technologiesThis lack of harmonization means that identical AI- generated outputs can be treated differently in different jurisdictions. As a result, the regulation of AI becomes difficult.

2.1 Foundation of Intellectual Property Laws

The foundational concepts of intellectual property are based on anthropocentric assumptions. The doctrines of authorship and inventorship presuppose human cognitive ability and legal responsibility for the work they generate. Under the doctrine of Copyright, the concept of authorship has been directly tied to human creativity. An author refers to a person who creates the work. For most types of work, it clearly refers to the human creator only. It requires a qualifying level of originality, while patent law requires inventive step and novelty traceable to human ingenuity (Ricketson & Ginsburg, 2022). Understanding the philosophy of these doctrines, they are based on the utilitarian theory given by Jeremy Bentham, which grants legal rights over creation of the mind, balancing the rewards given to the creator with public access. Provided that by such creation of mind, some good must be provided to the public as well as the inventor or author. Such a framework, given by the statutes and administrative frameworks of different countries, has the effect that authors and inventors must be natural persons. This, as a consequence, creates difficulties when outputs are originated from systems rather than minds. Natural persons are capable of holding rights, liabilities, and property, and these entities can be held accountable. There is a status of artificial legal persons given to corporations to enjoy legal personhood that enables them to hold rights and be held accountable (Kouroupis, K. & Efthymiou, I., 2026). But extending such status to AI systems raises complexities and questions about accountability, remedies, and incentives (Abbott, 2020). Therefore, courts and administrative bodies treat AI as merely an instrument, not as an independent body subject to rights.

2.2 Evolution of AI

The advancement of AI from a rule-based tool to a generative and deep-learning system has significant implications for intellectual property. Earlier, AI was used as a computer-assisted system which can process information and perform automated tasks only under human control and direction. With its development, AI has taken the form of a modern generative tool. It can produce texts, images, music, and technical solutions with high degrees of novelty (Goodfellow et al., 2014). This advancement blurs the line between assistance and autonomous generation, which makes it harder to allocate rights exclusively to humans.

A recent case that reflects these issues is the set of DABUS patent applications. In these applications, Dabus, an AI, is termed to be an inventor. There were divergent views given by various patent authorities and courts. The European Patent Office and many national offices refused to accept a machine as an inventor, while some jurisdictions engaged more substantively with the question. These cases serve as an example of how traditional doctrines interact with technological novelty to produce inconsistent legal outcomes and difficulty in the regulation of AI-driven innovation.

2.3 Existing Scholarly Debates

"Artificial intelligence has emerged as an engine that is driving the Fourth Industrial Revolution. The era of traditional machine learning is getting transformed into the modern era of machine learning."

The European Union emphasize on a human-centric approach to the regulation of Artificial Intelligence, particularly in the context of intellectual property. The EU AI Act does not confer the status of an artificial legal person to AI systems. "AI is specified as a transformative technology that must be governed with sector-specific regulations" (European Commission, 2020). The emphasis in the literature of the EU on the obligations of transparency, dataset governance, rights holders, and sustainable enforcement tools, rather than authors or inventors of AI systems, suggests that the EU does not recognize an AI system as a potential author or inventor.

The United States employs a doctrine-based and case-specific approach to regulate intellectual property. There is a release of formal written guidelines to clarify that AI-assisted inventions may meet the patentability requirements of the U.S. Patent Law, provided that the invention is made with the significant contribution of a natural person. Overall, the U.S. approach prioritizes flexibility and innovation while focusing on adapting new rules and regulations for emerging AI- related disputes (U.S. Patent and Trademark Office, 2024).

China takes a narrower approach to artificial intelligence use. It focuses on using a practical and innovative approach to regulation through detailed guidelines. To that end, the China National Intellectual Property Administration has established guidelines on how to apply for patents related to AI and how to determine copyrightability for AI-generated works (China National Intellectual Property Administration [CNIPA], n.d.). In the context of Chinese policy, AI is viewed as a technology and still requires a human creator in order to obtain a property right under existing IP laws. China's primary regulatory concerns are to create a well-defined and legally sound process for AI applicants, as well as to deploy AI technology as quickly and efficiently as possible (China National Intellectual Property Administration, 2025).

In India, the current policy regarding AI regulation in Intellectual property is in the initial stages. Currently, India does not have any binding legislation regarding the creation of Generative AI Copyright through the DPIIT's Working Paper on Generative AI & Copyright, released in 2025 some of the issues raised by India in this discussion include the concern that the use of copyrighted works without permission to create AI models is harmful to the authors of those works and creates uncertainty of ownership of the output generated. Most of the literature on this topic within India supports the notion that current laws relating to IP do not give adequate consideration to AI-produced works. As such, there is reluctance to create specific rights for AI without first achieving a broader level of agreement among stakeholders (Department for Promotion of Industry and Internal Trade, 2025).

3. CHALLENGES TO EXISTING IP FRAMEWORKS

3.1 Challenges to Copyright

Copyright laws are based on the idea that human creativity is the basis for copyright protection. Copyright law requires that the author take steps to create an original work and, as such, has to think about how they want their work to look, as well as to create a work with meaning. International copyright systems explicitly embed human authorship into their statutory framework. Copyright law gives authors the right to decide how their creation is going to be used by others, and protects them if someone else uses their work without permission. The Copyright Office of the United States Government reasserted that copyright protects only "the product of human creativity" (U.S. Copyright Office, 2023). This position was not newly invented for AI and traces back to judicial precedents like Burrow-Giles Lithographic Co. v. Sarony (1884). It was held in this judgment that copyright protects works reflecting "original intellectual conception" by an author. The modern policy guidance applies this same principle to generative systems, emphasising that copyright cannot recognise authorship where "no human Mind" produces the expression (U.S. Copyright Office, 2023).

Today, Copyright laws do not grant protection to works created by AI. AI-generated works, therefore, have no copyright protection in most jurisdictions and, as such, may be treated as works that are automatically considered to be in the public domain. The requirement of human authorship thus represents one of the main barriers that limit AI works from falling within the Copyright framework. Whether this barrier develops into something that allows for attributions of AI artwork as a legitimate form of authorship, thus allowing AI works to enter into the copyright system, will depend on how various legal frameworks define what is meant by authorship. In the present state of Copyright, however, "author" is very much based on the human experience and creative process, and therefore, AI creativity does not have a defined place within existing copyright models.

3.1.2 Issues with AI-Generated Artistic and Literary Works

Copyright law is in direct opposition to the concept of the creative contributions of Artificial Intelligence when it comes to art and literature. As a result of AI generative product development, it is now possible to create poetry, novels, artwork, music, vocal performances, and countless other visual art forms that are indistinguishable from the creative contributions made by a human artist or author. Many of these new forms of creative work have already been used for commercial purposes in areas such as publishing, advertising, video games, and social media, resulting in numerous disputes in the real world regarding who owns the rights to the creative work and how much credit should be given to those involved in the creation of the creative work.

One serious concern regarding generative AI's ability to create works that violate copyright law is the ability to recreate the distinctive voice or likeness of another human being. This issue was brought to light in and others the "case of Arijit Singh

v. Codible Ventures LLP and others", where a copyright was brought against an AI-created song that used the voice and name of an Indian singer, resulting in the Mumbai Court granting interim relief to the plaintiff against the use of his name, image, and voice without his permission (High Court, 2023). The Court found that the AI-created mimicry was an infringement against the personality and publicity rights of the singer. Similarly, Asha Bhosle, a well-known Indian singer, sought to protect her likeness and voice from being replicated or imitated by AI without her consent. She ultimately received interim protection from the Mumbai Court due to the fact that the court found that generative AI poses new threats to an individual's creative identity and financial interests (Times of India, 2024). In Thaler v. Perlmutter, the U.S. Copyright Office ruled that it was unable to register a piece of art created by an AI system autonomously. The ruling declared that copyright law "requires a human author" (U.S. Copyright Office Review Board, 2022). Rather, in Zarya of the Dawn, a graphic novel that included AI images, the U.S. Copyright Office accepted only the human-authored text and selection arrangement for registration, while specifically rejecting the images generated by an AI system for copyright protection (U.S. Copyright Office, 2023).

Many of the traditional copyright doctrines struggle with how they handle the cloning of voices and stylistic imitations or synthetic performances, where one is profiting off another person's reputation without actually duplicating their artistic expression in any way. Many of the same issues exist in literary genres as well, where the number of online sales of AI-generated books is on the rise, in turn creating concerns about offences of plagiarism, authorship, and displacement of markets with AI-generated books. Writers believe that by using AI-generated text trained upon already protected literary works, there is a loss of motivation for new authorship. To avoid any potential copyright infringement, the courts and copyright offices are cautious concerning the issue of human creation. This will only confirm such works where there is evidence of human creativity contributing to the creation of that work.

3.2 Challenges to Patent Law

3.2.1 Inventive Step, Novelty, and AI Assistance

Patent law has a different structural issue with respect to artificial intelligence than does copyright. Patent law is based on identifying an inventor at the time the application is filed. It defines inventorship in terms of conception, which has long associated with the mental activities of an individual. This definition of inventorship becomes more tenuous when considering that AI-based research will substantially contribute to the creation of inventions. It is especially in fields where AI suggests potential solutions to researchers who then conform, select, or refine these suggestions to develop a final product. Therefore, there is a tension between the notion that an AI-generated invention can be "new" and the issue of how to assign inventorship, ownership, and liability with respect to inventive contributions made using AI.

Inventive step and novelty are required for patentability, and the patent system currently assesses these against the standard of a "person skilled in the art." This standard is based on the assumption that innovation happens within the mental grasp of these professionals. However, AI allows for computational exploration of solution spaces at a much larger scale than a single individual can achieve. It is changing the discovery process and the time it takes to discover something. Accordingly, the amount that an individual can rely on their mental prowess for the invention is being reduced. However, the current patent law does not prohibit AI from assisting inventors. Instead, the patent law requires that any claims made in an application be based on the inventor's inventive contributions that derive from a natural person and that the claim be supported through a legally cognizable method of conception.

3.2.2 Patentability of AI-generated inventions

Creations generated by AI may also incorporate human imagination into the process, only occurring with some degree of distinction from their source of creation. In such cases, the patent system's orientation can produce a paradox. The inventions may satisfy substantive patentability indicators like novelty, utility, and non-obviousness, but still fail at the threshold requirement of valid inventorship. The challenge of AI-created design is further complicated by its intangible nature. Many international institutions affirm the view that most jurisdictions now require an inventor to be a natural person, based on the moral rights associated with personal inventors' work. It also involves the contractual relationships that exist between the inventor and the employer of the inventor in relation to assigning the patent rights to the employer. As a consequence, if there is no credible identification of a human inventor, then the technology is currently completely ineligible for patent protection. As a consequence, a business entity will not be able to benefit from the transparency associated with publicly reported patent applications. Therefore, they may ultimately rely more on maintaining confidentiality as opposed to reporting the novelty of AI-generated technology. This undermines the traditional patent bargain of exclusivity in exchange for public knowledge about the new technology. There is also a concern that a business entity may unintentionally mischaracterize inventorships by not reporting its AI-generated technology, causing an increase in legal compliance administrative costs.

3.2.3 DABUS as a Stress Test for Patent Doctrine

The DABUS patent application is a series of patent applications based on inventions that were allegedly created independently by an artificial intelligence system called DABUS, created by Stephen L. Thaler. The applicant has filed a patent application in various jurisdictions. It listed DABUS as the inventor, but not a human inventor. The intent of filing these applications was to determine if the current patent laws allow for legal recognition of an AI-based inventor. The applications represent a critical "litmus test" in exploring the conflicting nature of designating a machine, especially an artificial system, as an actual inventor. Patent authorities and the individuals who enforce patent laws face a significant dilemma related to these applications. Even though our patent statutes do not explicitly indicate that an invention must be produced by a human, there is an implied assumption that inventors are natural persons who can legally be recognized, held accountable, and have their rights transferred. There have been similar issues regarding the definition of an 'actual and first inventor' as outlined in the Patents Act of 1970 in India, through patent litigation and scholarly debate (Bansal, 2022).

The USPTO rejected the DABUS applications because the term inventor is defined within the U.S. patent statutes as "an individual, which has been interpreted to mean a natural person" (USPTO, 2020). Similarly, the "EPO legal board of appeal confirmed that, under the European Patent Convention, the inventor must be a natural person and upheld its decision to refuse publication of the designation of a machine inventor" (EPO, 2021). The UK Supreme Court supports the same conclusion as found by the EPO, stating that inventorship and ownership are not just descriptive terms of how an invention was made. They represent a legal status associated with the ability to create and transfer ownership rights (UK Supreme Court, 2023; WIPO Lex, 2023). All patent systems see inventorship as a legal status and associate ownership and liability with it.

3.3 Database Rights, Trade Secrets, and Model Ownership

The way AI is currently developed is moving away from focusing on isolated "works" and towards focusing on various resources that can be utilized to create new models. Resources that are needed to create new models include: training data, curated databases, model weights, fine-tuning procedures, evaluation benchmarks, and a method for deploying the model once finished. The sources and outputs of AI development do not fit neatly into the traditional copyright/patent categories. As a result, AI developers and the holders of these works can increasingly rely on different types of database rights, trade secrets, and new types of "model ownership" in order to control value in their respective AI life cycles. These issues are not theoretical;

they affect who is able to build models, who can audit the models, and who is able to compete in data and compute-intensive markets.

One of the most significant real-world challenges is the need for very large training corpora in order to create high-performing foundation models. Recent analyses of global policy have indicated that, as demand for training data has grown, there has been an increase in the use of web scraping methods to acquire this data, raising questions concerning intellectual property and related-rights issues (OECD, 2025). The EU allows for a type of protection for databases that is separate from copyright. This database right allows the individual who created a database to receive some rights for the investment put into the creation of the database. This includes rights to prohibit others from extracting or "re-using" the entire database or a "substantial part" of it.

3.3.1 Rights of Data Compilers Versus Developers

The second source of conflict relates to the conflicting interests of those who compile datasets, i.e., data compilers, and those who create models using compiled datasets, i.e., developers. "The Organization for Economic Cooperation and Development recognises that disputes between data compilers and developers are becoming more prevalent due to factors such as the economics of data collection, access, and use of data collected, and the failure to compensate the data compiler for their efforts" (OECD, 2025). In most cases, these conflicts produce inconsistent outcomes. Larger organizations can negotiate for licenses to use datasets or purchase datasets from proprietary sources. Smaller developers often rely on publicly available datasets, although the rights associated with these datasets may involve complex issues. In the context of the discussion by WIPO regarding the generation of artificial intelligence, WIPO highlights that the generated outputs from generative AI models are likely to be similar in character to the datasets used to train them (WIPO, 2024).

3.3.2 Transparency Versus Proprietary Algorithms and Trade Secrets

The issue of balancing transparency with the need to protect proprietary information has created a serious structural challenge for organizations in developing AIs. Training datasets, model weights, and fine-tuning strategies are typically protected by developers as proprietary assets, often referred to as trade secrets. Trade secrets are covered in the European Union under EU law as "undisclosed know-how and business information". Under "the Defend Trade Secrets Act of 2016, it pertains to the products or services of interstate or foreign commerce" (Defend Trade Secrets Act of 2016). These laws create an environment in which companies will frequently not disclose the sources of their training materials, the parameters of their models, and the measures they have undertaken to mitigate the potential of unsafe outcomes. They do this on the basis that such disclosures will diminish their competitive edge.

This legal framework reflects what happens in practice. Many companies will not share their training datasets, the model parameters, and the safety processes carried out, as they feel that doing so would harm their competitive advantage. However, there is now a growing demand from regulators, researchers, and users for increased transparency to assess for bias, safety, or risk from potential infringements by an AI tool's model. Specifically, WIPO is concerned about a growing possibility that the use of generative AI tools may inadvertently expose corporate trade secrets (WIPO, 2023). As such, database rights and trade secret laws are often a substitute for copyright or patent protection. The existence of strong secrecy makes it more difficult to audit AI systems, providing less opportunity for validating the training methods of AI systems, and ultimately creates market power concentration in the hands of a few entities only. The resulting challenges will not only have legal consequences, but they will ultimately shape how AI systems are developed and used, and who can actively participate in innovation and who will be excluded from innovations involving AI systems.

3.4 Economic and Policy Consequences

The way in which artificial intelligence interacts with existing jurisdictions creates uncertainty for businesses when determining how to best protect themselves under IP laws. An intellectual property regime that has not been developed to accommodate generative AI will create a greater incentive for firms to use secrecy as opposed to sharing information with the public. It will further undermine the public purpose of the intellectual property regime (Samuelson, 2023). A lack of transparency with datasets, model architectures, and training methods limits cumulative innovation and restricts outside evaluations of the quality of the innovation. Additionally, an overly broad set of protections from IP rights could also create significant problems for businesses. Granting broad proprietary rights to AI-generated outputs, training outputs and inputs, or system interfaces could allow for dominant firms to continue to exist and create high barriers to market entry for potential new companies (Lemley and Casey, 2023). Significant reliance on data and computational resources is necessary for the development of AI, so high levels of protection can exacerbate market concentrations. Thus, the intellectual property rights would function less to incentivise innovation and more as a means of excluding others from the marketplace.

This set of risks brings forth a continuing tension in global patenting policy: "under protection" versus "over protection." If an AI-created invention or other output does not fall within current IP law, firms will not have confidence that they will receive exclusive rights and, therefore, will invest less in socially beneficial innovative products (Abbott, 2020). On the other end of the spectrum, if AI-generated outputs are granted the same level of protection as human-generated outputs, then AI-generated creations could lead to monopolies over automating creativity at levels of sophistication that transcend human talent. The question should not be whether or not IP exists for AI-created innovations. Rather, it should be how to regulate IP so that innovators will be motivated through incentives, while preventing market distortions. This is especially relevant for startups and smaller developers, as large corporations will be able to absorb the cost of legal uncertainty, negotiate licensing agreements, or have a competitive advantage in utilising trade secrets. Innovation ecosystems rely on ongoing access to shared sources of information, and if shared informational resources are enclosed, the capacity for distributed innovation will be diminished in the long run (Frischmann, Selinger, and Kidd, 2019).

There are, of course, many implications for culture and distribution that stem from AI's impact on how we create. AI-based systems, when employed in the context of creative work, provide a vehicle through which economic power can be centralised on tech platforms while simultaneously extracting value from cultural labour in the form of ownership and control (Katyal, 2023). If IP laws do not provide coherence with these dynamics, they enable and sustain patterns of extraction while failing to protect a sustainable creative economy. Collectively, these economic and policy implications will require a more holistic approach when developing responses to the challenges associated with AI.

4. COMPARATIVE ANALYSIS ACROSS JURISDICTIONS

AI has a global impact, while IP laws are only enforced in specific areas. Because of this contradiction, the treatment of products created with AI or with AI assistance differs from country to country. Most jurisdictions enforce their intellectual property rights based on some degree of human involvement or oversight. However, there is a wide variation between jurisdictions on how to regulate transparency in AI, resolve disputes over the use of training data sets, and determine what constitutes "human Contribution." This section compares IP law in the EU, USA, UK, SA, and India. It shows how the different doctrines provide different incentives for innovation across borders.

4.1 European Union

The EU's AI regulatory framework is centred on how it will regulate AI, as opposed to how it will create legal entities in AI. This concept underpins both the EU's governance of AI and its governance of intellectual property. The "EU Artificial Intelligence Act (Regulation (EU) 2024/1689)" does not regulate intellectual property directly. However, it has an indirect impact on IP law. The Act also imposes transparency and documentation requirements on general AI models. The Act specifically sets out an obligation for the model developers to act within the constraints set forth in EU copyright law and creates an obligation for developers to make publicly available a summary of the content of their training datasets.

Under patent law, the EU has declared that an AI cannot be named as an inventor. Indeed, in the DABUS matter, the European Patent Office concluded that a person who is an "inventor" under the European Patent Convention must be a natural person with the capacity to hold legal rights and obligations. The decision upheld the reasoning that to be an inventor under the convention is a title or status afforded to a person. It cannot be based on a factually correct description of the inventive process, but rather is the result of a legal interpretation of the law (European Parliament, 2025). In recent studies released by the EU Parliament and the EUIPO, it has been reported that current areas for improvement in AI lie more in the legality of large-scale data collections for training AI than it does in determining legal ownership of AI-generated products (European Patent Office, 2021).

Overall, the EU framework reinforces human oversight in AI-assisted creations while using transparency obligations to rebalance interests between AI developers and rights holders.

4.2 United States

As it relates to AI regulation, the USA employs a "decentralized" and "guidance-driven" method. Rather than enacting all encompassing piece of AI regulation, AI regulations in the USA utilise a combination of judicial decisions and administrative guidance to adapt existing law regarding intellectual property. "In order for an AI-assisted invention to be considered patentable, inventorship must be attributed to the natural persons who contribute significantly" (USPTO, 2024a). In this regard, the guidance identifies AI as an advanced tool and requires applicants to indicate the specific human contributions to claims. This approach allows for the establishment of continuity within existing patent doctrine while also providing practical guidelines for innovators. A framework is provided for overall governance of AI in the USA. It offers recommendations on creating documentation for AI design, training data, and risk mitigation, which could be valuable during IP disputes, particularly during evaluation (NIST, 2023). In terms of copyright law, the USA copyright law is firmly anchored in human authorship. As stated by the US Copyright Office, works produced by AI are not subject to copyright protection. Copyright protection will extend only to the elements contributed by humans, such as selection, coordination, or modification (U.S. Copyright Office, 2023a). Overall, the U.S. approach preserves doctrinal flexibility while emphasising human contribution.

4.3 United Kingdom

The mixed model present in the UK relates to the manner in which copyright law has statutorily recognised computer-generated works, while simultaneously applying a strict rule to patent inventorship. This means that computer-generated works are protected under copyright, and have authorship attributed to the persons who make the arrangements. Whereas, "inventorship in patents remains confined to natural persons, as affirmed by the UK Supreme Court in Thaler v. Comptroller-General" (UK Supreme Court, 2023). The Court found that an inventor must be a natural person based on the definitions of inventorship and ownership as delineated in UK laws.

This decision harmonises the inventorship doctrine of the UK with that of the EPO and USPTO, and reaffirms UK patent law as consistent with international patent law.

4.4 South Africa

The position of South Africa in the discussions surrounding artificial intelligence is distinct and not always well understood overall. The course of action that has been fully completed with a patent application listing DABUS as an inventor was the only instance that took place in South Africa. This, however, must be taken into consideration with extreme caution, because it does not mean that the courts or legislative authorities made any kind of legal ruling that supports AI inventorship. The patent registration process operated by South Africa is formal, i.e., non-examining (Companies and Intellectual Property Commission, 2021). This means that the registration is determined by meeting basic formalities, and there is no substantive examination to ensure that there is not an inventor other than the one claimed. Applications that meet the minimum formal requirements to qualify as patents will be registered. This registration does not mean that there has been an assessment of the novelty, inventive step, or legal validity of the inventor. Therefore, any grant for the DABUS patent was not subject to judicial scrutiny or administrative interpretation as it relates specifically to whether an AI system can be considered as "inventor" under any patent law.

There is no reasoning provided regarding inventorship, legal personhood, or liability, as was done in the case of the European Patent Office, U.S. courts, and UK authorities. Thus, the outcome is more indicative of procedural attributes than a particular policy position. Additionally, the majority of academic commentary views the South African grant as an institutional aberration based on differences in patent administration, rather than as an indication of normative acceptance of AI inventorship (Abbott, 2020). As such, while the South African case has significance as a symbol of acceptance of AI inventorship, it does not provide direction for the international unification of patent laws related to AI Inventorship, and it should not be seen as a model for reform.

4.5 India

The human-based structure of Indian intellectual property laws is consistent and continues as such with copyright and patent legislation. These legislations presume human authorship/inventorship, creating its very foundation upon ideas conceived by mental activity and subsequently transferred to another individual. These doctrines work under the assumption that an individual exists as a natural person (Bansal, 2022). Currently, there exists no binding AI law in India. "Despite the recent interactions by policymakers regarding these issues, the emphasis is to address the considerations regarding the use of training data and lack of clarity regarding ownership while deferring the immediate need for statutory reform" (DPIIT 2021). The lack of adequate text and data mining exceptions, in addition to the absence of wide-ranging fair use protections, creates an uncertain legal environment for the developers of AI and the protected right holders of Copyright.

5. PROPOSED INTELLECTUAL PROPERTY ALLOCATION MODELS FOR AI-GENERATED OUTPUTS

The comparative analysis in the prior sections shows a clear agreement among many jurisdictions that AI systems are not legally classified as entities capable of owning IP. However, several jurisdictions have also shown that AI generates a wide range of expression and creativity, and therefore has the potential to have a large impact on how intellectual property law is treated. Because of this broad acceptance of this gap, a number of schools of thought have been developed to suggest alternate methods for allocating rights in relation to human responsibility and adapting intellectual property law.

This chapter identifies four of the most widely discussed allocation methods in contemporary IP legal literature-

1. Human oversight model
2. Input control model
3. Shared/joint authorship model

4. Default public domain model

5.1 Human Oversight Model

In this model, the allocation of ownership rights of assets created by an AI system is divided between the individuals who guided or helped to develop the output generated by the system, and the developer of the system. It attributes all human authorship or invention of the creation/invention to the best or most qualified person, directing, guiding, or shaping how the AI system performs. This model is consistent with current copyright and patent laws around the world.

The primary benefit of the human oversight model is its ability to maintain continuity with established legal rules and regulations. Since the courts and the intellectual property agency can independently enforce established legal frameworks, this results in much improved legal certainty in jurisdictions. The ability to identify and hold individuals accountable for the rights and obligations they create allows the human oversight model to provide greater incentive for actual human involvement in the development of outputs, versus having all these functions completely automated.

However, there have been significant challenges in determining the extent to which the human supervision model will remain viable if AI systems are allowed to operate independently at the highest level of autonomy. This has resulted in an evidence void when it comes to proving the level of human support required under this model. If proven, this may result in variable results or manipulation of claims for human involvement to facilitate obtaining an intellectual property right (Abbott, 2020).

5.2 Input-Control Model

According to the input control model of AI Rights, those who control the inputs to an AI are allocated the intellectual property rights in connection to the associated outputs. The input control model considers those making the key investments along with the organisational decisions regarding the AI as the holders of IPR for the AI outputs, as opposed to an end-user triggering a result via the AI. The input control model acknowledges the current economic environment of AI development, where developing value through data collection and system design has considerably more economic value than developing the moment-by-moment creative output from the AI.

One benefit of the Input-Control Model of AI Ownership is that it is economically realistic. The input control model provides predictability into the outcomes of ownership by firms that undertake significant investments into the development of AI products, and it is consistent with existing models of allocating ownership, as seen in database rights, trade secret and contract. For enterprises developing AI products, developing clarity around commercialisation and planning investments based on the input control model of ownership is highly beneficial.

The downside of the input control model is that it may create a concentration of IPR among large technology companies that maintain control over their infrastructure or data inputs to an AI. This concentration could limit entrance into the AI market for new innovators and limit the diversity of the creative AI ecosystem. In addition, increased reliance on the input control model may weaken the expressive rationale of copyright law and sever the connection between ownership rights and the creator's intent or contribution. Therefore, policy research has expressed the concern that increased reliance on input control ownership could create a state of using IPR law as a means for dominating the market rather than fostering innovation (OECD, 2025; Lemley & Casey, 2023).

5.3 Shared or Joint Authorship Models

The concept of shared joint authorship models is to provide an avenue for AI-assisted production to enable multiple authors and co-inventors for human collaboration through recognition of the many types of human contributors, including, but not limited to, programmers, curators, system designers, and prompt or output generators. Both the terms of AI-human collaboration are present in the literature, yet current legal statutes do not allow for AI's legal personhood and do not allow for shared human and AI authorship.

The most significant advantage of joint authorship is to the extent that it is inclusive and representative of the dispersed nature of AI development, and to avoid the monopolistic control of a single entity within an increasingly networked Creative Web. By recognising that shared contribution exists, it may help to mitigate disputes regarding attribution.

However, Joint Authoring does pose considerable administrative and financial burdens associated with shared ownership and the increased risk of complications in licensing and cross-licensing. The complexities of co-owned inventors, due to the nature of invention law requiring at least one contribution to the conception of each patent claim by all joint inventors, occur as the sheer volume of joint inventorship requires even greater economies of scale to enable compliance, creating challenging circumstances. Courts have reiterated that inventorship status is a legal status and not a policy statement regarding responsibility or accountability before a court, as such, limiting further development of the practical application (UK Supreme Court, 2023).

5.5 Public Domain Default Model

According to the public domain default approach, the work produced by autonomous AI systems, with no substantial human contribution, is not entitled to intellectual property protection. Instead, that work is immediately subject to the public domain. The public domain default model is premised on the belief that intellectual property law exists primarily to create incentives to create, share, and publish new creative works by human beings. Thus, providing intellectual property rights to only the automated products of computer programs is misaligned with that purpose.

The public domain default model will allow for the broadest distribution and re-use, and therefore, the greatest levels of external innovation. Public domain defaults result in no ambiguity over the definition of authorship and are consistent with the United States' current copyright practices, denying protection for works produced without a human author (U.S. Copyright Office, 2023). From the perspective of competition, a public domain default model will limit the ability of individuals or organisations to monopolise the automated productivity of computers.

However, utilising a public domain default model may discourage firms from investing in AI-generated content, products, or services. Firms may be forced to rely more heavily on trade secrets, contractual provisions, and technological protection measures, resulting in less public transparency and accessibility to that knowledge. Additionally, it has been noted by other scholars that public domain defaults may allow the largest platforms that can monetise AI services to benefit disproportionately by leveraging their economies of scale, not necessarily through ownership of intellectual property (Samuelson, 2023; Frischmann et al., 2019).

5.6 Normative Implications for IP Reform

The selection of an allocation model encompasses greater consequences on innovation, market structure, and cultural Production. Excessive protection rewards existing market leaders while insufficient protection may disincentivize investments that generate social benefits. Comparative evidence indicates that reforms should be undertaken carefully, with reliance on empirical research and international cooperation instead of sweeping philosophical changes. Ultimately, the challenge facing Intellectual Property Law is not identifying machines as the creators but clarifying how human accountability, authority, and involvement are determined in machine-mediated creations.

6. REAL-WORLD REGULATORY IMPACT

Legal uncertainty regarding the ownership of creative works, inventions, and data used to train AI is currently influencing how people within the creative industries, tech companies, and marketplaces operate. The lack of established doctrines will not only be difficult to interpret; however, it will also have an impact on investment decisions, the way that organisations license their products, and how they compete. Throughout this section, I will examine how the AI-IP relationship manifests in practice, as well as how current doctrinal decisions affect innovation ecosystems.

6.1 Impact on creative industries

Generative A.I. is one of the earliest industries to be affected by the disruptions caused by A.I., as AI systems have now started to produce music, images, literature, and audiovisual content, and compete directly with human creators in these areas. As a result, many times copyright laws do not extend to protect works created or generated by A.I. systems, so businesses have turned to other means in order to protect and regulate the use of A.I.-created works, which include contractual controls and platform governance (Guadamuz, 2017).

This change will have a different impact on the creative industry; while large platforms/corporations can use licensing and technical restrictions to manage and/or protect A.I.-created material, individual and/or small studios do not have the same leverage, and studies on cultural markets indicate that there is an increasing risk of capitalising creative value on behalf of the intermediary platforms, rather than the creators themselves (Kretschmer et al., 2021). Therefore, the outcome of the absence of copyright protection for A.I.-created work does not result in open access to the work to the general public and may lead to providing private sector mechanisms to replace public sector protections.

6.2 Impact on technology firms and startups

New technologies are changing the way businesses develop products that rely on patents. Uncertainty about AI (artificial intelligence) and who can claim to be an inventor under current law is affecting the way companies approach innovation.

Under patent law, inventors are generally required to have an actual idea and to have made a clear contribution to the development of an invention. AI can discover inventions through algorithms and techniques such as computer simulations, which do not lend themselves to easily assigning inventorship. Academic research suggests that when companies cannot clearly identify inventors of an AI-generated invention, they often turn to trade secrets rather than disclosing their invention via a patent (Burk, 2019).

This change in strategy has a systemic impact on innovation and technology development. A patent's purpose is to help share technical knowledge and provide technical information to the public. Conversely, trade secrets are designed to protect and keep the company's technical knowledge private. Trade secrets provide greater protection to larger and more established companies – those who have the resources to develop and maintain extensive confidential business systems – compared to startup companies and small research-oriented companies.

As artificial intelligence becomes more extensively used in the research and development of new products and processes, these structural problems may limit cumulative innovation.

Uncertainty also affects investment decision-making. Venture capitalists use intellectual property (IP) as an important measure of perceived value in a company. If a company cannot provide reliable protection for its AI-based output, it will face challenges in obtaining financial backing, unless its products have advanced commercial viability (Rai, 2021). Therefore, uncertainty regarding AI inventorship will have far-reaching consequences for businesses seeking to raise capital.

6.3 Market Structure and Power Concentration

The way AI and IP regulations affect the real world is heavily dependent on market structure. Scholars who study digital marketplaces link the marketplace structure's degree of control over data, algorithms, and platforms to the amount of traditional IP protection given to companies (Cohen 2019).

This structural change presents a competition risk. Without the protection of IP rights for AI–based content, dominant businesses can profit off providing services at scale, through network effects, and through contractual barriers to entry. Smaller companies do not have access to the same mechanisms as larger companies. Studies of law and political economies indicate that these mechanisms can reinforce the power of "informational capitalism," in which control over the systems of knowledge outweighs ownership of particular IP rights (Zuboff 2019).

From the perspective of governance, there is a paradox. Although eliminating the IP rights for AI-generated outputs may help avoid forming formal monopolies, it can create informal monopolies through private control. In evaluating the consequences of AI–IP laws upon businesses, there is a need to examine not only the means of distributing rights but also the balance of market power and the availability of institutions to enforce those rights.

6.4 Cross- Border Effects and Global Innovation

The territorial nature of IP law presents an additional challenge when considering how AI is regulated internationally. Companies train AI in one country, deploy it in another, and monetise it in several locations around the world. Different national regulatory frameworks for authorship/inventorship create tension in both the licensing and enforcement of AI use. As a result of this fragmentation of regulatory frameworks for IP globally, scholars analysing the "international governance of IP have indicated that the fragmentation leads to increased compliance costs associated with the regulation of AI and the creation of opportunities for companies"(Gervais, 2022).

In addition, developing economies find themselves in a particularly difficult situation because, without clearly articulated and predictable rules regarding the interaction between AI and IP, there is a strong likelihood that they will be excluded from global value chains, thus forcing them to create a dependency on large multinational platforms for their work. Comparative studies show that countries with high levels of regulatory uncertainty tend to concentrate their technological power, thereby reinforcing existing global technological hierarchies rather than disrupting them (Dinwoodie & Dreyfuss, 2018).

These cross-border effects highlight that any evaluation of AI and IP regulation is not possible to do simply on a domestic legal level; rather, the actual effects of AI/IP regulation will depend on the manner in which individual nations interact with the broader global innovation systems in which they participate.

6.5 Synthesis: Practical Lessons

A global trend exists in many different industries, with most legal systems making clear their commitment to the protection of human-centred IP laws. However, the entities within the marketplace continue to adapt their ip practices through licensing agreements, non-disclosure agreements, and other methods of contracting/enterprise governance developed under the auspices of Covered Services. As a result of this growing utilisation of training data, we see a transfer of power from the creators and small innovators to the data-rich intermediaries established by platform operators.

The current trend of AI on IP continues to increase as it expands its impact on IP development, but also changes the manner in which people work in creative sectors, influence innovation and enhances competitiveness in the global marketplace. Thus, identifying and understanding these impacts will help drive reforms aimed at recognising and managing the impact of technology on the development of balanced and inclusive opportunities for both technological development and inclusive innovation.

7. SOLUTIONS AND RECOMMENDATIONS

The previous analysis has shown that the current intellectual property structures that govern creativity and invention through artificial intelligence (AI) are inadequate for allowing AI-created works without violating fundamental principles of the law. In other words, the challenge is more than simply inconsistencies in legal theory; it is also the potential for regulatory inaction to lead to private control, secrecy, and the concentration of markets. The following section presents targeted solutions that will enhance human responsibility while enhancing legal clarity, stimulating innovation, and increasing cross-border collaboration.

First, instead of simply redefining an author or inventor as a machine/technology entity, legislators and the courts should provide clear definitions of what it means for a human to have contributed meaningfully (in other words, what is sufficient human involvement). Scholars suggest that by focusing on the decision-making authority, the creative selection, and who is liable for the work produced, creates a more coherent way to allocate authorship than using abstract and/or ambiguous concepts such as creativity (Margoni, 2018). Providing clear guidelines will reduce the opportunity for strategic over-claiming of human input and increase predictability for creators and developers.

Second, policymakers should avoid providing broad proprietary rights over the machine-generated output of AI systems and instead assist in supporting accountability and transparency through the establishment of policies encouraging their development. Open-access legal theory provides a warning about providing exclusivity rights to machines generating data, which may create 'circle-jerk' monopolies of machine-generated knowledge (Grimmelmann, 2016). A default position that allows for fully autonomous output to go unprotected while still giving an individual the right to protect their human-assisted contribution would maintain the incentive system within the IP area without producing the contrary effect.

Third, harmonisation of soft laws via institutional coordination is one way of achieving this objective. Research comparing national responses to international intellectual property governance shows that fragmented national responses have disproportionately impacted smaller innovators and developing countries (Yu, 2022). The use of non-binding principles, created through collaboration and academic dialogue, will not only guide national reconstructions, but they will also provide flexibility to individual countries based on their context. An example of an area where such coordination would be of value is in the ability to create uniformity in relation to the documentation standards that define how to attribute ownership of ideas and the expectations of the level of human responsibility for AI-mediated creations.

Finally, the empirical evidence regarding the effects of AI on innovation incentives must form the basis of all reform efforts and not be based upon speculative assumptions. Open-access studies have shown that the effects of AI on innovation incentives differ depending on the sector and the degree of automation (Bessen et al., 2021). Therefore, a prerequisite to any successful reform will be to clarify how human agency, control, and responsibility are identified in relation to AI mediation.

8. CONCLUSION

AI has disrupted the creative aspect of humans and intellectual property. This has created pressure on intellectual property as now they need to be evolved. There is always a consistent and certain notion that only humans can be the natural creators. But now, in the era of technology, AI can generate stories, art, and generate new invention ideas like a human. There is now a challenge for the legislative bodies to create and adapt to new laws without challenging the authenticity, quality, and credibility of such protected property. This will help in incentivizing creativity, attracting investors, maintaining fair competition, and ensuring that knowledge remains accessible.

On the other hand, the rapid development of generative AI systems has brought the uncertainty of the law regarding when machine outputs were created through human thought. This uncertainty hinders commercialization, and disputes arise as to ownership, licensing, and liability. It is known that uncertainty regarding standards leads to development or commercialization shifting to jurisdictions that are perceived to be more lenient, creating further global inconsistency in standards.

Looking at the way forward, regulators must strike a balance that provides sufficient legal clarity on what constitutes intellectual property and affords protection to work created by AI and machines (Efthymiou, I-P, and Kouroupis,.K., 2026). Future changes should be made after keeping in mind that the focus should be on making it clear that when humans give their contribution to any work, more transparency must be there on how AI systems are developed and used. It must be ensured that data is used responsibly. There is a need for clear laws and rules to be established for the handling of training data and copyrighted materials. Good licensing agreements must be enabled for the development of AI for good purposes, which can make the ethical use of AI easy and clear. There is a need for nations to collaborate and come to a uniformity on this. Organizations such as WIPO were established to help such discussions lead to coherence in the way we define law, enforce rules, and recognize rights. Therefore, we want our law to incentivize a situation in which AI and people can collaborate effectively, with humans remaining in control and conferring benefits to those who use AI instruments. With a few clever pre-emptive adjustments to the intellectual property law, we can keep this body of law viable for

REFERENCES

Abbott, R. (2020). *The reasonable robot: Artificial intelligence and the law*. Cambridge University Press., https://papers.ssrn.com/sol3/papers.cfm?abstract_id=3611370 DOI: 10.1017/9781108631761

Almada, M. (2025). Artificial intelligence regulation and governance in the European Union. *Computer Law & Security Review*, *49*, 105758. DOI: 10.1016/j.clsr.2024.105758

Bansal, R. (2022). Artificial intelligence and intellectual property rights in India. *Journal of Intellectual Property Rights*, *27*(3), 169–180. https://or.niscpr.res.in/index.php/JIPR/article/view/1205

Bessen, J. E., Impink, S. M., & Seamans, R. (2021). The impact of artificial intelligence on innovation. *Research Policy*, *50*(9), 104279. DOI: 10.1016/j.respol.2021.104279

Burk, D. L. (2019). Law and artificial intelligence. *University of Illinois Law Review*, *2019*(4), 1957–1996.

China National Intellectual Property Administration. (n.d.). Guidelines for patent applications related to artificial intelligence (trial). https://www.cnipa.gov.cn/

Cohen, J. E. (2019). *Between truth and power: The legal constructions of informational capitalism*. Oxford University Press. DOI: 10.1093/oso/9780190246693.001.0001

Companies and Intellectual Property Commission. (2021). *Patent journal: Acceptance of patent application 2021/03242* (South Africa).

Defend Trade Secrets Act of. 2016, Pub. L. No. 114–153, 130 Stat. 376. https://www.congress.gov/114/plaws/publ153/PLAW-114publ153.pdf

Department for Promotion of Industry and Internal Trade. (2025). *Working paper on generative artificial intelligence and copyright*. Government of India. https://www.dpiit.gov.in

Dinwoodie, G. B., & Dreyfuss, R. C. (2018). *A neofederalist vision of TRIPS*. Oxford University Press.

Efthymiou, I.-P., & Kouroupis, K. (Eds.). (2026). *Legal and Regulatory Impacts on AI Development*. IGI Global Scientific Publishing., DOI: 10.4018/979-8-3373-7996-8

European Commission. (2020). *White paper on artificial intelligence: A European approach to excellence and trust*. https://commission.europa.eu

European Commission. (2023). *Proposal for a regulation laying down harmonised rules on artificial intelligence (AI Act)*. https://eur-lex.europa.eu

European Parliament. (2025). *Generative AI and copyright: Legal challenges and policy options*. https://www.europarl.europa.eu

European Patent Office. (2021). *Decision J 0008/20 (Designation of inventor/ DABUS)*. https://www.epo.org

Frischmann, B., Selinger, E., & Kidd, I. (2019). *Re-engineering humanity*. Cambridge University Press.

Gervais, D. J. (2022). Artificial intelligence and international intellectual property law. *Vanderbilt Journal of Transnational Law*, *55*(2), 301–356.

Goodfellow, I., Bengio, Y., & Courville, A. (2014). Generative adversarial nets. *Advances in Neural Information Processing Systems*, 27. https://papers.nips.cc/paper/2014/file/5ca3e9b122f61f8f06494c97b1afccf3-Paper.pdf

Grimmelmann, J. (2016). There's no such thing as a computer-authored work—and it's a good thing, too. *Columbia Journal of Law & the Arts*, 39(3), 403–420. https://papers.ssrn.com/sol3/papers.cfm?abstract_id=2736569

Guadamuz, A. (2017). Artificial intelligence and copyright. *WIPO Journal*, *9*(2), 169–186.

Katyal, S. K. (2023). Private accountability in the age of artificial intelligence. *UCLA Law Review. University of California, Los Angeles. School of Law*, *70*, 1–60.

Kouroupis, K., & Efthymiou, I. (Eds.). (2026). *Legal Challenges of AI Across Interdisciplinary Sectors*. IGI Global Scientific Publishing., DOI: 10.4018/979-8-3373-5452-1

Kretschmer, M.. (2021). Creativity, artificial intelligence, and the future of copyright. *Journal of the Copyright Society*, *68*(3), 353–392.

Kriz, D. (2023). Artificial intelligence and the future of creative production. *Technology Innovation Management Review*, *13*(2), 22–35.

Lemley, M. A., & Casey, B. (2023). Remedies for AI harms. *Texas Law Review*, *101*(6), 1191–1266.

Margoni, T. (2018). *The harmonisation of authorship in EU copyright law*. CREATe Working Paper. https://zenodo.org/record/1256996

McCarthy, J., Minsky, M., Rochester, N., & Shannon, C. (1955). *A proposal for the Dartmouth summer research project on artificial intelligence*. Dartmouth College.

National Institute of Standards and Technology. (2022). *Artificial intelligence definition and characteristics*. https://www.nist.gov

National Institute of Standards and Technology. (2023). *AI risk management framework (AI RMF 1.0)*. https://www.nist.gov

Organisation for Economic Co-operation and Development. (2025). *Intellectual property issues in artificial intelligence trained on scraped data*. https://www.oecd.org

Presno Linera, M., & Meuwese, A. (2025). AI regulation and constitutional design in the EU. *European Constitutional Law Review*, *21*(1), 1–26.

Rai, A. K. (2021). Intellectual property and artificial intelligence. *Chicago-Kent Law Review*, *96*(1), 1–26.

Ricketson, S., & Ginsburg, J. (2022). *International copyright and neighbouring rights* (3rd ed.). Oxford University Press. DOI: 10.1093/oso/9780198801986.001.0001

Roberts, H., Cowls, J., Morley, J., Taddeo, M., Wang, V., & Floridi, L. (2021). The Chinese approach to artificial intelligence governance. *AI & Society*, *36*(1), 59–77. DOI: 10.1007/s00146-020-00992-2

Samuelson, P. (2023). Generative AI and copyright. SSRN *Electronic Journal*. https://papers.ssrn.com/sol3/papers.cfm?abstract_id=4374012 DOI: 10.1126/science.adi0656

Schuett, J. (2019). A legal definition of artificial intelligence. *Stanford Technology Law Review : STLR : An Online High-Technology Law Journal from Stanford Law School*, *22*(1), 1–50.

Senftleben, M. (2021). Copyright, creativity, and innovation. *Amsterdam Law School Research Paper*. https://ssrn.com/abstract=3032406

Thaler v. Comptroller-General of Patents, Designs and Trade Marks, [2023] UKSC 49. https://www.supremecourt.uk

Thaler v. Perlmutter, 999 F.3d ___ (D.C. Cir. 2025).

Times of India. (2024). *Bombay HC grants interim relief to Asha Bhosle against AI misuse*. https://timesofindia.indiatimes.com

United Kingdom Supreme Court. (2023). Thaler v. Comptroller-General of Patents, Designs and Trade Marks [2023] UKSC 49. https://www.supremecourt.uk/cases/docs/uksc-2021-0201-judgment.pdf

United States Patent and Trademark Office. (2024). *Inventorship guidance for AI-assisted inventions*. https://www.uspto.gov

U.S. Copyright Office. (2023). *Copyright registration guidance: Works containing AI-generated material.* https://www.copyright.gov/ai

World Intellectual Property Organization. (2023). *Generative AI: Navigating intellectual property.* https://www.wipo.int

World Intellectual Property Organization. (2024). *Generative AI and intellectual property.* https://www.wipo.int

Yu, P. K. (2022). Artificial intelligence, big data, and intellectual property. *Vanderbilt Journal of Entertainment and Technology Law, 24*(4), 1–54.

Chapter 10

The Urgency of Regulating Copyright of Artificial Intelligence Works With the Fair Use Principle From the Perspective of Development Law Theory

Junaidi Junaidi
https://orcid.org/0000-0002-2948-696X
Sjakhyakirti University, Indonesia

ABSTRACT

As artificial intelligence (AI) develops rapidly, new copyright issues are emerging, particularly concerning the limitations of fair use of AI-generated works. Using copyrighted works without permission is illegal. Legal studies that adopt a normative approach focus on the applicable legal principles to deepen understanding of the current legal basis. Although Indonesia has copyright laws, there are currently no specific provisions to regulate limitations. Comprehensive regulations are therefore needed to control the application of AI in creative activities and enforce the principle of fair use. Based on the theory of development law, AI can be viewed as a tool for updating legislation, providing a foundation for legal and social innovation that aligns with the values of the Indonesian state philosophy, Pancasila

DOI: 10.4018/979-8-2600-0000-7.ch010

INTRODUCTION

Thanks to increasingly rapid technological advancements, human life is becoming more dynamic and practical. These advancements enable people to communicate more quickly using various modern tools and applications. In the current era of the 5.0 industrial revolution, elements focused on cyber-physical systems have diminished, and the focus is now on the role of humans as the core of civilization. Information and communication technology (ICT) infrastructure is used by humans in their daily lives. This method does not only focus on the relationship between machines and machines or robotics, but also on the relationship between humans and machines (human to machine) and vice versa (Ramli, 2021). Advances in computer and communication technology have enabled the development of various computer network systems around the world. Computer network applications are not only used in laboratories, universities, or scientific research activities. Additionally, various businesses, banking, government, military and defense institutions, legal systems, and the judiciary have adopted computer networks (Junaidi, 2020). This is what creates Artificial Intelligence (AI). Artificial intelligence is software that has been trained and programmed to function as a machine with human-like thinking capabilities (Fadillah, 2023).

Artificial intelligence (AI) has transformed many areas, including graphic design (Rochim, 2024; Saputra et al., 2023). AI enables designers and artists to produce work using methods that were previously unthinkable, such as automated design, instant image processing and using algorithms to generate unique visuals (Putro et al., 2023). AI also enriches the creative process in graphic arts by providing advanced tools for image manipulation and the creation of adaptive and interactive artworks (Anantrasirichai & Bull, 2022; Bonadio & McDonagh, 2020). Additionally, AI enables human-machine collaboration, allowing artists to use AI as a tool to expand their imagination and create more complex and innovative artworks (Anifa, 2024). Since the turn of the millennium, artificial intelligence has developed capabilities such as facial recognition, speech understanding and the processing of large amounts of data. According to the European Parliament, artificial intelligence is defined as a system that performs tasks mimicking human intelligence, such as receiving and processing information, learning from its environment, and performing specific tasks independently. Ultimately, artificial intelligence mimics the way the human brain works (Maidanyk, 2021).

Significant investments in artificial intelligence research and the development of AI-based products have been made by large technology companies such as Google, Facebook and Amazon, driving progress in this technology (Hildebrandt, 2020). Although technology is intended to make human life easier, its advancement has instead caused many problems, especially those related to the law. Because artificial intelligence (AI) systems can produce works such as text, images, or music with only the instructions given, artificial intelligence (AI) allows humans to produce something similar to human thinking without having to have special skills. This development has two sides. Technology can help many industries, but it can also be misused to commit violations (Rama, et al, 2023).

Artificial Intelligence (AI) has become an effective tool for transforming how intellectual property (IP) is created, managed, and utilised. Of course, the changes and advancements in this technology bring new challenges and opportunities for inventors, both in the business world and in other fields related to intellectual property. On the other hand, artificial intelligence (AI) helps inventors create new forms of intellectual property and also improves efficiency and facilitates the exploitation of other intellectual property. However, complex legal and ethical issues related to intellectual property ownership, such as ownership, patentability, copyright infringement, and data security, may also arise with artificial intelligence (Singh & Singh, 2023). The issue of copyright for works created by artificial intelligence has become a significant subject of debate. Furthermore, Spindler (2019) argues that using artificial intelligence in the painting process positions artificial intelligence as the true creative agent, rather than the human artist.

According to Kumari and Sharma (2020), Soraya and Althafuzzfar (2024) and Fadillah (2024), intellectual property rights (IPRs) are the legal protections granted to individuals or groups for the results of their innovations, inventions or creative works. Intellectual property law gives creators exclusive control over the use and distribution of their artistic, literary, and musical works. Copyright law prohibits the unauthorized preparation, modification, or use of a work (Vakili, 2025). Conversely, patents grant inventors the right to protect their technological inventions, preventing others from making, using, or selling those inventions without permission (Lindsey et al., 2022). Additionally, IPR include moral rights, which enable creators to maintain an emotional and intellectual connection to their work even when third parties receive rights to benefits such as copyright. This ensures recognition of the artwork and protects its integrity from changes or uses that could harm the creator's reputation.

In the United States, there is a copyright case involving OpenAI and the Authors Guild. The case is being heard in the United States District Court, where the Authors Guild is working with famous authors such as George RR Martin, John Grisham, Jodi Picoult, and Shanbhag Lang, all of whom represent professional writers whose livelihoods depend heavily on their work. The plaintiffs claim that OpenAI and the institutions supporting it have violated copyright laws by training Large Language Models (LLMs) using copyrighted works repeatedly. The plaintiffs assume that their work has been used to train data, improving the quality of tools like ChatGPT. It is claimed that ChatGPT can provide summaries, analyses, and detailed summaries of the work, which would not be possible if the material had not been understood by the LLM. The plaintiffs argue that OpenAI had ample opportunity to obtain licences for their work and should have done so promptly (Saber, 2024).

In November 2023, the Beijing Internet Court ruled on the landmark case of LI v. LIU, the first case in China to involve copyright for images created by artificial intelligence, setting an important precedent in intellectual property rights. This decision sparked numerous comments from both the artificial intelligence industry and the general public. Mr LI filed a copyright infringement lawsuit against Ms LIU, a Baijiahao blogger. The court ruled that Ms. LIU had infringed the copyright of images created through artificial intelligence, including the right to distribute them via information networks. According to the court, images created through artificial intelligence can be copyrighted as long as an element of human creativity is involved in the process of creating the image. The court found the defendant guilty of copyright infringement for the image in question (Song, 2023).

To protect ideas and creativity in the fields of art, culture, and education, copyright laws are necessary. Copyright laws protect intellectual property rights, particularly copyrights granted to creators and other copyright holders. Copyright protects creative works such as books, music, visual art, films, and technology. Essentially, copyright law protects economic and moral rights (Puspasari & Sardjono, 2023). Copyright forms the foundation of the intellectual property protection system and is designed to protect human original works. Copyright gives creators the authority to control how their work is used and profited from, including moral and economic rights. Furthermore, copyright is essential in fostering innovation, encouraging creators and establishing an environment that promotes technological and cultural advancement (Ramadika & Wijaya, 2025). Without copyright protection for artificial intelligence content, monetary income could shift from human creators to the owners or managers of artificial intelligence and platforms (Israhadi, 2023). This shift is likely to have a significant impact on the income of writers, artists, and other creators. This could lead to economic inequality and a decline in the value placed on human creativity (Crawford & Schultz, 2024; Dwivedi et al., 2023). Therefore, in this new era, it is crucial to find ways to balance transformative technological advancements like GenAI with copyright protection and fair compensation for content creators (Lee et al., 2024; Liu et al., 2023; Lucchi, 2023).

Copyright owners are protected by copyright law from third parties who wish to copy their work. This ensures that the original form of the work remains true to the creator's expression (WIPO, 2016). In the WIPO Study on Limitations and Exceptions to Copyright and Related Rights in the Digital Environment, held in Geneva from 23 to 27 June 2003, the objective of the study was to enhance or strengthen the principles of exceptions and limitations in copyright law (Ricketson, 2003). When interpreting the doctrine of fair use, which permits the use of copyrighted works for purposes such as commentary, criticism, education or research, the legal system must determine where the line is drawn between the essential elements of the creative process that result in new and innovative works. This could undermine the legal basis of copyright (Amankwah-Amoah et al., 2024; Appel et al., 2023; Crawford & Schultz, 2024; Gans, 2024; Jodha & Bera, 2023).

According to a legal doctrine known as the fair use principle, people can use copyrighted works for certain recognized purposes, such as education, research, analysis, review, or news reporting, without seeking permission from the copyright owner (Habibi, et al, 2025). This doctrine, which first appeared in the American legal system, has become an important basis for balancing copyright protection with the interests of society. The principle of fair use is particularly important in the development of technology because it provides greater access to copyrighted works that are necessary for generating new innovations (Rejeki, et al, 2023).

Current regulations indicate that Indonesia's copyright law system must be changed to meet the needs of the information technology industry. In today's digital age, competition in the field of technology is increasingly fierce, and the ability to gain access to technology and information is crucial to gaining an advantage. Copyright regulations that are too strict in situations like this can hinder innovation because they limit access to the resources needed to create new goods or services. On the other hand, overly lenient policies can harm creators or copyright holders by reducing their incentive to create something new. Therefore, it is crucial to strike a balanced equilibrium between protecting copyright and fulfilling public interests through the application of wise principles for fair use.

Regulations governing the use of artificial intelligence in the creation of works are becoming increasingly important. Legal uncertainty regarding copyright for works produced by artificial intelligence can lead to significant legal disputes, especially in the digital sector. The principle of fair use is applied in this situation to regulate the use of creative works that use artificial intelligence technology without infringing on the copyright of the original creator.

The problems studied are, *firstly* the urgency of copyright regulation of artificial intelligence with the application of fair use and *secondly* the application of fair use to artificial intelligence in copyright law: a development law perspective.

RESEARCH METHOD

Normative legal research is the type of research conducted by the author. Normative legal research focuses on doctrines and principles in the field of law. Research on legal principles is a type of research that aims to identify the legal foundations or doctrines of positive law, often referred to as doctrinal research (Muhaimin, 2020).

Normative legal research uses various models to gather information from different perspectives on the legal issues being studied, as well as to find solutions or the best solutions. The approaches used in this research include the legal approach, the conceptual approach, and the comparative approach.

1. The statute approach is a research method that evaluates legal products by examining relevant regulations to assess whether the research is in line with legal products (Muhaimin, 2020). In this study, various theories and concepts related to copyright, fair use, and artificial intelligence (AI) will be analyzed. Law No. 28 of 2014 concerning Copyright is one of the laws relevant to the subject of this study.
2. Ascertains how these principles relate to the issues under study, particularly in cases where there are no specific regulations governing them, a conceptual approach is used (Muhaimin, 2020). This study uses a conceptual approach to examine how Law No. 28 of 2014 on Copyright applies the principle of fair use to works created by artificial intelligence. This is because there is no specific law regulating the use of works created by artificial intelligence. The purpose of this approach is to explore the relationship between the doctrine and concepts of copyright law when assessing the validity of its use.
3. Comparative approach, analyzing Indonesian laws with laws in other countries to find similarities and differences (Muhaimin, 2020). In this study, we will examine how the principle of fair use is applied to works produced by artificial intelligence in accordance with Law No. 28 of 2014 concerning Copyright, as well as applicable regulations in other countries. By making this comparison, we will identify the advantages, disadvantages, and differences of the current regulations, and this will help us find a more effective legal model to protect copyright.

RESULT AND DISCUSSION

The Urgency of Copyright Regulation of Artificial Intelligence with the Application of Fair Use

According to Erdozain (2019), advances in AI have made it possible to create many digital products at very low production cost. Consequently, conventional copyright law systems are struggling to protect the rights of creators. Important questions about copyright ownership, the identity of legal actors and legal liability in the event of infringement arise from AI's ability to produce works independently. While some argue that artificial intelligence can be held legally liable (Naqvi, 2020), this idea contradicts the Indonesian legal system, which only recognises humans and legal entities as legal subjects. Therefore, neither the copyright owner nor the party directly responsible for copyright infringement can be considered to be AI.

It is clear that products created by artificial intelligence must meet the requirements set by copyright law in order to be protected. It should be noted that the two requirements are that there must be a 'work' that can be protected and that the work must be 'original' (Iaia, 2022). National, European, or international lawmakers generally do not define the term 'work.' However, most people agree that active human participation is necessary to provide legal protection for copyrighted works. According to a number of authors, this clearly stems from the emphasis on copyright law focusing on creators, protecting their creative efforts (Gervais, 2020). This requires the ability to make conscious choices, which in turn requires a 'mind,' 'feelings,' or 'soul,' as mentioned earlier in Kazuo Ishiguro's poetic description (Keller, 2023). In other words, this is in line with the statement made by the United States Copyright Office that, in order to obtain copyright protection, a work must be created by a human being (U.S. Copyrights, 2021). The decision is similar to the case where a photograph taken by a monkey cannot be considered a work that can be registered (U.S. Copyrights, 2021).

For copyright protection, a work must be both original and exist in physical form (Iaia, 2022). Originality is always associated with human intellectual contribution in both international and national law. This aligns with the United States Copyright Office's policy, which clearly states that works eligible for copyright must be created by humans, as demonstrated by the Monkey Selfie Photo case (U.S. Copyright Office, 2021). Articles 1(2) and 1(3) of Indonesian Law No. 28 of 2014 indirectly state that works created entirely by artificial intelligence (AI) cannot be considered creations, as the creator must be human.

Artificial intelligence (AI) can reproduce existing works independently and often without the consent of the copyright owner (Daffa, 2024). The opaque nature of AI, known as the 'black box problem', makes it more difficult to verify the data used in the training process. These characteristics further increase the likelihood of widespread and persistent copyright infringement. However, expanding the definition of creator to include those who develop or own AI raises many questions. This is primarily because the output generated by AI can differ from what its creators intended (Raharjo, 2023).

Overall, regulations regarding copyright limitations are an important component of copyright law because they recognize the interests of the public as users of creative works (Ricketson & Ginsburg, 2022). Historically, copyright law has been accepted since the Berne Convention for the Protection of Literary and Artistic Works in 1886.

However, the legal provisions regarding the rights of creators were still lacking in the Berne Convention. As a result, the Berne Convention has been amended seven times (Shefan, 2023). During the process of change, one of the rules that greatly influenced the enforcement of copyright law at that time was the 1928 amendment in Rome. This amendment established additional provisions that limited the rights of creators and copyright holders to produce their works and communicate about them (Adu & Walt, 2021).

It is essential to understand the philosophical concepts underlying copyright protection before applying fair use practices. Copyright protection is based on three main philosophical premises. First, to preserve the interests of authors or copyright holders in their creative works; second, to provide legal recognition of individual works; and third, to provide financial incentives to encourage the creation of new works. Each of these general methods can describe the copyright system specifically or as a whole. However, the non-physical aspects of copyright are the most important (Longan, 2023). According to Indonesian copyright law, fair use means using someone else's work for educational, research, or study purposes, as long as it is not used for commercial purposes. In order to maintain ethical use, it is very important to cite the source of the work. As a result, this action is not considered a copyright infringement (Rahmayanti, et al, 2024).

According to the author, the doctrine of fair use or fair treatment is a principle that requires artists to accept or grant permission for their works to be used by the public for various purposes. One condition is that the work remains the property of its original creator. This recognition is demonstrated by acknowledging and mentioning the creator. In short, cite the source of the work used. Plagiarism is considered a legal violation under this principle. In copyright law, the principle of fair use allows the use of artistic works without the creator's permission provided certain conditions are met. The purpose of use, the characteristics of the work, the amount of the work used, and its impact on the potential market for the original work are among the criteria considered (Aini & Wauran, 2021).

The Standing Committee on Copyright and Related Rights (SCCR) has established a classification of rules regarding copyright limitations and exceptions, which includes:

1. 'Provisions that do not cover, or allow for exceptions to, the protection of certain categories of works or materials.' In the context of analysis, this can be referred to as a 'limitation' on protection, meaning that protection is not required for certain types of subjects.
2. Rules that grant immunity from actions suspected of infringing copyright, which may be used for specific purposes, such as reporting news or teaching. These provisions grant immunity from actions suspected of infringing copyright for

specific types of use, such as for news reporting or educational purposes, or when certain conditions are met. Because it allows for the removal of liability that would otherwise arise, this can be referred to as 'permitted use' or an exception to protection.

3. Regulations that allow the use of copyrighted works in conjunction with the provision and/or payment of compensation to the creator and/or copyright holder. 'Subject to the conditions allowing certain use of copyrighted material, provided that compensation is paid to the copyright owner.' Most people refer to this license as a 'compulsory license' or 'binding license,' and the specific provisions allowing it can be found in Article 11 bis (2) and 13, as well as the Annex to the Paris Act of Berne. Provided that certain conditions are met, this type of license may be granted.

In Indonesia, the concept of fair use only applies to works for which the copyright is still valid. As long as certain requirements are met, people can use such works without the owner's permission. However, according to Article 42 of the Copyright Act, certain types of works are excluded from the scope of fair use. These include minutes of meetings of state institutions, legal regulations, speeches by state or public officials, court verdicts and decisions of arbitration bodies. Additionally, works whose copyright has expired, and works that do not meet the requirements under Articles 1(2) and 1(3) of the Copyright Act, cannot utilise the fair use provisions (Aurelia, et al., 2025).

Article 46 of the Indonesian Copyright Law recognises the principle of fair use, whereby one copy of a published work may be used for personal purposes without the creator's or copyright owner's permission (Fadli et al., 2023). This rule enables people to complete more work, provided they remain within reasonable limits. However, this usage has some limitations. Firstly, reproduction can be considered non-infringing if it meets certain requirements relating to education, research, scientific writing, national security, public interest, government administration, legal and judicial processes, or performances without charge. Secondly, reproduction can be carried out using digital media or digital copies for storage purposes. Thirdly, reproduction can be carried out using digital media or digital copies for storage purposes (Manullang, et al, 2023).

The principles set out in Section 107 of the United States Copyright Act cannot be applied directly in Indonesia. Unlike the United States, Indonesia has a limited copyright restriction system (Putri, 2023). Nevertheless, international regulations governing copyright restrictions still adhere to the three-step test of the Berne Convention:

1. Only certain situations may be restricted;

2. These restrictions must not effectively prevent the use of copyright;
3. The rights of the creator must not be affected by the restrictions.

These three tests form the legal basis for limiting the use of works related to the development of artificial intelligence (AI), including the training of large-scale AI models.

In fact, these copyright restrictions are intended to strike a fair balance between the interests of creators and users of copyrighted works. Each country may enact copyright laws to prevent the use of copyrighted works without obtaining the creator's consent or compensation. The Berne Convention for the Protection of Literary and Artistic Works stipulate that its member states must grant permission for the reproduction of literary and artistic works under certain conditions, provided that such reproduction does not interfere with the general use of the work and does not prejudice the legitimate interests of the author (Junaidi, 2025).

Article 107 sets out four non-exclusive factors that courts must consider when deciding whether the use of a copyrighted work is protected as 'fair use'. According to this provision, the use of a work is not considered copyright infringement if it is fair, which is considered an exception to the copyright holder's exclusive rights. Several factors must be considered to determine whether a particular use is reasonable in a given situation. These factors include (Aurelia, et al, 2025):

1. The target audience and usage characteristics, including whether it is for educational purposes or business interests, should be considered; This research shows that, in fair use assessments, courts are now placing more emphasis on the purpose and characteristics of the first factor's use. This is particularly evident when transformative use becomes a pivotal consideration in digital copyright disputes (Aufderheide & Jaszi, 2018). Courts usually permit the widespread use of copyrighted works if the subsequent use is not expressive and has a different purpose to that of the original work. According to this perspective, copyright protection indirectly limits the market's acquisition of works, neglecting the utilisation of the data contained within them (Gorwa, 2024). Over the past decade, an analysis of court decisions indicates that judicial interpretation has relaxed the requirements for the second and third factors but tightened scrutiny of the fourth factor, particularly in relation to advances in digital technology.
2. Characteristics of copyrighted works; Regarding the second factor, the unique characteristics of copyrighted works particularly the level of creativity are often overlooked when discussing non-commercial reproduction. According to Netanel (2008), the most basic explanation for this is that AI and search engines only process 'raw data' or 'patterns', which have lost their artistic value. Consequently, the creative aspect inherent in the original work becomes irrelevant. This cre-

ates a conceptual vacuum whereby creative works that should be protected by copyright law become irrelevant when evaluated by machines.

3. The ratio of copyrighted works used in relation to the total number of works; and The quantity and value of the shares used have changed significantly. In order to optimise the learning process in the digital world, technologies such as artificial intelligence require a comprehensive dataset of works. The legal recognition that comprehensive duplication of work for transformation purposes represents a significant reinterpretation of this factor (Ding & Yang, 2025). A comprehensive legal analysis shows that technological needs have outweighed the third factor. This demonstrates that the principle of legal compliance has evolved beyond the original regulations.
4. The impact of usage on the market prospects or value of copyrighted works Research has shown that the development of generative artificial intelligence poses the most significant economic threat. According to the fair use doctrine, the impact on the potential market is one of its key components. Besides copyright infringement, artificial intelligence is transforming the market in unprecedented ways. According to recent legal precedents, fair use analysis in the context of large language model (LLM) training is extremely difficult and highly dependent on the specific details of each case (Didsbury & Zhu, 2025). Artificial intelligence (AI) is often described as a fundamental change. However, the latest AI developments demonstrate its potential for direct market substitution (Didsbury & Zhu, 2025). A transformative paradox occurs when the use of AI is considered financially innovative (De La Durantaye, 2025). In such situations, creators must demonstrate the value of their work, which could diminish the financial incentive within the creative industry, has created a dilemma known as the 'transformative paradox'. This dilemma arises because, while a use may be conceptually considered transformative, if its output directly competes with and replaces the original work, the fair use claim could be rejected (Eyo-Udo et al., 2025). Consequently, it was agreed that plaintiffs in future cases would need to present concrete evidence of copyright-infringing output and its systemic market impact. This shifts the burden of proving financial loss to the creators of the works.

Although there are already rules regarding the four components that determine fair use, these are not consistently applied in court decisions. This is because these factors are open to interpretation based on the specific details of each case. Consequently, it is difficult to determine the relationship between them. Consequently, clear boundaries regarding the scope of fair use are difficult to establish. However, there was agreement on the 'transformative' test, which stems from the first fair use factor. According to this test, derivative works have an element of creativity or novelty. A work is considered transformative if it makes a new contribution, has a different purpose or features, and alters the original expression, meaning or message, resulting in a new artistic identity. Therefore, this test also considers the creator's intentions: did they simply want to create something new, or were they merely imitating previous works? Courts often determine reasonable use by identifying transformative components (Aurelia, et al, 2025).

The existence of the internet has made the use of artificial intelligence easier in various aspects of society; however, this has also caused problems in the legal sector (Pirmansyah, et al, 2021). Artificial intelligence has solved ethical and technical problems, as well as replacing markets. The research findings directly contradict the utilitarian theory of copyright, which states that people are motivated to create. According to this theory, legal uncertainty and high court costs reduce the desire to create (Findings). Furthermore, the concept of 'transformative use', a key component of the artificial intelligence Fair Use defence, is jeopardised by AI's capacity to memorise data (Ramadhani, 2025). Generative AI models can accurately or nearly accurately replicate copyrighted training data when asked to do so (Craig & Kerr, 2025). Having clear regulations on artificial intelligence is essential to ensure legal certainty. Often, current legal regulations do not change as quickly as society progresses, especially in this era of rapid advancement. This situation can lead to a lack of regulations (Junaidi & Fadzil, 2024). Legislation and regulations related to protection against discrimination when developing artificial intelligence systems Compliance with current laws is essential for the implementation of artificial intelligence. Legal research and the involvement of legal professionals may be necessary to ensure that systems are created and operated in accordance with applicable regulations (Junaidi, et al, 2024). Most people believe that large data sets created by artificial intelligence (AI) have the ability to provide more insightful responses than human reasoning can (Junaidi, 2023).

Currently, the Fair Use Doctrine focuses more on creating a balance of actual rights than on providing technological assistance to specific parties. Those with resources use them to fight lawsuits in order to maintain the mass reproduction of materials, while individual creators face economic hardship. If the law cannot address the systemic impact of the abundance of content generated by artificial intelligence, the sustainability of the creative industry will be threatened. Therefore, rather than simply interpreting the law on a case-by-case basis, a more comprehensive solution is needed that provides fair compensation to creators. This could be achieved by requiring licensing or amending laws to consider the economic impact of large-scale artificial intelligence systems (Morganti & Valdes, 2025). The proposed new framework emphasizes licensing obligations and fair compensation, aligning with international legal standards such as the mandatory collective rights management standard for non-commercial use. It provides a model that can be applied globally in today's digital age.

This measure is most effective in preventing artificial intelligence copyright infringement related to commercialization issues amid legal uncertainty. As a result, Indonesia must review its copyright laws and manage artificial intelligence comprehensively. According to the document, the testing procedure must be carried out in three stages (Murfianti, 2020):

1. Artworks may only be reproduced in certain situations;
2. Reproduction may be carried out without interfering with legitimate interests or in an exploitative manner;
3. Only if it does not interfere with the rights and interests of copyright holders or creators.

Legal uncertainty arises due to ambiguities in Indonesia's Copyright Law, particularly with regard to artificial intelligence. Therefore, the regulations must be amended to:

1. clarify that AI is a tool and not a legal entity with rights and responsibilities;
2. establish regulations governing the use of copyrighted works for AI instruction;
3. establish a fair licensing and compensation system.

This approach is consistent with development law theory, which considers law to be a tool for shaping society and ensuring that the rights of creators and the public interest are proportionally protected.

The Application of Fair Use to Artificial Intelligence in Copyright Law: A Development Law Perspective

With the use of modern technology, artificial intelligence has great potential to revolutionize the world. It is essential to ensure that the application of artificial intelligence is carried out morally and with the participation of all relevant parties. Using artificial intelligence to prepare for future challenges (Junaidi, 2025). Thanks to advances in artificial intelligence, there are many opportunities for change in various aspects of life, such as the social, economic and legal spheres. It is crucial that AI is developed based on ethics, accountability, and broad public participation (Roberts et al., 2021; Qian et al., 2024). Therefore, the state has a constitutional responsibility to ensure that the use of AI does not conflict with the public interest or violate fundamental rights, including intellectual property rights. Gordon (2021) also proposed that legislation should address issues such as machine bias, legal decision-making processes and legal accountability relating to this technology. Creative works depend on the characteristics formed by the human mind's ability to create, feel, and desire (Luhur, 2024). This principle is very important in the use of artificial intelligence because works created by artificial intelligence can be accessed quickly after the user gives a command or instruction. As long as the works produced by artificial intelligence are in a tangible format, such as writing, images, or music, and are accessible to the public, they can be considered to meet the criteria for protection under copyright law.

Essentially, the purpose of copyright law is to protect creative works and grant their creators full rights. According to the Berne Convention and the TRIPS Agreement, copyright arises automatically when a work is created (Pasechnyk, 2022). However, Article 1, paragraph 2 of Law No. 28 of 2014 states that the creator must be an individual or group of individuals. Therefore, even though artificial intelligence (AI) can produce publishable works such as writing, images or music, these are not automatically protected by copyright because the process must involve human intelligence to a significant degree. According to this law, copyright is automatically owned once a work is created without the need for official registration (Chai, et al, 2022).

The issue of copyright ownership of works produced by artificial intelligence is one of the main problems facing copyright recognition. Explicitly, the Copyright Law (UUHC) does not recognize artificial intelligence as a creator or copyright owner. A creator is defined as an ‘individual’ or ‘group of individuals’ who create a unique work, according to Article 1, paragraph 2 of the Copyright Law (Andani & Santoso, 2025). The limitation on copyright known as fair use reflects the social function of copyright, namely fair use or fair behavior. Certain countries allow the reproduction of works without it being considered a violation of copyright (Hanson, et al, 2022), but this must be done in a fair and reasonable manner.

According to the fair use principle, there are four things that can make an activity considered fair use. These consist of (U.S. Copyright, 2021):

1. The purpose and character of the use of the work (the purpose and character of the use, including whether it is for commercial or non-profit educational purposes);
2. The character of the use of the copyrighted work;
3. The quantity and duration of the use of the copyrighted work; and
4. The effect of the use on the market prospects and value of the copyrighted work

The court will evaluate whether the actions in a case can be considered fair use based on the four components of fair use mentioned earlier. According to the doctrine of fair use, an individual or group may use a work of art without seeking permission from the creator for specific purposes, such as news reporting, commentary or criticism, research, or other educational purposes.

According to the author, in Indonesia, copyright law does not grant legal status or copyright to artificial intelligence (AI). The main issue that arises is therefore not 'AI rights', but rather how copyrighted works can be used in the development and training of artificial intelligence. The principle of 'fair use', consisting of four considerations common in the US legal system, cannot be automatically applied in Indonesia. Indonesia has restrictive copyright limitations and must pass the three-step test of the Berne Convention, which includes the following:

1. Use is only permitted in certain situations;
2. This use must not interfere with the normal exploitation of the copyrighted work;
3. This use must not unreasonably prejudice the legitimate interests of the author.

One of the measures taken by the state to maintain legal order is the application of the fair use principle, which limits creative works produced by artificial intelligence. The state has the right to make regulations, which are then regulated in legislation. Generally, law is the result of government policies regulated by various laws. In this case, law functions as both an object of development and a subject. Part of the task of legal social oversight is to create order and truth in society (Hainadri, 2021).

Mochtar Kusumaatmadja created a theory of development law based on this idea. This theory shows a pattern of cooperation involving all stakeholders in the social community. In the process, Mochtar Kusumaatmadja emphasised that pragmatic goals (for development) are very important, as demonstrated by Roescoe Pound and Eugen Ehrlich. This highlights the connection between Laswell and McDougal's statements, which stress that legal researchers and legal practitioners must collaborate to develop legal theory (theory about law) with practical applications. Mochtar Kusumaatmadja effectively transformed the understanding of law from being a tool for building society (Yudha, et al, 2024).

Mochtar Kusumaatmadja believes that the concept of law as a tool for social change is broader in Indonesia than in the United States. In the United States, legal reform is more focused on legislation, although court decisions remain important. Mochtar explained that this approach is based on adapting to the Indonesian context. He also stated that Laswell's policy-oriented theory and Northrop's cultural philosophy had influenced this approach; both theories view law as a tool for renewal (Fadillah, 2022). The ideal law in Indonesia is one that is both relevant and accepted by society. At the heart of this theory is the idea that, as well as helping to realize and implement the values present in society, law serves to protect and maintain the progress that has been achieved. Therefore, since developing societies tend to experience rapid transformation, the law must support the dynamics of social change.

According to Mochtar Kusumaatmadja's theory of development law, to improve the quality of society, law must act as an agent of social transformation. In Indonesia, a country that is still developing, the role of law is not limited to enforcing certainty and order. On the contrary, law is expected to be a tool for social engineering, helping society to behave in accordance with the goals of development and modernization. This allows social change and progress to occur in a structured and controlled manner (Herlinda & Darmawan, 2022).

Mochtar Kusumaatmadja believes that law helps maintain social stability. Because of its function as an instrument of order, law tends to be conservative, meaning it seeks to maintain the existing state of affairs. Because all achievements must be maintained, preserved, and stored, this function is important for all societies, including developing ones. However, the law doesn't just need to be conservative for a rapidly developing society. Additionally, the law must have the capacity to help and encourage change. According to the conventional perspective, law only serves to maintain static order. They believe that law does not play a significant role in reform (Yudha, et al., 2024).

The core of this idea is that in the process of development and revitalization, order and discipline are expected, and even highly necessary. In the context of development and renewal, law as a norm is also expected to guide human actions in the desired direction. Therefore, unwritten legal regulations are needed, which must be in line with the prevailing social norms (See, 2020). Mochtar Kusumaatmadja (1970) defines law as a form of social regulation alongside religion, tradition and social norms. Like other social rules, law serves as a moral guide for individuals. However, it is different in that it is binding and coercive, and can be enforced through a formal process. Furthermore, specific provisions exist regarding the types, techniques and tools used for legal enforcement. Various changes that have been made to the legal system demonstrate this. One example is the legislation that was created to meet the demands of development and social justice. As power is coercive, it is an essential element of a legal system. However, this power must also be regulated by law. As French philosopher Blaise Pascal said, 'Justice without power is powerless; power without justice is tyranny.' This illustrates the relationship between law and power (Jerphagnon, 2025).

According to the author, this theory essentially shows that the law helps to preserve what has already been achieved. Law is also a representation and implementation of the values that develop within the structure of society. As society is undergoing rapid change, the government must support the process of social transformation. According to Mochtar Kusumaatmadja's theory of development law, law should help transform society for the better. There is a possibility that law will function as a 'tool of social engineering.' Regulations are intended to help human activities achieve their goals in the process of development and improvement (Kusumaatmadja, 1986).

In further explanation, he said that the main purpose of law is not only to create order. There are other purposes of law to achieve justice that differ according to context and scale, in accordance with the development of society in a particular place and time. Therefore, although legal certainty is very important, we also need to understand and study the concept of justice that develops in society (Kusumaatmadja, 1990). According to Prof. Mochtar Kusumaatmadja (1995), based on the idea of 'law as a tool for social engineering,' there are two main dimensions in development law theory. The first is the order or regularity required in the development process, which is considered desirable and even necessary. The second dimension is the role of law as a regulatory tool or a means for development, serving to guide development.

Roscoe Pound's view of 'law as a means of social engineering', which was popular in the United States, is similar (Shidarta, 2006). However, Mochtar Kusumaatmadja redefined law as a tool for societal progress (Kusumaatmadja, 1986). This is based on the belief that law can function as a regulator in the development process and be used to encourage progress and renewal in human behavior (Kusumaatmadja, 1986). According to Mochtar, the concept of law as a means is more comprehensive than the concept of law as a tool for several reasons (Shidarta, 2006):

1. Lawmaking plays a greater role in legal reform in Indonesia than jurisprudence does in the United States, where court decisions are prioritized.
2. The view of law as a 'tool' can produce consequences similar to the 'legalism' system that existed during the Dutch colonial era, which Indonesian society may reject due to its memories of colonialism.
3. The concept of law as a tool for societal renewal has been demonstrated through the application of international law, even before it was officially recognised as the basis for national legal policy.

The role of law as a tool for social engineering is very important for renewing society, and Indonesia, as a developing country, is in dire need of a good legal system (Gunawan & Arumbinang, 2023). In a community, the two main objectives of law are security and justice. Law not only aims to maintain order, but also to realise justice in various forms and qualities appropriate to society and the times. Mochtar Kusumaatmadja said that law not only helps maintain order but also helps renew society. This concept stems from the idea that law must be used to ensure that change and progress in society occur systematically (Marif & Nurhaedah, 2024). Thus, legislation can serve as a catalyst and catalyst for the revitalization of Indonesian society, using law as a proactive and active tool for renewing society rather than merely as a tool for social change.

Mochtar Kusumaatmadja's legal reforms focus on change through lawmaking or legislation (*Gesetzgebungslehre*). Consequently, any reform efforts must be grounded in legal principles rather than other disciplines such as sociology. He believes that current legislation should play a more active role in driving social change and helping the government to achieve its national development goals. Therefore, adequate regulation is crucial in development law, so that law can function as a tool to improve society especially with regard to achieving development and effective law enforcement must support this.

According to Kusumaatmadja's theory of digital development, the role of law must evolve from merely regulating society to forming an essential foundation of the digital world. This approach focuses not only on applying rules to regulate social behavior, but also on the importance of law in establishing social structures and technologies that ensure fair access, data protection and the equitable use of algorithms. Without significant changes from this perspective, Indonesian law could become an outdated collection of rules irrelevant to today's digital society (Risnandar & Zaenudin, 2025). As part of development efforts, changes to Indonesian law require a shift from using law as a policy tool to upholding the rule of law based on the constitution, where law serves as a guarantee of justice for all segments of society, not just a means to achieve specific development goals. Three main components can contribute to reconstructing this legal system: lawmaking involving community participation; judicial processes supporting progress; and responsive public service administration (Risnandar & Zaenudin, 2025).

Based on Mochtar Kusumaatmadja's views, it can be concluded that a legal theory of development is necessary, one that views law as a tool for creating social change rather than merely maintaining order. The laws required in relation to the development of artificial intelligence (AI) are therefore as follows:

1. Predicting and overcoming the impact of technology;
2. Providing certainty to all parties;
3. Balancing technological progress with copyright protection.

Therefore, copyright and AI regulations should be considered a strategy for engineering a society that encourages innovation while ensuring fairness and legal certainty.

The author argues that it is crucial to emphasize the existence of agreements and contracts governing interactions between developers, users, and artificial intelligence, as the implementation of development law theory can only be achieved through the process of lawmaking, court rulings, or both. The aim is to ensure that all parties understand their rights and responsibilities regarding the works created. Ultimately, it is essential to educate people about copyright and the laws related to artificial intelligence use. Governments, schools, and relevant organizations should conduct training courses explaining copyright, the risks of infringement, and the responsibilities associated with artificial intelligence use. By increasing knowledge about these issues, it is hoped that an ecosystem will be created that supports innovation and copyright protection while reducing the likelihood of future conflicts.

According to the author, in the future, products created by Artificial intelligence can be legally recognized if they meet certain conditions. First and foremost, the results produced by Artificial intelligence require human intelligence as a component. Second, artificial intelligence products must be completely original. Third, the work is not just a prototype or an unfinished idea, but has been created in a digital form that is accessible to the public. Finally, products created by Artificial intelligence fall under the category of intellectual property rights (IPR) protection, similar to works of art. If all four main criteria are met in an artificial intelligence-created work, the creator or rights holder may claim economic rights over the artificial intelligence-generated product.

In order to ensure that the law keeps pace with the digital age without compromising human copyright, the following measures must be taken, based on the theory of development law:

1. It must be reaffirmed that artificial intelligence (AI) acts only as an instrument and not as the source of the work.
2. Clear regulations are needed to govern the use of copyrighted works during the AI training process.
3. A fair licensing and compensation system is needed.
4. The general public must be educated about copyright and AI.

This approach guarantees that the law remains relevant in the computer and internet era while safeguarding the rights of human creators.

CONCLUSION

Artificial intelligence (AI) raises many copyright issues, particularly with regard to ownership of works, copyright holders, and the permitted use of protected works. As copyright protection requires human creativity and thought, AI cannot currently be considered a creator or granted legal status under Indonesian law. Furthermore, Indonesia lacks specific regulations governing the use of copyrighted works in the development and training of AI, creating legal uncertainty for creators and technology developers. Unlike in the United States, however, they do not apply the broad principle of fair use.

According to Mochtar Kusumaatmadja's theory of development law, the law should be a tool for social transformation capable of keeping pace with technological developments while upholding justice and legal certainty. Consequently, Indonesian copyright regulations must be amended to clarify that AI is merely a tool, restrict the use of copyrighted works for the development of AI, and establish a fair licensing and compensation system for creators. In line with the objectives of national legal development and the noble values of Pancasil, these amendments are expected to strike a balance between protecting creators' rights and encouraging technological innovation.

REFERENCES

Adu, T. L., & Van der Walt, T. B. (2021). The level of awareness and understanding of copyright laws and policies among academic librarians in Ghana. *Journal of Academic Librarianship, 47*(3), 102317. DOI: 10.1016/j.acalib.2021.102317

Aini, F. N., & Wauran, I. (2021). Pemenuhan Prinsip Fair Use dalam Cover Lagu Berdasar Hukum Hak Cipta Indonesia. *Jurnal Ilmiah Kebijakan Hukum, 15*(1), 111–132. Advance online publication. DOI: 10.30641/kebijakan.2021.V15.111-132

Amankwah-Amoah, J., Abdalla, S., Mogaji, E., Elbanna, A., & Dwivedi, Y. K. (2024). The impending disruption of creative industries by generative AI: Opportunities, challenges, and research agenda. *International Journal of Information Management, 79*(102759), 102759. Advance online publication. DOI: 10.1016/j.ijinfomgt.2024.102759

Anantrasirichai, N., & Bull, D. (2022). Artificial intelligence in the creative industries: A review. *Artificial Intelligence Review, 55*(1), 589–656. DOI: 10.1007/s10462-021-10039-7 PMID: 41625196

Andani, A.P., & Santoso, B. (2025), "Perlindungan Hukum Terhadap Pelanggaran Karya Cipta yang Dihasilkan Melalui AI dan dikomersilkan Berdasarkan Undang-Undang Nomor 28 Tahun 2014 tentang Hak Cipta", *JIHHP: Jurnal Ilmu Hukum, Humaniora dan Politik*, 5(3), 2274-2281, DOI: 10.38035/jihhp.v5i3.4039

Anifa, Y. (2024). Dampak Kecerdasan Buatan dalam Mempertimbangkan Aspek Sosial dan Budaya dalam Desain Grafis. *VCoDe: Visual Communication Design Journal, 3*(2), 145–156. DOI: 10.26887/vcode.v3i2.4256

Appel, G., Neelbauer, J., & Schweidel, D. A. (2023). Generative ai has an intellectual property problem. Harvard Business Review. Retrieved from https://hbr.org/2023/04/generative-ai-has-an-intellectual-property-problem

Aufderheide, P., & Jaszi, P. (2018). *Reclaiming fair use: How to put balance back in copyright* (2nd ed.). University of Chicago Press. DOI: 10.7208/chicago/9780226374222.001.0001

Aurelia, C. N.. (2025). Enhancing Fair Use in Protecting Appropriated Artworks: A Comparative Analysis of Safeguarding Indonesian Copyright Law. *Lex Scientia Law Review, 9*(1), 1181–1222. DOI: 10.15294/lslr.v9i1.20570

Bonadio, E., & McDonagh, L. (2020). Artificial intelligence as producer and consumer of copyright works: evaluating the consequences of algorithmic creativity. *Intellectual Property Quarterly, 2020*(2), 112-137. https://ssrn.com/abstract=3617197

Chai, Y. X., Ren, S. J. F., & Zhang, J. Q. (2022). The impacts of digital content piracy and copyright protection policies when consumers are loss averse. *Journal of Industrial and Management Optimization, 18*(5), 3587–361. DOI: 10.3934/jimo.2021127

Craig, C., & Kerr, I. (2025). The Death of The AI Author. In *Robot Law* (Vol. II, pp. 250–285). Edward Elgar Publishing., DOI: 10.4337/9781800887305.00014

Crawford, K., & Schultz, J. (2024). The Work of Copyright Law in the Age of Generative AI. *Grey Room, 94*(94), 56–62. DOI: 10.1162/grey_a_00389

Daffa, B.M. (2024). "Aspek Hukum Penggunaan Metode Stable Diffusion Oleh Artificial Intelligence Terhadap Suatu Ciptaan Berdasarkan Undang-Undang Nomor 28 Tahun 2014". *COMSERVA: Jurnal Penelitian dan Pengabdian Masyarakat,* 3(11), 4434-4442, DOI: 10.59141/comserva.v3i11.1254

De la Durantaye, K. (2025). Control and Compensation. A Comparative Analysis of Copyright Exceptions for Training Generative AI. *IIC; International Review of Industrial Property and Copyright Law, 56*(4), 737–770. DOI: 10.1007/s40319-025-01569-6

Didsbury, H., & Zhu, X. A. (2025). Transformative training: an analysis of AI training data and fair use in *Authors Guild v. OpenAI Inc. Publishing Research* 4: e002. DOI: 10.48130/pr-2025-0001

Ding, T., & Yang, L. (2025). Intellectual Property Protection and Corporate Digital Transformation: An Empirical Analysis from The Perspectives of Intellectual Property Protection and Digital Governance. *International Review of Economics & Finance, 100*(104125), 1–11. DOI: 10.1016/j.iref.2025.104125

Dwivedi, Y. K., Kshetri, N., Hughes, L., Slade, E. L., Jeyaraj, A., Kar, A. K., Baabdullah, A. M., Koohang, A., Raghavan, V., Ahuja, M., Albanna, H., Albashrawi, M. A., Al-Busaidi, A. S., Balakrishnan, J., Barlette, Y., Basu, S., Bose, I., Brooks, L., Buhalis, D., & Wright, R. (2023). Opinion Paper: "So what if ChatGPT wrote it?" Multidisciplinary perspectives on opportunities, challenges and implications of generative conversational AI for research, practice and policy. *International Journal of Information Management, 71*(102642), 1–63. DOI: 10.1016/j.ijinfomgt.2023.102642

Erdozain, J. C. (2019). How will the IP world respond to the rise of AI. Managing Intell. Prop., 281, 56. https://www.managingip.com/article/2a5bqo2drurt0bwowm4jf/how-will-the-ip-world-respond-to-the-rise-of-ai

Eyo-Udo, N. L., Apeh, C. E., Bristol-Alagbariya, B., Udeh, C. A., & Ewim, C. P.-M. (2025). International Trade Law in the Modern World: A Review of Evolving Practices and Agreements. *International Journal of Multidisciplinary Research and Growth Evaluation*, *6*(1), 840–847. DOI: 10.54660/.IJMRGE.2025.6.1.840-847

Fadillah, N. (2022). Tinjauan Teori Hukum Pembangunan Mochtar Kusumaatmadja dalam Undang-Undang Ibu Kota Negara (IKN). *Supremasi Hukum: Jurnal Kajian Ilmu Hukum*, *11*(1), 45–65. DOI: 10.14421/sh.v11i1.2559

Fadillah, R. N. (2024). Perlindungan Hak Atas Kekayaan Intelektual Artificial Intelligence (AI) dari Perspektif Hak Cipta dan Paten. *Das Sollen: Jurnal Kajian Kontemporer Hukum dan Masyarakat*, 2(2), 1-25. http://dx.doi.org/DOI: 10.11111/dassollen.xxxxxxx

Fadli, M. A., Hamidi, M. P., Edwin, F. A., & Aritonang, R. G. K. (2023). Let's Play Content as a Fiduciary Collateral under Indonesian Law: Potential Challenges. Yuridika, 38(3), 481-498.

Gans, J. S. (2024). Copyright policy options for generative artificial intelligence (No. w32106). National Bureau of Economic Research. Retrieved from https://www.nber.org/papers/w32106

Geiger, C. (2024). Elaborating a Human Rights-Friendly Copyright. *IIC; International Review of Industrial Property and Copyright Law*, *55*(6), 1129–1164. DOI: 10.1007/s40319-024-01481-5

Gervais, D. J. (2020). "The machine as author". *Iowa Law Review,* 105(5), 2053, https://ilr.law.uiowa.edu/print/volume-105-issue-5/the-machine-as-author dan lihat juga P Bernt Hugenholtz and João Pedro Quintais, Copyright and Artificial Creation: Does EU Copyright Law Protect AI-Assisted Output? *IIC - International Review of Intellectual Property and Competition Law* (2021), DOI: 10.1007/s40319-021-01115-0

Gordon, J. S. (2021). AI and law: Ethical, legal, and socio-political implications. *AI & Society*, *36*(2), 403–404. DOI: 10.1007/s00146-021-01194-0

Gorwa, R. (2024). *The Politics of Platform Regulation: How Governments Shape Online Content Moderation.* (Oxford Studies in Digital Politics). New York, NY: Oxford University Press. DOI: 10.1093/oso/9780197692851.001.0001

Gunawan, Y., & Arumbinang, M. H. (2023). The Climate Change Litigation Based Human Rights Approach in Corporations: Prospects and Challenges. *Journal of Human Rights, Culture and Legal System*, *3*(2), 288–307. DOI: 10.53955/jhcls.v3i2.116

Habibi, M.. (2025). Tinjauan Yuridis tentang Doktrin Fair Use dalam Undang-Undang Hak Cipta Indonesia dan Implikasinya terhadap Pengembangan Software. *Rewang Rencang: Jurnal Hukum Lex Generalis*, *6*(4), 1–18. DOI: 10.56370/jhlg.v6i4.860

Hainadri, H. D. (2021). Hukum Sebagai Sarana Pembaharuan Dalam Masyarakat (*law as a tool of social engineering*). *Datin Law Jurnal*, *2*(1), 66–75. DOI: 10.36355/dlj.v2i1.566

Hanson, N. A., Permata, R. R., & Ramli, T. S. (2022). Protection of Video Copyrights on the Youtube Platform in Fair Use Principles. *Jurnal Legalitas*, *15*(2), 108–121. DOI: 10.33756/jelta.v15i2.15448

Herlindah, H., & Darmawan, Y. (2022). Development Legal Theory and Progressive Legal Theory: A Review, in Indonesia's Contemporary Legal Reform. *Peradaban Journal of Law and Society*, *1*(1), 15–27. DOI: 10.59001/pjls.v1i1.22

Hildebrandt, M. (2020). The Artificial Intelligence of European Union Law. *German Law Journal*, *21*(1), 74–79. DOI: 10.1017/glj.2019.99

Iaia, V. (2022). To be, or not to be … original under copyright law, that is (one of) the main questions concerning AI-produced works. *GRUR International: Journal of European and International IP Law*, *71*(9), 793–812. DOI: 10.1093/grurint/ikac087

Israhadi, E. I. (2023). The Impact of Developments in Artificial Intelligence on Copyright and other Intellectual Property Laws. *Journal of Law and Sustainable Development*, *11*(11), 1–22. DOI: 10.55908/sdgs.v11i11.1965

Jerphagnon, L. (2025). Blaise Pascal French philosopher and scientist. https://www.britannica.com/biography/Blaise-Pascal

Jodha, D., & Bera, P. (2023). Copyright issues in the era of Ai-A critical analysis. *Res Militaris*, *13*(3), 1737–1748. https://resmilitaris.net/uploads/paper/4b260afa4323eb024822083157506b97.pdf

Junaidi, J. (2020). Kajian Yuridis Pasal 362 Kitab Undang-Undang Hukum Pidana dan Pasal 30 Undang-Undang Nomor 11 tahun 2008 tentang Informasi dan Transaksi Elektronik Terhadap Pencurian Internet Wifi. *Jurnal Wajah Hukum*, *4*(2), 325–330. DOI: 10.33087/wjh.v4i2.190

Junaidi, J. (2023). "Artificial Intelligence in The Implementation of Criminal Case ettlement Law Using Restorative Justice Based on Progressive Law". *ICL-C Proceedings of the First International Cyber Law Conference*, ICL-C 2023, 198-204, DOI: 10.4108/eai.11-11-2023.2351325

Junaidi, J. (2025), "Transformation of Literary Culture in the Digital Era Through the Use of Artificial Intelligence as a Form of Copyright Protection", *in Role of AI in Producing, Evaluating, and Interpreting Literary Materials, IGI Global Scientific Publishing*, 177-204, DOI: 10.4018/979-8-3373-1057-2

Junaidi, J. (2025). Education Reform in the Application of Artificial Intelligence in Supporting Learning Media: AI and Education for Sustainable Development Goals (SDGs). In *AI, Policy, and the Future of Human-Centered Education* (pp. 57–84). IGI Global Scientific Publishing., DOI: 10.4018/979-8-3373-5781-2.ch003

Junaidi, J., & Fadzil, R. M.Rozlinda Mohamed Fadzil. (2024). Urgency of Legal Personal Data Protection in E-Commerce Transactions Involving Artificial Intelligence. *Walisongo Law Review (Walrev)*, *6*(2), 96–108. DOI: 10.21580/walrev.2024.6.2.25548

Junaidi, J., Pujiyono, P., & Fadzil, R. M. (2024). Legal Reform of Artificial Intelligence's Liability to Personal Data Perspectives of Progressive Legal Theory. *Journal of Law and Legal Reform*, *5*(2), 587–612. DOI: 10.15294/jllr.vol5i2.3437

Keller, P. (2023), A first look at the copyright relevant parts in the final AI Act compromise. Kluwer Copyright Blog. https://copyrightblog.kluweriplaw.com/2023/12/11/a-first-look-at-the-copyrightrelevant-parts-in-the-final-ai-act-compromise/

Kumari, N., Sharma, R., & Priyanka, . (2020). Intellectual property right (IPR). *Asian Journal of Pharmacy and Technology*, *10*(4), 250–254. DOI: 10.5958/2231-5713.2020.00042.2

Kusumaatmadja, M. (1970). *Fungsi dan Perkembangan Hukum dalam Pembangunan Nasional*. Fakultas Hukum, Universitas Padjadjaran.

Kusumaatmadja, M. (1970). *Fungsi dan perkembangan hukum dalam pembangunan nasional. Lembaga Penelitian Hukum dan Kriminologi*. Fakultas Hukum, Universitas Padjadjaran.

Kusumaatmadja, M. (1976). *Hukum, masyarakat dan pembinaan hukum nasional*. Binacipta.

Kusumaatmadja, M. (1986). *Pembinaan Hukum Dalam Rangka Pembangunan Hukum Nasional*. Lembaga Penelitian Hukum Dan Kriminologi Fakultas Hukum Universitas Padjajaran Cetakan Kedua.

Lee, K., Cooper, A. F., & Grimmelmann, J. (2024). "Talkin''Bout AI Generation: Copyright and the Generative-AI Supply Chain (The Short Version)". *In Proceedings of the Symposium on Computer Science and Law*, 48-63. DOI: 10.1145/3614407.3643696

Lindsey, T.. (2022). *Hak Kekayaan Intelektual, Suatu Pengantar Edisi Kedua.* Alumni.

Liu, Y. C., Kuo, C. H., Wang, G. H., ... (2023). "Dialogues of Liu, Kuo & Wang creation: Collaborative content generation by the human author and ChatGPT and its impact on the evolving intellectual property landscape". In E. Y. Li, et al. (Eds23). *Proceedings of The International Conference on Electronic Business*, 23, 166-185. ICEB'23, Chiayi, Taiwan, October 19-23, 2023.

Longan, M. (2023). A System Out of Balance: A Critical Analysis of Philosophical Justifications for Copyright Law Through the Lenz of Users' Rights. *University of Michigan Journal of Law Reform. University of Michigan. Law School, 56*(3), 779–826. DOI: 10.36646/mjlr.56.3.system

Lucchi, N. (2023). ChatGPT: A case study on copyright challenges for generative AI systems. *European Journal of Risk Regulation, 15*(3), 1–23. DOI: 10.1017/err.2023.59

Luhur, W. H. (2024). Penerapan Teknologi Kecerdasan Buatan dalam Proses Penciptaan Karya [IKJ]. *Seminar Nasional Institut Kesenian Jakarta, 2*, 325–335. DOI: 10.52969/seminarikj.v2i.77

Maidanyk, L. (2021). Artificial Intelligence and Sui Generis Right: A Perspective for Copyright of Ukraine. *Justice in Eastern Europe, 3*(11), 144–154. DOI: 10.33327/AJEE-18-4.3-n000076

Manullang, S.O., et al., (2023), "Limits of the Concept of Fair Use in Law Number 28 of 2014 Concerning Copyright," *Jurnal Hukum dan HAM Wara Sains,* 2(3), 181-187, DOI: 10.58812/jhhws.v2i03.251

Marif, M., & Nurhaedah, N. (2024). Teori Hukum Pembangunan dalam Eksistensi Mediasi Penal di Tingkat Penuntutan dalam Mewujudkan Penegakan Hukum Yang Berkeadilan. *Indonesian Journal of Legality of Law, 6*(2), 217–224. DOI: 10.35965/ijlf.v6i2.4529

Morganti, P., & Valdes, R. C. (2025). Toward Fair Futures? Digital Transformation and Income Disparities in the Era of Industry 5.0. Springer Books. In Teran-Bustamante, A., & Salazar, M. V. (Eds.), *Digital Transformation and Innovation in Organizations* (pp. 45–76). Springer., DOI: 10.1007/978-3-031-90794-4_3

Muhaimin, (2020). *Metode Penelitian Hukum,* Mataram: Mataram University Press.

Murfianti, F. (2020). Hak Cipta dan Karya Seni di Era Digital. *Acintya Jurnal Penelitian Seni Budaya, 12*(1), 44–58. DOI: 10.33153/acy.v12i1.3147

Nafila Andriana Putri, N. A. (2023). Copyright Protection for Internet Memes: The Doctrine of Fair Use in Indonesia. *JIPRO: Journal of Intellectual Property*, *6*(2). Advance online publication. DOI: 10.20885/jipro.vol6.iss2.art3

Naqvi, Z. (2020). Artificial intelligence, copyright, and copyright infringement. Marq. Intell. Prop. L. Rev., 24, 15. https://scholarship.law.marquette.edu/iplr/vol24/iss1/4/

Netanel, N. W. (2008). *Introduction, Copyright's Paradox*. Oxford University Press, UCLA School of Law Research Paper No. 08-06, Available at *SSRN*: https://ssrn.com/abstract=1099457

Pasechnyk, O. (2022). International aspects of intellectual property rights protection. *Baltic Journal of Economic Studies*, *8*(5), 146–157. DOI: 10.30525/2256-0742/2022-8-5-146-157

Puspasari, A. T., & Sardjono, A. (2023). Pembatasan Hak Cipta Terkait Remix Lagu Berdasarkan Doktrin Fair Use dan Undang- Undang Nomor 28 Tahun 2014 Tentang Hak Cipta. *Technology and Economics Law*, *2*(2), 265–281. DOI: 10.21143/TELJ.vol2.no2.1040

Putro, H. P.. (2023). *Development Of Artificial Intelligence Applications (Studi Kasus & Implementasi AI Menggunakan Berbagai Bahasa Pemrograman*. Jambi: PT. Sonpedia Publishing Indonesia. Qian, Y., Siau, K.L. & Nah, F.F., "Societal impacts of artificial intelligence: Ethical, legal, and governance issues". *Societal Impacts*, *3*(1000040), 1–5. DOI: 10.1016/j.socimp.2024.100040

Raharjo, B. (2023). *Teori Etika Dalam Kecerdasan Buatan (AI)*. Penerbit Yayasan Prima Agus Teknik.

Rahmayanti, S. D., Rismana, V., Salsabila, Y., & Syechnas, N. Z. Z. (2024). Pertarungan Hukum Hak Cipta di Era Digital: Studi Kasus dalam Konteks Indonesia. *Jurnal Dunia Ilmu Hukum*, *2*(1), 16–21. DOI: 10.59435/jurdikum.v2i1.372

Rama, B. G. A., Prasada, D. K., & Mahadewi, K. J. (2023). Urgensi Pengaturan *Artificial Intelligence* (AI) dalam Bidang Hukum Hak Cipta di Indonesia. *Jurnal Rechten*, *12*(2), 209–224. DOI: 10.56013/rechtens.v12i2.2395

Ramadhani, M. (2025). The fair use doctrine in the digital age: Reassessing the balance between creators' rights and public access. *Journal of Law Science*, *7*(4), 733–742. DOI: 10.35335/jls.v7i4.6749

Ramadika, I. K. G. A., & Wijaya, I. K. K. A.I Ketut Gede Adi RamadikaI Ketut Kasta Arya Wijaya. (2025). Legality of Copyright Protection on *Artificial Intelligence* Works. *Pemuliaan Keadilan*, *2*(1), 105–114. DOI: 10.62383/pk.v2i1.460

Ramli, A. M. (2021). "Pentingnya Bangsa Ini Mulai Adaptif Dengan Industri 5.0", https://www.republika.co.id/berita/qynv2n371/pentingnya-bangsa-ini-mulai-adaptif-dengan-industri-50

Redi Pirmansyah, R., Junaidi, J., & Merta, M. M. (2021). OSC 2021-FHISIP Universitas Terbuka, Indonesia Theme 1: "*Law Transformation the Law of Personal Data Protection in The Pandemic Time Covid-19 On Electronic System*". https://www.cips-indonesia.org/post/siaran-pers-kebocoran-data-ingatkan-pentingnya-legislasi-ruu-perlindungan-data-pribadi

Rejeki, H.P., Iriyanti & Dadang. (2023). "Pelanggaran Hak Ciptapada Penjualan Buku Hasil Bajakan Berdasarkan Undang-Undang Nomor 28 tahun 2014 tentang Hak Cipta", *Jurnal Surya Kencana Dua: Dinamika Masalah Hukum dan Keadilan*, 10(1), 52-68, DOI: 10.32493/SKD.v10i1.y2023.32254

Ricketson, S. (2003). *SCCR/9/7: WIPO Study on Limitations and Exceptions of Copyright and Related Rights in The Digital Environment.*

Ricketson, S., & Ginsburg, J. (2022). *International Copyright and Neighbouring Rights: The Berne Convention and Beyond (3rd edn).* Oxford Law Pro. 891, DOI: 10.1093/oso/9780198801986.001.0001

Risnandar, A. & Zaenudin, K.M.I.S., "Development and Evolution of Indonesian Law from the Perspective of Development Law Theory", *Rechtsvinding*, 3(2), 9-24. https://doi.org/DOI: 10.59525/rechtsvinding.v3i2.893

Roberts, H., Cowls, J., Morley, J., Taddeo, M., Wang, V., & Floridi, L. (2021). The Chinese approach to artificial intelligence: An analysis of policy, ethics, and regulation. *AI & Society*, *36*(1), 59–77. DOI: 10.1007/s00146-020-00992-2

Rochim, A. A. (2024). Kecerdasan Buatan: Resiko, Tantangan dan Penggunaan Bijak pada Dunia Pendidikan. *Antroposen: Journal of Social Studies and Humaniora*, *3*(1), 13–25. DOI: 10.33830/antroposen.v3i1.6780

Saber, A., Shaheen, N., & Giammalva, S. (2024). "6 AI Cases And What They Mean For Copyright Law," Law360 (New York, 2024), https://www.crowell.com/a/web/7QtNejMH1FSM1n5Ddt6cdU/6-ai-cases-and-what-they-mean-for-copyright-law.pdf

Saputra, A. M. A., et al. (2023). *Teknologi Informasi: Peranan TI dalam berbagai Bidang*. Jambi: PT. Sonpedia Publishing Indonesia.

See, B. R. (2020). Peranan Hukum sebagai Sarana Pembaharuan Masyarakat Menuju Masyarakat Indonesia yang Sadar Hukum. *Jurnal Hukum Caraka Justitia*, *1*(1), 42–50. DOI: 10.30588/jhcj.v1i1.700

Shefan, A. (2023). Copyright and Its Exceptions and Limitations in Ukraine. In Richter, H. (Ed.), *Competition and Intellectual Property Law in Ukraine. MPI Studies on Intellectual Property and Competition Law* (Vol. 31, pp. 507–536). Springer., DOI: 10.1007/978-3-662-66101-7_20

Shidarta. (2006). *Karakteristik Penalaran Hukum Dalam Konteks Ke-Indonesiaan.* Jakarta: CV Utomo.

Singh, S., & Singh, A. (2023). "Intellectual Property Rights and Artificial Intelligence: Contemporary Convergence and Probable Challenges", *2023 10th IEEE Uttar Pradesh Section International Conference on Electrical, Electronics and Computer Engineering (UPCON)*, Gautam Buddha Nagar, India, 1279-1286, DOI: 10.1109/UPCON59197.2023.10434754

Song, S. (2023). *China's First Case on Copyrightability of AI-Generated Picture.* https://www.kwm.com/cn/en/insights/latest-thinking/china-s-first-case-on-copyrightability-of-ai-generated-picture.html

Soraya, J., & Althafzufar, M. A. (2024). Intellectual Property Rights Protection for Actors in the Creative Economy Based on Intellectual Property Rights Law Number 28 of 2014 Concerning Copyright. *Realism: Law Review*, *2*(1), 39–53. DOI: 10.71250/rlr.v2i1.30

Spindler, G. (2019), "Copyright Law and Artificial Intelligence," *IIC - International Review of Intellectual Property and Competition Law* 50, 1049-1051, https://doi.org/https://dx.doi.org/10.18196/iclr.v5i2.17927

United States Copyright Office. (2021). *Copyright Law of the United States: and Related Laws Contained in Title 17 of the United States Code*, (P. 19-20).

U.S. Copyright Office Guidelines. (2021) Compendium of U.S. Copyright Office practices, 3rd edn. https://www.copyright.gov/comp3/docs/compendium.pdf

Vakili, S. (2025). A Study on the Intellectual Property Challenges of AI-Based Algorithms in International Law. *Legal Studies in Digital Age*, *4*(2), 1–8. DOI: 10.61838/kman.lsda.134

World Intellectual Property Organization. (2016). *Understanding Copyright and Related Rights.* (P. 5).

Yudha, F. M.. (2024). Kajian Deskriptif Analitis tentang Teori Hukum Pembangunan Prof. Dr. Mochtar Kusumaatmadja, S.H., LLM. *Ensiklopedia of Journal*, *7*(1), 362–368. DOI: 10.33559/eoj.v7i1.2627

Compilation of References

. Dastin, J. (2018, October 11). Amazon scraps secret AI recruiting tool that showed bias against women. Reuters.

. Hilty, R., Hoffmann, J., & Scheuerer, S. (2020). *Intellectual property justification for artificial intelligence.*

. Kaur, J., 2016. Intellectual Property Law in times of artificial intelligence: is it a misnomer to consider the Bot a possible IP right holder? Journal of legal studies and research, 2(6)

. Kossow, N., & Transparency International. (2021). Algorithmic transparency and accountability: Tools for practitioners. Transparency International Knowledge Hub.

. Laukyte, M., & Lucchi, N. (2022). Creative AI: The complex relationship between human inventiveness and intellectual property. *BioLaw Journal – Rivista di Bio-Diritto*, (3), 169–183.

. The Copyright, Designs and Patents Act, 1988, s.178

. Vesala, J. (2023). Developing artificial intelligence-based content creation: Are EU copyright and antitrust law fit for purpose? *IIC – International Review of Intellectual Property and Competition Law, 54*(3), 351–380. https://doi.org/DOI: 10.1007/s40319-023-01301-2

Aayog, N. I. T. I. (2018). *National strategy for artificial intelligence: AI for all.* Government of India. https://www.niti.gov.in

Aayog, N. I. T. I. (2021). *Responsible AI for all: Operationalizing principles for responsible AI.* Government of India. https://www.niti.gov.in

Abbott, R. (2020). *The reasonable robot: Artificial intelligence and the law.* Cambridge University Press., https://papers.ssrn.com/sol3/papers.cfm?abstract_id=3611370 DOI: 10.1017/9781108631761

Action, C. O. S. T. (2020). *Global digital human rights network* (CA19143). European Cooperation in Science and Technology. https://www.cost.eu/actions/CA19143/

Adam, M., & Hockuard, C. (2023). *Artificial intelligence, democracy and elections*. European Parliamentary Research Service. https://www.europarl.europa.eu/RegData/etudes/BRIE/2023/751478/EPRS_BRI(2023)751478_EN.pdf

Adepoju, D. A., & Adepoju, A. G. Establishing Ethical Frameworks for Scalable Data Engineering and Governance in AI-Driven Healthcare Systems.

Adu, T. L., & Van der Walt, T. B. (2021). The level of awareness and understanding of copyright laws and policies among academic librarians in Ghana. *Journal of Academic Librarianship*, *47*(3), 102317. DOI: 10.1016/j.acalib.2021.102317

AI Now Institute. (2023). *Algorithmic accountability: Moving beyond audits*. AI Now Institute.

Aini, F. N., & Wauran, I. (2021). Pemenuhan Prinsip Fair Use dalam Cover Lagu Berdasar Hukum Hak Cipta Indonesia. *Jurnal Ilmiah Kebijakan Hukum*, *15*(1), 111–132. Advance online publication. DOI: 10.30641/kebijakan.2021.V15.111-132

Akerkar, R. (2026). Blueprints for Intelligence: Building Robust AI Systems. In *Artificial Intelligence: Transcending Traditional Paradigms* (pp. 447-491). Cham: Springer Nature Switzerland. DOI: 10.1007/978-3-031-91084-5_12

Alanoca, S., Gur-Arieh, S., Zick, T., & Klyman, K. (2025, June). Comparing Apples to Oranges: A Taxonomy for Navigating the Global Landscape of AI Regulation. In *Proceedings of the 2025 ACM Conference on Fairness, Accountability, and Transparency* (pp. 914-937). DOI: 10.1145/3715275.3732059

Al-Emran, M., Abu-Hijleh, B., & Alsewari, A. A. (2025). Examining the impact of Generative AI on social sustainability by integrating the information system success model and technology-environmental, economic, and social sustainability theory. *Education and Information Technologies*, *30*(7), 9405–9426. DOI: 10.1007/s10639-024-13201-0

Ali, A. (2025). Ethical Considerations and Bias in AI Applications in Healthcare. In *Artificial Intelligence in Healthcare Information Systems—Security and Privacy Challenges* (pp. 129–146). Springer Nature Switzerland., DOI: 10.1007/978-3-031-84404-1_7

Ali, H., & Aysan, A. F. (2025). Ethical dimensions of generative AI: A cross-domain analysis using machine learning structural topic modeling. *International Journal of Ethics and Systems*, *41*(1), 3–34. DOI: 10.1108/IJOES-04-2024-0112

Al-Kfairy, M., Mustafa, D., Kshetri, N., Insiew, M., & Alfandi, O. (2024, September). Ethical challenges and solutions of generative AI: An interdisciplinary perspective. In *Informatics* (Vol. 11, No. 3, p. 58). Multidisciplinary Digital Publishing Institute. DOI: 10.3390/informatics11030058

Almada, M. (2025). Artificial intelligence regulation and governance in the European Union. *Computer Law & Security Review*, *49*, 105758. DOI: 10.1016/j.clsr.2024.105758

Almeida, D., Shmarko, K., & Lomas, E. (2022). The ethics of facial recognition technologies, surveillance, and accountability in an age of artificial intelligence: A comparative analysis of US, EU, and UK regulatory frameworks. *AI and Ethics*, *2*(3), 377–387. DOI: 10.1007/s43681-021-00077-w PMID: 34790955

Alvarado, R. (2023). AI as an epistemic technology. *Science and Engineering Ethics*, *29*(5), 32. DOI: 10.1007/s11948-023-00451-3 PMID: 37603120

Alvarez, J. M., Colmenarejo, A. B., Elobaid, A., Fabbrizzi, S., Fahimi, M., Ferrara, A., Ghodsi, S., Mougan, C., Papageorgiou, I., Reyero, P., Russo, M., Scott, K. M., State, L., Zhao, X., & Ruggieri, S. (2024). Policy advice and best practices on bias and fairness in AI. *Ethics and Information Technology*, *26*(2), 31. DOI: 10.1007/s10676-024-09746-w

Alvarez, R. M., Hall, T. E., & Trechsel, A. H. (2009). Internet Voting in Comparative Perspective: The Case of Estonia. *PS, Political Science & Politics*, *42*(3), 497–505. DOI: 10.1017/S1049096509090787

Al-Zain, M. A., Abdulaziz, S. A., & Al Mubarak, M. (2025). Ethics of artificial intelligence and robotics. In *Ethical Standards for Technological and Business Education Sustainability* (pp. 85–98). Emerald Publishing Limited. DOI: 10.1108/978-1-83608-036-720251006

Amankwah-Amoah, J., Abdalla, S., Mogaji, E., Elbanna, A., & Dwivedi, Y. K. (2024). The impending disruption of creative industries by generative AI: Opportunities, challenges, and research agenda. *International Journal of Information Management*, *79*(102759), 102759. Advance online publication. DOI: 10.1016/j.ijinfomgt.2024.102759

Amin, P. (2024). Disinformation and the Impact on Democracy. In Hewage, C., Yasakethu, L., & Jayakody, D. N. K. (Eds.), *Data Protection*. Springer., DOI: 10.1007/978-3-031-76473-8_12

Anand, A. J., & Bhambri, P. (Eds.). (2025). *Handbook of AI-driven threat detection and prevention: A holistic approach to security*. CRC Press., DOI: 10.1201/9781003521020

Anantrasirichai, N., & Bull, D. (2022). Artificial intelligence in the creative industries: A review. *Artificial Intelligence Review, 55*(1), 589–656. DOI: 10.1007/s10462-021-10039-7 PMID: 41625196

Andani, A.P., & Santoso, B. (2025), "Perlindungan Hukum Terhadap Pelanggaran Karya Cipta yang Dihasilkan Melalui AI dan dikomersilkan Berdasarkan Undang-Undang Nomor 28 Tahun 2014 tentang Hak Cipta", *JIHHP: Jurnal Ilmu Hukum, Humaniora dan Politik*, 5(3), 2274-2281, DOI: 10.38035/jihhp.v5i3.4039

Anderljung, M., Hazell, J., & von Knebel, M. (2025). Protecting society from AI misuse: When are restrictions on capabilities warranted? *AI & Society, 40*(5), 3841–3857. DOI: 10.1007/s00146-024-02130-8

Andrada, G., Clowes, R. W., & Smart, P. R. (2023). Varieties of transparency: Exploring agency within AI systems. *AI & Society, 38*(4), 1321–1331. DOI: 10.1007/s00146-021-01326-6 PMID: 35035112

Angwin, J., Larson, J., Mattu, S., & Kirchner, L. (2016). *Machine bias: There's software used across the country to predict future criminals. And it's biased against blacks*. ProPublica.

Anifa, Y. (2024). Dampak Kecerdasan Buatan dalam Mempertimbangkan Aspek Sosial dan Budaya dalam Desain Grafis. *VCoDe: Visual Communication Design Journal, 3*(2), 145–156. DOI: 10.26887/vcode.v3i2.4256

Appel, G., Neelbauer, J., & Schweidel, D. A. (2023). Generative ai has an intellectual property problem. Harvard Business Review. Retrieved from https://hbr.org/2023/04/generative-ai-has-an-intellectual-property-problem

Arsenault, A. C., & Kreps, S. (2022). *AI and International Politics*. The Oxford Handbook of AI Governance.

Aufderheide, P., & Jaszi, P. (2018). *Reclaiming fair use: How to put balance back in copyright* (2nd ed.). University of Chicago Press. DOI: 10.7208/chicago/9780226374222.001.0001

Auld, G., Casovan, A., Clarke, A., & Faveri, B. (2022). Governing AI through ethical standards: Learning from the experiences of other private governance initiatives. *Journal of European Public Policy, 29*(11), 1822–1844. DOI: 10.1080/13501763.2022.2099449

Aurelia, C. N.. (2025). Enhancing Fair Use in Protecting Appropriated Artworks: A Comparative Analysis of Safeguarding Indonesian Copyright Law. *Lex Scientia Law Review, 9*(1), 1181–1222. DOI: 10.15294/lslr.v9i1.20570

Ayaa, M. M., Peprah, W. K., Mensah, M. O., Owusu-Sekyere, A. B., & Daniel, B. (2022). Influence of heuristic techniques and biases in investment decision-making: A conceptual analysis and directions for future research. *International Journal of Academic Research in Business & Social Sciences, 12*(5), 1252. DOI: 10.6007/IJARBSS/v12-i5/13339

Ayres, I., Bankman, J., & Hemel, D. J. (2025). (forthcoming). The Law and Economics of Guilt and Shame. *The University of Chicago Law Review. University of Chicago. Law School.*

Ayub, T. (2024). Regulating AI: Balancing Innovation, Ethics, and Public Policy A Critical Analysis of AI Governance and Policy Approaches. Ethics, and Public Policy A Critical Analysis of AI Governance and Policy Approaches (December 11, 2024).

Baggini, J., & Fosl, P. S. (2024). *The ethics toolkit: A compendium of ethical concepts and methods.* John Wiley & Sons.

Bajdor, P., & Bhambri, P. (2024). The principles of sustainability. In Bhambri, P., & Bajdor, P. (Eds.), *Handbook of technological sustainability: Innovation and environmental awareness* (pp. 3–17). CRC Press., DOI: 10.1201/9781003475989-2

Bajgar, O., & Horenovsky, J. (2023). Negative human rights as a basis for long-term AI safety and regulation. *Journal of Artificial Intelligence Research, 76*, 1043–1075. Advance online publication. DOI: 10.1613/jair.1.14020

Balasubramaniam, N., Kauppinen, M., Hiekkanen, K., & Kujala, S. (2022, March). Transparency and explainability of AI systems: ethical guidelines in practice. In *International working conference on requirements engineering: foundation for software quality* (pp. 3-18). Cham: Springer International Publishing. DOI: 10.1007/978-3-030-98464-9_1

Balasubramaniam, N., Kauppinen, M., Rannisto, A., Hiekkanen, K., & Kujala, S. (2023). Transparency and explainability of AI systems: From ethical guidelines to requirements. *Information and Software Technology, 159*, 107197. DOI: 10.1016/j.infsof.2023.107197

Banh, L., & Strobel, G. (2023). Generative artificial intelligence. *Electronic Markets, 33*(1), 63. DOI: 10.1007/s12525-023-00680-1

Bansal, R. (2022). Artificial intelligence and intellectual property rights in India. *Journal of Intellectual Property Rights, 27*(3), 169–180. https://or.niscpr.res.in/index.php/JIPR/article/view/1205

Bao, Y., Gong, W., & Yang, K. (2023). A literature review of human–AI synergy in decision making: From the perspective of affordance actualization theory. *Systems, 11*(9), 442. DOI: 10.3390/systems11090442

Barocas, S., Hardt, M., & Narayanan, A. (2019). *Fairness and machine learning*. fairmlbook.org

Barocas, S., & Selbst, A. D. (2016). Big data's disparate impact. *California Law Review, 104*(3), 671–732.

Batakis, D., Chatzivasileiou, S., & Efthymiou, I. P. (2025). Artificial Intelligence and its impact on healthcare.

Batsurovska, I. B., & Kurylen, V. M. (2024). Models and algorithms of machine learning in computer systems of artificial intelligence: New trends and prospects for development. *Modern Aspects of Sciences, XLVII*, 462–473.

Baumann, J., & Loi, M. (2023). Fairness and risk: An ethical argument for a group fairness definition insurers can use. *Philosophy & Technology, 36*(3), 45. DOI: 10.1007/s13347-023-00624-9 PMID: 37346393

BBC News. (2025a). Facebook and Instagram get rid of fact checkers. https://www.bbc.com/news/articles/cly74mpy8klo

Becerra, M., & Waisbord, S. R. (2021). The curious absence of cyber nationalism in Latin America: Lessons for the study of digital sovereignty and governance. *Communication and the Public, 6*(1–4), 67–79. DOI: 10.1177/20570473211046730

Becker, S. J., Nemat, A. T., Lucas, S., Heinitz, R. M., Klevesath, M., & Charton, J. E. (2023). A Code of Digital Ethics: Laying the foundation for digital ethics in a science and technology company. *AI & Society, 38*(6), 2629–2639. DOI: 10.1007/s00146-021-01376-w

Beier, K. (2025). Virtue Ethics and AI. *A Companion to Applied Philosophy of AI*, 178-191.

Bender, E. M., Gebru, T., McMillan-Major, A., & Shmitchell, S. (2021). On the dangers of stochastic parrots: Can language models be too big? In *Proceedings of the 2021 ACM Conference on Fairness, Accountability, and Transparency* (pp. 610–623). ACM. DOI: 10.1145/3442188.3445922

Benjamin, R. (2019). *Race After Technology: Abolitionist Tools for the New Jim Code*. Polity Press.

Benkler, Y., Faris, R., & Roberts, H. (2018). *Network propaganda: Manipulation, disinformation, and radicalization in American politics*. Oxford University Press. DOI: 10.1093/oso/9780190923624.001.0001

Bennett, W. L., & Livingston, S. (2018). The disinformation order: Disruptive communication and the decline of democratic institutions. *European Journal of Communication, 33*(2), 122–139. DOI: 10.1177/0267323118760317

Bessen, J. E., Impink, S. M., & Seamans, R. (2021). The impact of artificial intelligence on innovation. *Research Policy, 50*(9), 104279. DOI: 10.1016/j.respol.2021.104279

Bhambri, P., & Kant, S. (2025a, February 1–2). A taxonomy of bias in machine learning: Classification, sources, and implications for ethical AI [Paper presentation]. Lincoln-SPAST Global Sustainability Programme (SGS-24), 1st International Conference of L-GPR Program, Lincoln University College, Malaysia. SPAST Proceedings, 1(2). https://spast.org/techrep/article/view/5387/724

Bhambri, P., & Kant, S. (2025b, April 26–27). Ethical AI systems: A comprehensive framework for bias mitigation and fairness in machine learning [Paper presentation]. Lincoln-SPAST Global Sustainability Programme (SGS-24), 2nd International Conference of L-GPR Program, Lincoln University College, Malaysia. SPAST Proceedings, 1(2). https://spast.org/techrep/article/view/5385/698

Bhambri, P. (2024). Enhancing cybersecurity with SuperHyperSoft computing approaches. In Smarandache, F., & Majumder, P. (Eds.), *Modern SuperHyperSoft computing trends in science and technology* (pp. 127–148). IGI Global., DOI: 10.4018/979-8-3693-6875-6.ch005

Binns, R. (2018). Fairness in machine learning: Lessons from political philosophy. *Proceedings of the 1st Conference on Fairness, Accountability and Transparency*, 149–159.

Birkstedt, T., Minkkinen, M., Tandon, A., & Mäntymäki, M. (2023). AI governance: Themes, knowledge gaps and future agendas. *Internet Research, 33*(7), 133–167. DOI: 10.1108/INTR-01-2022-0042

Blanchard, A., Thomas, C., & Taddeo, M. (2025). Ethical governance of artificial intelligence for defence: Normative tradeoffs for principle to practice guidance. *AI & Society, 40*(1), 185–198. DOI: 10.1007/s00146-024-01866-7 PMID: 41640455

Bodig, M., & Ződi, Z. (2024). Is Code Destined to Supersede Law? Behavioural Control by Code and Legal Theory. Behavioural Control by Code and Legal Theory (June 20, 2024).

Bodini, M. (2024). Generative Artificial Intelligence and Regulations: Can We Plan a Resilient Journey Toward the Safe Application of Generative Artificial Intelligence? *Societies (Basel, Switzerland), 14*(12), 268. DOI: 10.3390/soc14120268

Bommasani, R., Hudson, D. A., Adeli, E., ... (2021). On the opportunities and risks of foundation models. *arXiv preprint.* https://arxiv.org/abs/2108.07258

Bonadio, E., & McDonagh, L. (2020). Artificial intelligence as producer and consumer of copyright works: evaluating the consequences of algorithmic creativity. *Intellectual Property Quarterly, 2020*(2), 112-137. https://ssrn.com/abstract=3617197

Bostrom, N. (2014). *Superintelligence: Paths, dangers, strategies.* Oxford University Press.

Bradshaw, S., & Howard, P. N. (2019). *The global disinformation disorder: 2019 global inventory of organised social media manipulation.* Computational Propaganda Research Project. https://comprop.oii.ox.ac.uk

Brindha, R., Pongiannan, R. K., Bharath, A., & Sanjeevi, V. K. S. M. (2025). Introduction to Multimodal Generative AI. In *Multimodal Generative AI* (pp. 1–36). Springer Nature Singapore., DOI: 10.1007/978-981-96-2355-6_1

Bringas Colmenarejo, A., Nannini, L., Rieger, A., Scott, K. M., Zhao, X., Patro, G. K., & Kinder-Kurlanda, K. (2022, July). Fairness in agreement with European values: An interdisciplinary perspective on AI regulation. In *Proceedings of the 2022 AAAI/ACM Conference on AI, Ethics, and Society* (pp. 107-118). DOI: 10.1145/3514094.3534158

Brookings Institution. (2019). *Algorithmic bias detection and mitigation: Best practices and policies to reduce consumer harms.* Brookings Institution.

Brown, I., & Marsden, C. T. (2023). *Regulating code: Good governance and better regulation in the information age.* MIT press.

Brownsword, R. (2019). Law, technology and society: Re-imagining the regulatory environment. *Law, Innovation and Technology, 11*(1), 1–26.

Brown, T. B., Mann, B., & Ryder, N.. (2020). Language models are few-shot learners. *Advances in Neural Information Processing Systems, 33*, 1877–1901.

Bryson, J. J., Diamantis, M. E., & Grant, T. D. (2017). Of, for, and by the people: The legal lacuna of synthetic persons. *Artificial Intelligence and Law, 25*(3), 273–291. DOI: 10.1007/s10506-017-9214-9

Budziszewska, A., & Potluka, O. (2024). The use of AI and technology by civil society organizations and its international implications. In *The Routledge Handbook of Artificial Intelligence and Philanthropy* (pp. 94–106). Routledge. DOI: 10.4324/9781003468615-8

Buolamwini, J., & Gebru, T. (2018). Gender shades: Intersectional accuracy disparities in commercial gender classification. *Proceedings of Machine Learning Research, 81*, 77–91.

Burk, D. L. (2019). Law and artificial intelligence. *University of Illinois Law Review, 2019*(4), 1957–1996.

Burrell, J. (2016). How the machine "thinks": Understanding opacity in machine learning algorithms. *Big Data & Society, 3*(1), 1–12. DOI: 10.1177/2053951715622512

Busuioc, M. (2021). Accountable artificial intelligence: Holding algorithms to account. *Public Administration Review, 81*(5), 825–836. DOI: 10.1111/puar.13293 PMID: 34690372

Calo, R. (2016). Robotics and the lessons of cyberlaw. *California Law Review, 103*(3), 513–563.

Camilleri, M. A. (2023). Artificial intelligence governance: Ethical considerations and implications for social responsibility. *Expert Systems: International Journal of Knowledge Engineering and Neural Networks, 41*(7), e13406. Advance online publication. DOI: 10.1111/exsy.13406

Cantero Gamito, M., & Marsden, C. T. (2024). Artificial intelligence co-regulation? The role of standards in the EU AI Act. *International Journal of Law and Information Technology, 32*, eaae011. DOI: 10.1093/ijlit/eaae011

Capital.gr. (2023, October 19). High Level Advisory Committee for Artificial Intelligence (AI) established. https://www.capital.gr/english/3744823/high-level-advisory-committee-for-artificial-intelligence-ai-established/

Capraro, V., Lentsch, A., Acemoglu, D., Akgun, S., Akhmedova, A., Bilancini, E., Bonnefon, J.-F., Brañas-Garza, P., Butera, L., Douglas, K. M., Everett, J. A. C., Gigerenzer, G., Greenhow, C., Hashimoto, D. A., Holt-Lunstad, J., Jetten, J., Johnson, S., Kunz, W. H., Longoni, C., & Viale, R. (2024). The impact of generative artificial intelligence on socioeconomic inequalities and policy making. *PNAS Nexus, 3*(6), pgae191. Advance online publication. DOI: 10.1093/pnasnexus/pgae191 PMID: 38864006

Carrillo, M. (2020). Artificial intelligence: From ethics to law. *Telecommunications Policy, 44*(6), 101937. DOI: 10.1016/j.telpol.2020.101937

Cath, C. (2018). Governing artificial intelligence: Upholding human rights & dignity. European Parliament Research Service (EPRS) Study.

Caton, L., & Hwang, Y. (2025). Ethical Implications and Governance of AI in Research. In *Institutional guide to using AI for research* (pp. 167–183). Springer Nature Switzerland. DOI: 10.1007/978-3-031-94809-1_8

Caton, S., & Haas, C. (2020). Fairness in machine learning: A survey. *ACM Computing Surveys, 54*(6), 1–35. DOI: 10.1145/3457607

Cave, S., Nyrup, R., Vold, K., & Weller, A. (2019). Motivations and risks of machine ethics. *Proceedings of the IEEE, 107*(3), 562–574. DOI: 10.1109/JPROC.2018.2865996

Cech, F. (2021). Mechanisms for algorithmic accountability through the lens of actor and forum agency. *Government Information Quarterly, 38*(4), 101615. DOI: 10.1016/j.giq.2021.101615

Cetinkaya, N. E., & Krämer, N. (2025). Between transparency and trust: Identifying key factors in AI system perception. *Behaviour & Information Technology*, •••, 1–15. DOI: 10.1080/0144929X.2025.2533358

Chai, Y. X., Ren, S. J. F., & Zhang, J. Q. (2022). The impacts of digital content piracy and copyright protection policies when consumers are loss averse. *Journal of Industrial and Management Optimization, 18*(5), 3587–361. DOI: 10.3934/jimo.2021127

Challoumis, C. (2024). Behavioral Economics Concepts and the QE Method. [IJMMU]. *International Journal of Multicultural and Multireligious Understanding, 11*(10), 166–212. DOI: 10.18415/ijmmu.v11i10.6138

Chatzivasileiou, D., Psomiadi, A., Efthymiou-Egleton, T. W., & Kassar, L. (2024). AI, International Relations & Religion. *Journal of Politics and Ethics in New Technologies and AI, 3*(1), e37109–e37109. DOI: 10.12681/jpentai.37109

Chekhratova, O., & Pohorielova, T. (2024). Ethical research in the age of artificial intelligence.

Chellakkannu, S. (2024, April). Embracing generative AI in design: practical implementations. In *International Conference on Artificial Intelligence and its Applications in the Age of Digital Transformation* (pp. 189-206). Cham: Springer Nature Switzerland. DOI: 10.1007/978-3-031-71429-0_15

Chemnad, K., & Othman, A. (2024). Digital accessibility in the era of artificial intelligence—Bibliometric analysis and systematic review. *Frontiers in Artificial Intelligence*, *7*, 7. DOI: 10.3389/frai.2024.1349668 PMID: 38435800

Chen, J., Sun, J., & Wang, G. (2022). From unmanned systems to autonomous intelligent systems. *Engineering (Beijing)*, *12*, 16–19. DOI: 10.1016/j.eng.2021.10.007

Chen, P., Wu, L., & Wang, L. (2023). AI fairness in data management and analytics: A review on challenges, methodologies and applications. *Applied Sciences (Basel, Switzerland)*, *13*(18), 10258. DOI: 10.3390/app131810258

Cheong, I., Caliskan, A., & Kohno, T. (2023). Is the US Legal System Ready for AI's Challenges to Human Values? *arXiv preprint arXiv:2308.15906.*

Cheong, B. C. (2024). Transparency and accountability in AI systems: Safeguarding wellbeing in the age of algorithmic decision-making. *Frontiers in Human Dynamics*, *6*, 1421273. DOI: 10.3389/fhumd.2024.1421273

Cheong, I., Caliskan, A., & Kohno, T. (2025). Safeguarding human values: Rethinking US law for generative AI's societal impacts. *AI and Ethics*, *5*(2), 1433–1459. DOI: 10.1007/s43681-024-00451-4 PMID: 40352581

China National Intellectual Property Administration. (n.d.). Guidelines for patent applications related to artificial intelligence (trial). https://www.cnipa.gov.cn/

Choi, I., & Zu, J. (2025). Generating language assessment content free from representational harms. *Language Testing*, *42*(4), 539–560. DOI: 10.1177/02655322251349560

Choung, H., David, P., & Seberger, J. S. (2023). A multilevel framework for AI governance. *arXiv preprint arXiv:2307.03198.*

Citron, D. K., & Chesney, R. (2019). Deep fakes: A looming challenge for privacy, democracy, and national security. *California Law Review*, *107*, 1753–1819.

Citron, D. K., & Pasquale, F. (2014). The scored society: Due process for automated predictions. *Washington Law Review (Seattle, Wash.)*, *89*(1), 1–33.

Clift, S. L. (2004) E-Government and Democracy: Representation and Citizen Engagement in the Information Age. Research provided for the UN UNPAN/DESA for the 2003 World Public Sector Report. Available at: https://www.publicus.net/articles/cliftegovdemocracy.pdf

Cobbe, J., Lee, M. S. A., & Singh, J. (2021). Reviewable automated decision-making: A framework for accountable algorithmic systems. *Computer Law & Security Review*, *41*, 105561.

Cohen, J. E. (2019). *Between truth and power: The legal constructions of informational capitalism*. Oxford University Press. DOI: 10.1093/oso/9780190246693.001.0001

Companies and Intellectual Property Commission. (2021). *Patent journal: Acceptance of patent application 2021/03242* (South Africa).

Council of Europe. (2017). Recommendation CM/Rec(2017)5 of the Committee of Ministers to member states on standards for e-voting. https://search.coe.int/cm?i=0900001680726f6f

Council of Europe. Committee on Artificial Intelligence (CAI). https://www.coe.int/en/web/artificial-intelligence/cai

Council of Europe. Framework Convention on Artificial Intelligence and Human Rights, Democracy and the Rule of Law. https://www.coe.int/en/web/artificial-intelligence

Craig, C., & Kerr, I. (2025). The Death of The AI Author. In *Robot Law* (Vol. II, pp. 250–285). Edward Elgar Publishing., DOI: 10.4337/9781800887305.00014

Crawford, K., & Paglen, T. (2019). Excavating AI: The politics of images in machine learning datasets. https://excavating.ai

Crawford, K. (2021). *Atlas of AI: Power, politics, and the planetary costs of artificial intelligence*. Yale University Press.

Crawford, K., & Schultz, J. (2024). The Work of Copyright Law in the Age of Generative AI. *Grey Room*, *94*(94), 56–62. DOI: 10.1162/grey_a_00389

Creemers, R. (2023). China's governance of artificial intelligence: A regulatory overview. *Journal of Cyber Policy*, *8*(1), 1–24.

Cruz-Aguilar, M. A. (2025). The epistemic revolution of AI: Reconfiguring the foundations of scientific knowledge. *AI & Society*, •••, 1–17. DOI: 10.1007/s00146-025-02658-3

Cupać, J., Schopmans, H., & Tuncer-Ebetürk, İ. (2024). Democratization in the age of artificial intelligence: Introduction to the special issue. *Democratization*, *31*(5), 899–921. DOI: 10.1080/13510347.2024.2338852

Curtis, P. (2023, July 28). Artificial intelligence is powering politics—but it could also reboot democracy. *The Guardian*. https://www.theguardian.com/commentisfree/2023/jul/28/artificial-intelligence-powering-politics-reboot-democracy

Cuzzolin, F., & Sultana, M. (2024). *Epistemic Uncertainty in Artificial Intelligence*. Springer Nature Switzerland., DOI: 10.1007/978-3-031-57963-9

Da Silva, M., Horsley, T., Singh, D., Da Silva, E., Ly, V., Thomas, B., Daniel, R. C., Chagal-Feferkorn, K. A., Iantomasi, S., White, K., Kent, A., & Flood, C. M. (2022). Legal concerns in health-related artificial intelligence: A scoping review protocol. *Systematic Reviews*, *11*(1), 123. DOI: 10.1186/s13643-022-01939-y PMID: 35715812

Dabis, A., & Csáki, C. (2024). AI and ethics: Investigating the first policy responses of higher education institutions to the challenge of generative AI. *Humanities & Social Sciences Communications*, *11*(1), 1–13. DOI: 10.1057/s41599-024-03526-z

Daffa, B.M. (2024). "Aspek Hukum Penggunaan Metode Stable Diffusion Oleh Artificial Intelligence Terhadap Suatu Ciptaan Berdasarkan Undang-Undang Nomor 28 Tahun 2014". *COMSERVA: Jurnal Penelitian dan Pengabdian Masyarakat*, 3(11), 4434-4442, DOI: 10.59141/comserva.v3i11.1254

DALY, E. & MAY, J. R. (). DIGNITY UNDER LAW: A GLOBAL HANDBOOK FOR CIVIL SOCIETY.

Danaher, J. (2024). Generative AI and the future of equality norms. *Cognition*, *251*, 105906. DOI: 10.1016/j.cognition.2024.105906 PMID: 39067317

De Cremer, D., Narayanan, D., Nagpal, M., McGuire, J., & Schweitzer, S. (2024). AI Fairness in Action: A human-computer perspective on AI fairness in organizations and society. *International Journal of Human-Computer Interaction*, *40*(1), 1–3. DOI: 10.1080/10447318.2023.2273673

De Gregorio, G., & Dunn, P. (2022). The European risk-based approaches: Connecting constitutional dots in the digital age. *Common Market Law Review*, *59*(2), 473–500. DOI: 10.54648/COLA2022032

De la Durantaye, K. (2025). Control and Compensation. A Comparative Analysis of Copyright Exceptions for Training Generative AI. *IIC; International Review of Industrial Property and Copyright Law*, *56*(4), 737–770. DOI: 10.1007/s40319-025-01569-6

de Laat, P. B. (2021). Companies committed to responsible AI: From principles towards implementation and regulation? *Philosophy & Technology*, *34*(4), 1135–1193. DOI: 10.1007/s13347-021-00474-3 PMID: 34631392

De Silva, D., & Alahakoon, D. (2022). An artificial intelligence life cycle: From conception to production. *Patterns (New York, N.Y.)*, *3*(6), 100489. DOI: 10.1016/j.patter.2022.100489 PMID: 35755876

De Stefano, V. (2020). Algorithmic management and collective bargaining. *Comparative Labor Law & Policy Journal*, *41*(1), 1–14.

Defend Trade Secrets Act of. 2016, Pub. L. No. 114–153, 130 Stat. 376. https://www.congress.gov/114/plaws/publ153/PLAW-114publ153.pdf

Denia, E. (2025). AI narratives model: Social perception of artificial intelligence. *Technovation, 146*, 103266. DOI: 10.1016/j.technovation.2025.103266

Department for Promotion of Industry and Internal Trade. (2025). *Working paper on generative artificial intelligence and copyright*. Government of India. https://www.dpiit.gov.in

Dewasiri, N. J., & Singh, R. (2025). Advancing Climate Neutrality Through Sustainable Production and Consumption: Pathways to a Resilient Future. In *Climate Neutrality Through Smart Eco-Innovation and Environmental Sustainability* (pp. 3–19). Springer Nature Switzerland. DOI: 10.1007/978-3-031-83250-5_1

Diareme, K. C., Liapakis, A., & Efthymiou, I. (2022). Big Data, Sentiment Analysis, and Examples during the COVID-19 Pandemic. *HAPSc Policy Briefs Series, 3*(2).

Diareme, K. C., Liapakis, A., & Efthymiou, I. P. (2022). Data, sentiment analysis, and examples during the COVID-19 pandemic. *Hellenic Archives of Psychiatry and Clinical Psychology in Behavioral Sciences, 33779*(2), 21–30. Advance online publication. DOI: 10.12681/hapscpbs.33779

Didsbury, H., & Zhu, X. A. (2025). Transformative training: an analysis of AI training data and fair use in *Authors Guild v. OpenAI Inc. Publishing Research* 4: e002. DOI: 10.48130/pr-2025-0001

Ding, T., & Yang, L. (2025). Intellectual Property Protection and Corporate Digital Transformation: An Empirical Analysis from The Perspectives of Intellectual Property Protection and Digital Governance. *International Review of Economics & Finance, 100*(104125), 1–11. DOI: 10.1016/j.iref.2025.104125

Dinwoodie, G. B., & Dreyfuss, R. C. (2018). *A neofederalist vision of TRIPS*. Oxford University Press.

Domingos, P. (2015). *The master algorithm*. Basic Books.

Doshi-Velez, F., & Kim, B. (2017). Towards a rigorous science of interpretable machine learning. *arXiv preprint*. https://arxiv.org/abs/1702.08608

DRCF / UK regulators. (2021). Auditing algorithms: The existing landscape, role of regulators and future outlook. UK Government. https://www.gov.uk/government/publications

Dunleavy, P., & Margetts, H. (2025). Data science, artificial intelligence and the third wave of digital era governance. *Public Policy and Administration*, *40*(2), 185–214. DOI: 10.1177/09520767231198737

Dwivedi, Y. K., Kshetri, N., Hughes, L., Slade, E. L., Jeyaraj, A., Kar, A. K., Baabdullah, A. M., Koohang, A., Raghavan, V., Ahuja, M., Albanna, H., Albashrawi, M. A., Al-Busaidi, A. S., Balakrishnan, J., Barlette, Y., Basu, S., Bose, I., Brooks, L., Buhalis, D., & Wright, R. (2023). Opinion Paper: "So what if ChatGPT wrote it?" Multidisciplinary perspectives on opportunities, challenges and implications of generative conversational AI for research, practice and policy. *International Journal of Information Management*, *71*(102642), 1–63. DOI: 10.1016/j.ijinfomgt.2023.102642

Dworkin, R. (1977). *Taking rights seriously*. Harvard University Press.

Ebers, M. (2025). Truly risk-based regulation of artificial intelligence how to implement the EU's AI Act. *European Journal of Risk Regulation*, *16*(2), 684–703. DOI: 10.1017/err.2024.78

Edeji, O. C. (2025). Intersections of the Right to Education and Human Dignity in International Human Rights Law: A Purpose-Based Analysis. *Laws*, *14*(3), 33. DOI: 10.3390/laws14030033

Edwards, L., & Veale, M. (2017). Slave to the algorithm? Why a "right to an explanation" is probably not the remedy you are looking for. *Duke Law & Technology Review*, *16*, 18–84. DOI: 10.31228/osf.io/97upg

Efthymiou, I. P. (2024). Exploring the frontier: New and future trends in research. In B. Orlando (Ed.), Innovation capabilities and entrepreneurial opportunities of smart working (pp. 48–68). IGI Global Scientific Publishing. DOI: 10.4018/978-1-7998-8797-3.ch003

Efthymiou, I. P. (2024). Human wellbeing in the age of artificial intelligence.

Efthymiou, I. P. (2026). Bridging the Divide: Addressing the Leadership Gap. In *Bridging the Leadership Gap: Policies and Workplace Equality* (pp. 103-138). IGI Global Scientific Publishing.

Efthymiou, I.-P. (2024). AI, ethics and the future of work: What's left for us? Independently published.

Efthymiou-Egleton, I. (2016). *Trends in health care: A global challenge*. Xlibris Corporation.

Efthymiou–Egleton, T. W., Sidiropoulos, S., Spanos, E., & Stougiannou, E. (2020). Big Data and Democracy. *HAPSc Policy Briefs Series*, *1*(2), 18–25. DOI: 10.12681/hapscpbs.26473

Efthymiou, I. (2025). Diversity, Inclusion, and Equity in Gig Economies. In Doon, R. (Ed.), *Sustainability and Adaptability of Gig Economies in Global Business* (pp. 231–260). IGI Global Scientific Publishing., DOI: 10.4018/979-8-3693-9385-7.ch009

Efthymiou, I. (2026). Bridging the Divide: Addressing the Leadership Gap. In Montaudon-Tomas, C., & Pinto-López, I. (Eds.), *Bridging the Leadership Gap: Policies and Workplace Equality* (pp. 103–138). IGI Global Scientific Publishing., DOI: 10.4018/979-8-3373-6345-5.ch004

Efthymiou, I. P. (2025). The role of e-government and e-governance in modern societies. In Wongmahesak, K., Ahmad, J., & Mustanir, A. (Eds.), *Public governance practices in the age of AI* (pp. 45–68). IGI Global Scientific Publishing., DOI: 10.4018/979-8-3693-9286-7.ch003

Efthymiou, I. P. (2026). AI in epidemiology: Predicting the unpredictable leveraging AI to detect outbreaks and predict disease. In *Innovation and transformation of public health* (pp. 281–308). IGI Global Scientific Publishing.

Efthymiou, I. P. (2026). Ethical Considerations and Regulatory Standards for Adversarial Defense. In Nowroozi, E., Taheri, R., & Cordeiro, L. (Eds.), *Adversarial Example Detection and Mitigation Using Machine Learning*. Springer., DOI: 10.1007/978-3-031-99447-0_2

Efthymiou, I. P., & Egleton, T. E. (2023). Artificial intelligence for sustainable smart cities. In *Handbook of research on applications of AI, digital twin, and internet of things for sustainable development* (pp. 1–11). IGI Global. DOI: 10.4018/978-1-6684-6821-0.ch001

Efthymiou, I. P., Egleton, T. E., & Psomiadi, A. (2025). The Role of Artificial Intelligence in Humanitarian Aid for Achieving Sustainable Development Goals. In *Cases on AI-Driven Solutions to Environmental Challenges* (pp. 43–72). IGI Global Scientific Publishing. DOI: 10.4018/979-8-3693-7483-2.ch003

Efthymiou, I. P., Vozikis, A., & Sidiropoulos, S. (2019). Application of Sociocybernetic model in the field of Health Management. *International Journal of Scientific and Engineering Research*, *10*(5), 451–560.

Efthymiou, I., & Egleton, T. E. (2025). The Impact of Deep Fakes in Markets and Economies. In Gupta, G., Bohara, S., Kovid, R., & Pandla, K. (Eds.), *Deepfakes and Their Impact on Business* (pp. 19–50). IGI Global Scientific Publishing., DOI: 10.4018/979-8-3693-6890-9.ch002

Efthymiou, I.-P., & Kouroupis, K. (Eds.). (2026). *Legal and Regulatory Impacts on AI Development*. IGI Global Scientific Publishing., DOI: 10.4018/979-8-3373-7996-8

Elia, M., Ziethmann, P., Krumme, J., Schlögl-Flierl, K., & Bauer, B. (2025). Responsible AI, ethics, and the AI lifecycle: How to consider the human influence? *AI and Ethics*, *5*(4), 1–18. DOI: 10.1007/s43681-025-00666-z

Epstein, R. A. (2025). Nudges, preferences and competences: A critique of both neoclassical and behavioral economics. *Behavioural Public Policy*, *9*(3), 1–15. DOI: 10.1017/bpp.2024.54

Erdozain, J. C. (2019). How will the IP world respond to the rise of AI. Managing Intell. Prop., 281, 56. https://www.managingip.com/article/2a5bqo2drurt0bwowm4jf/how-will-the-ip-world-respond-to-the-rise-of-ai

EU High-Level Expert Group on Artificial Intelligence. (2019). *Ethics guidelines for trustworthy AI*. European Commission. https://digital-strategy.ec.europa.eu

Eubanks, V. (2018). *Automating inequality: How high-tech tools profile, police, and punish the poor*. St. Martin's Press.

Eubanks, V. (2018). *Automating Inequality: How High-Tech Tools Profile, Police, and Punish the Poor*. St. Martin's Press.

European Commission. (2019). Ethics guidelines for trustworthy AI. https://ec.europa.eu/futurium/en/ai-alliance-consultation.1.html

European Commission. (2020). *White paper on artificial intelligence: A European approach to excellence and trust*. https://commission.europa.eu

European Commission. (2021). *Proposal for a regulation laying down harmonised rules on artificial intelligence (Artificial Intelligence Act)*. COM(2021) 206 final.

European Commission. (2022). Eurobarometer on media and news. https://europa.eu/eurobarometer/surveys/detail/2832

European Commission. (2023). *Proposal for a regulation laying down harmonised rules on artificial intelligence (AI Act)*. https://eur-lex.europa.eu

European Commission. (2024). *Artificial Intelligence Act: Regulation of the European Parliament and of the Council on Artificial Intelligence*. Official Journal of the European Union.

European Data Protection Board. (2020). *Guidelines on processing personal data in the context of connected vehicles and mobility related applications.*

European Parliament. (2021, April 22). Digital transformation: Importance, benefits and EU policy. https://www.europarl.europa.eu/news/en/headlines/society/20210414STO02010/digital-transformation-importance-benefits-and-eu-policy

European Parliament. (2025). *Generative AI and copyright: Legal challenges and policy options*. https://www.europarl.europa.eu

European Patent Office. (2021). *Decision J 0008/20 (Designation of inventor/ DABUS)*. https://www.epo.org

European Union. (2024). Regulation (EU) 2024/1689 of the European Parliament and of the Council of 12 July 2024 laying down harmonised rules on artificial intelligence (Artificial Intelligence Act) and amending certain Union legislative acts. Official Journal of the European Union, L 234, 1–83. https://eur-lex.europa.eu/legal-content/EN/TXT/?uri=OJ:L:2024:234:TOC 23

Ewert, B., Loer, K., & Thomann, E. (2021). Beyond nudge: Advancing the state-of-the-art of behavioural public policy and administration. *Policy and Politics*, *49*(1), 3–23. DOI: 10.1332/030557320X15987279194319

Eyo-Udo, N. L., Apeh, C. E., Bristol-Alagbariya, B., Udeh, C. A., & Ewim, C. P.-M. (2025). International Trade Law in the Modern World: A Review of Evolving Practices and Agreements. *International Journal of Multidisciplinary Research and Growth Evaluation*, *6*(1), 840–847. DOI: 10.54660/.IJMRGE.2025.6.1.840-847

Fadillah, R. N. (2024). Perlindungan Hak Atas Kekayaan Intelektual Artificial Intelligence (AI) dari Perspektif Hak Cipta dan Paten. *Das Sollen: Jurnal Kajian Kontemporer Hukum dan Masyarakat*, 2(2), 1-25. http://dx.doi.org/DOI: 10.11111/dassollen.xxxxxxx

Fadillah, N. (2022). Tinjauan Teori Hukum Pembangunan Mochtar Kusumaatmadja dalam Undang-Undang Ibu Kota Negara (IKN). *Supremasi Hukum: Jurnal Kajian Ilmu Hukum*, *11*(1), 45–65. DOI: 10.14421/sh.v11i1.2559

Fadli, M. A., Hamidi, M. P., Edwin, F. A., & Aritonang, R. G. K. (2023). Let's Play Content as a Fiduciary Collateral under Indonesian Law: Potential Challenges. Yuridika, 38(3), 481-498.

Farina, M., Yu, X., & Lavazza, A. (2025). Ethical considerations and policy interventions concerning the impact of generative AI tools in the economy and in society. *AI and Ethics*, *5*(1), 737–745. DOI: 10.1007/s43681-023-00405-2

Feinberg, J., & Narveson, J. (1970). The nature and value of rights. *The Journal of Value Inquiry*, *4*(4), 243–257. DOI: 10.1007/BF00137935

Feldstein, S. (2024). Evaluating Europe's push to enact AI regulations: How will this influence global norms? *Democratization*, *31*(5), 1049–1066. DOI: 10.1080/13510347.2023.2196068

Ferrara, E. (2023). Fairness and bias in artificial intelligence: A brief survey of sources, impacts, and mitigation strategies. *Scientific Reports*, *13*, 21854. DOI: 10.1038/s41598-023-48938-0

Ferrari, F., Van Dijck, J., & Van den Bosch, A. (2025). Observe, inspect, modify: Three conditions for generative AI governance. *New Media & Society*, *27*(5), 2788–2806. DOI: 10.1177/14614448231214811

Feuerriegel, S., Hartmann, J., Janiesch, C., & Zschech, P. (2024). Generative ai. *Business & Information Systems Engineering*, *66*(1), 111–126. DOI: 10.1007/s12599-023-00834-7

Finnis, J. (1980). *Natural law and natural rights*. Oxford University Press.

Firmin, C., Lefevre, M., Langhoff, K., & Ruch, G. (2025). A line in the sand: Moving from surface improvement to foundational shifts to develop effective responses to extra-familial risks and harms. *British Journal of Social Work*, *55*(3), 1296–1316. DOI: 10.1093/bjsw/bcae193

Flew, T., Suzor, N., & Liu, B. R. (2013). Copyrights and copy fights: Copyright law and the digital economy. *International Journal of Technology Policy and Law*, *1*(3), 297–315. DOI: 10.1504/IJTPL.2013.057009

Floridi, L. (2021a). The European legislation on AI: A brief analysis of its philosophical approach. *Philosophy & Technology*, *34*(2), 215–222. DOI: 10.1007/s13347-021-00460-9 PMID: 34104628

Floridi, L. (2021b). *Ethics, governance, and policies in artificial intelligence*. Springer. DOI: 10.1007/978-3-030-81907-1

Floridi, L. (2023). AI as an agency without intelligence: On the philosophical foundations of artificial intelligence. *Philosophy & Technology*, *36*, 1–21.

Floridi, L., Cowls, J., Beltrametti, M., Chatila, R., Chazerand, P., Dignum, V., Luetge, C., Madelin, R., Pagallo, U., Rossi, F., Schafer, B., Valcke, P., & Vayena, E. (2018). AI4People—An ethical framework for a good AI society. *Minds and Machines*, *28*(4), 689–707. DOI: 10.1007/s11023-018-9482-5 PMID: 30930541

Fossa, F. (2025). Artificial intelligence and human autonomy: The case of driving automation. *AI & Society*, *40*(3), 1851–1862. DOI: 10.1007/s00146-024-01955-7

Fountain, J. E. (2001). *Building the virtual state: Information technology and institutional change*. Brookings Institution Press.

Franzoni, V. (2023, June). From black box to glass box: advancing transparency in artificial intelligence systems for ethical and trustworthy AI. In *International Conference on Computational Science and Its Applications* (pp. 118-130). Cham: Springer Nature Switzerland. DOI: 10.1007/978-3-031-37114-1_9

Frischmann, B., Selinger, E., & Kidd, I. (2019). *Re-engineering humanity*. Cambridge University Press.

Fry, H. (2018). *Hello world: Being human in the age of algorithms*. W. W. Norton & Company.

Futurism (2022), New European Political Party Is Led by an Artificial Intelligence, https://futurism.com/political-party-led-by-artificial-intelligence

Gans, J. S. (2024). Copyright policy options for generative artificial intelligence (No. w32106). National Bureau of Economic Research. Retrieved from https://www.nber.org/papers/w32106

Garbuio, M., & Lin, N. (2021). Innovative idea generation in problem finding: Abductive reasoning, cognitive impediments, and the promise of artificial intelligence. *Journal of Product Innovation Management*, *38*(6), 701–725. DOI: 10.1111/jpim.12602

Gebru, T., Morgenstern, J., Vecchione, B., ... (2018). Datasheets for datasets. *arXiv preprint*. https://arxiv.org/abs/1803.09010

Gebru, T., Morgenstern, J., Vecchione, B., Vaughan, J. W., Wallach, H., Daumé, H.III, & Crawford, K. (2021). Datasheets for datasets. *Communications of the ACM*, *64*(12), 86–92. DOI: 10.1145/3458723

Geiger, C. (2024). Elaborating a Human Rights-Friendly Copyright. *IIC; International Review of Industrial Property and Copyright Law*, *55*(6), 1129–1164. DOI: 10.1007/s40319-024-01481-5

Gervais, D. J. (2020). “The machine as author”. *Iowa Law Review,* 105(5), 2053, https://ilr.law.uiowa.edu/print/volume-105-issue-5/the-machine-as-author dan lihat juga P Bernt Hugenholtz and João Pedro Quintais, Copyright and Artificial Creation: Does EU Copyright Law Protect AI-Assisted Output? *IIC - International Review of Intellectual Property and Competition Law* (2021), DOI: 10.1007/s40319-021-01115-0

Gervais, D. J. (2022). Artificial intelligence and international intellectual property law. *Vanderbilt Journal of Transnational Law*, *55*(2), 301–356.

Gervais, D. J. (2023). Towards an effective transnational regulation of AI. *AI & Society*, *38*(1), 391–410. DOI: 10.1007/s00146-021-01310-0 PMID: 34776652

Ghose, A., Ali, S. A., & Deshmukh, S. (2024). Navigating the Legal and Ethical Framework for Generative AI: Fostering Responsible Global Governance. In *Exploring the Ethical Implications of Generative AI* (pp. 168-184). IGI Global Scientific Publishing. DOI: DOI: 10.4018/979-8-3693-1565-1.ch010

Ghosh, A., Saini, A., & Barad, H. (2025). Artificial intelligence in governance: Recent trends, risks, challenges, innovative frameworks and future directions. *AI & Society*, *40*(7), 1–23. DOI: 10.1007/s00146-025-02312-y

Gofen, A., Moseley, A., Thomann, E., & Kent Weaver, R. (2021). Behavioural governance in the policy process: Introduction to the special issue. *Journal of European Public Policy*, *28*(5), 633–657. DOI: 10.1080/13501763.2021.1912153

Goktas, P. (2024). Ethics, transparency, and explainability in generative ai decision-making systems: A comprehensive bibliometric study. *Journal of Decision Systems*, •••, 1–29. DOI: 10.1080/12460125.2024.2410042

Gonzalez-Sendino, R., Martínez-Rego, D., Alvares-Chillida, G., & Fontenla-Romero, Ó. (2024). Mitigating bias in artificial intelligence: Fair data generation using causal models. *Expert Systems with Applications*, *238*, 122064. DOI: 10.1016/j.eswa.2023.122064

Goodfellow, I., Bengio, Y., & Courville, A. (2014). Generative adversarial nets. *Advances in Neural Information Processing Systems*, 27. https://papers.nips.cc/paper/2014/file/5ca3e9b122f61f8f06494c97b1afccf3-Paper.pdf

Goodman, B., & Flaxman, S. (2017). European Union regulations on algorithmic decision-making and a “right to explanation”. *AI Magazine*, *38*(3), 50–57. DOI: 10.1609/aimag.v38i3.2741

Gordon, G., Rieder, B., & Sileno, G. (2022). On mapping values in AI Governance. *Computer Law & Security Review*, *46*, 105712. DOI: 10.1016/j.clsr.2022.105712

Gordon, J. S. (2021). AI and law: Ethical, legal, and socio-political implications. *AI & Society*, *36*(2), 403–404. DOI: 10.1007/s00146-021-01194-0

Gorwa, R. (2024). *The Politics of Platform Regulation: How Governments Shape Online Content Moderation*. (Oxford Studies in Digital Politics). New York, NY: Oxford University Press. DOI: 10.1093/oso/9780197692851.001.0001

Gorwa, R., Binns, R., & Katzenbach, C. (2020). Algorithmic content moderation: Technical and political challenges in the automation of platform governance. *Big Data & Society*, *7*(1). Advance online publication. DOI: 10.1177/2053951719897945

Gray, M. L., & Suri, S. (2019). *Ghost Work: How to Stop Silicon Valley from Building a New Global Underclass*. Houghton Mifflin Harcourt.

Grimmelmann, J. (2016). There's no such thing as a computer-authored work—and it's a good thing, too. *Columbia Journal of Law & the Arts*, 39(3), 403–420. https://papers.ssrn.com/sol3/papers.cfm?abstract_id=2736569

Guadamuz, A. (2017). Artificial intelligence and copyright. *WIPO Journal*, *9*(2), 169–186.

Guess, A. M., Nyhan, B., & Reifler, J. (2020). Exposure to untrustworthy websites in the 2016 US election. *Nature Human Behaviour*, *4*(5), 472–480. DOI: 10.1038/s41562-020-0833-x PMID: 32123342

Guess, A., Nagler, J., & Tucker, J. (2019). Less than you think: Prevalence and predictors of fake news dissemination on Facebook. *Science Advances*, *5*(1), eaau4586. Advance online publication. DOI: 10.1126/sciadv.aau4586 PMID: 30662946

Gunasekara, L., El-Haber, N., Nagpal, S., Moraliyage, H., Issadeen, Z., Manic, M., & De Silva, D. (2025). A systematic review of responsible artificial intelligence principles and practice. *Applied System Innovation*, *8*(4), 97. DOI: 10.3390/asi8040097

Gunawan, Y., & Arumbinang, M. H. (2023). The Climate Change Litigation Based Human Rights Approach in Corporations: Prospects and Challenges. *Journal of Human Rights, Culture and Legal System*, *3*(2), 288–307. DOI: 10.53955/jhcls.v3i2.116

Gupta, S., Kamboj, S., & Bag, S. (2023). Role of risks in the development of responsible artificial intelligence in the digital healthcare domain. *Information Systems Frontiers*, *25*(6), 2257–2274. DOI: 10.1007/s10796-021-10174-0

Gurumurthy, A., & Chami, N. (2022). AI and the Global South: Designing inclusive futures. *Development and Change*, *53*(4), 901–918.

Habermas, J. (1996). *Between facts and norms: Contributions to a discourse theory of law and democracy*. MIT Press. DOI: 10.7551/mitpress/1564.001.0001

Habibi, M.. (2025). Tinjauan Yuridis tentang Doktrin Fair Use dalam Undang-Undang Hak Cipta Indonesia dan Implikasinya terhadap Pengembangan Software. *Rewang Rencang: Jurnal Hukum Lex Generalis*, *6*(4), 1–18. DOI: 10.56370/jhlg.v6i4.860

Hacker, K., & van Dijk, J. (2000a). *Digital democracy: Issues of theory and practice*. Sage publications. DOI: 10.4135/9781446218891

Hacker, P. (2023). The European AI liability directives–Critique of a half-hearted approach and lessons for the future. *Computer Law & Security Review*, *51*, 105871. DOI: 10.1016/j.clsr.2023.105871

Hacohen, U. Y. (2023). User-based algorithmic auditing. *SSRN*. https://ssrn.com/abstract=4564776 DOI: 10.2139/ssrn.4540163

Hague, B. N., & Loader, B. D. (1999). *Digital democracy: Discourse and decision making in the information age*. Routledge.

Hainadri, H. D. (2021). Hukum Sebagai Sarana Pembaharuan Dalam Masyarakat (*law as a tool of social engineering*). *Datin Law Jurnal*, *2*(1), 66–75. DOI: 10.36355/dlj.v2i1.566

Hallamaa, J., & Kalliokoski, T. (2022). AI ethics as applied ethics. *Frontiers of Computer Science*, *4*, 776837. DOI: 10.3389/fcomp.2022.776837

Hanson, N. A., Permata, R. R., & Ramli, T. S. (2022). Protection of Video Copyrights on the Youtube Platform in Fair Use Principles. *Jurnal Legalitas*, *15*(2), 108–121. DOI: 10.33756/jelta.v15i2.15448

Hardt, M., Price, E., & Srebro, N. (2016). Equality of opportunity in supervised learning. *Advances in Neural Information Processing Systems*, •••, 29.

Harris, J., & Anthis, J. R. (2021). The moral consideration of artificial entities: A literature review. *Science and Engineering Ethics*, *27*(4), 53. DOI: 10.1007/s11948-021-00331-8 PMID: 34370075

Hartmann, D., De Pereira, J. R. L., Streitbörger, C., & Berendt, B. (2025). Addressing the regulatory gap: Moving towards an EU AI audit ecosystem beyond the AI Act by including civil society. *AI and Ethics*, *5*(4), 3617–3638. DOI: 10.1007/s43681-024-00595-3

Helberger, N., Karppinen, K., & D'Acunto, L. (2018). Exposure diversity as a design principle for recommender systems. *Information Communication and Society*, *21*(2), 191–207. DOI: 10.1080/1369118X.2016.1271900

Helbing, D., Frey, B. S., Gigerenzer, G., Hafen, E., Hagner, M., Hofstetter, Y., Van den Hoven, J., Zicari, R. V., & Zwitter, A. (2019). Will democracy survive big data and artificial intelligence? In *Towards digital enlightenment: Essays on the dark and light sides of the digital revolution* (pp. 73–98). Springer., DOI: 10.1007/978-3-319-90869-4_7

Herath Pathirannehelage, S., Shrestha, Y. R., & von Krogh, G. (2025). Design principles for artificial intelligence-augmented decision making: An action design research study. *European Journal of Information Systems*, *34*(2), 207–229. DOI: 10.1080/0960085X.2024.2330402

Herlindah, H., & Darmawan, Y. (2022). Development Legal Theory and Progressive Legal Theory: A Review, in Indonesia's Contemporary Legal Reform. *Peradaban Journal of Law and Society*, *1*(1), 15–27. DOI: 10.59001/pjls.v1i1.22

Hermann, E., & Puntoni, S. (2025). Generative AI in marketing and principles for ethical design and deployment. *Journal of Public Policy & Marketing*, *44*(3), 332–349. DOI: 10.1177/07439156241309874

Hernández, E. G. (2024). Towards an ethical and inclusive implementation of artificial intelligence in organizations: a multidimensional framework. *arXiv preprint arXiv:2405.01697.*

Hevelke, A., & Nida-Rümelin, J. (2015). Responsibility for crashes of autonomous vehicles: An ethical analysis. *Science and Engineering Ethics*, *21*(3), 619–630. DOI: 10.1007/s11948-014-9565-5 PMID: 25027859

Hildebrandt, M. (2015). *Smart technologies and the end(s) of law*. Edward Elgar Publishing. DOI: 10.4337/9781849808774

Hildebrandt, M. (2020). The artificial intelligence of European Union law. *German Law Journal*, *21*(1), 74–95. DOI: 10.1017/glj.2019.99

Hine, E., & Floridi, L. (2024). Artificial intelligence with American values and Chinese characteristics: A comparative analysis of American and Chinese governmental AI policies. *AI & Society*, *39*(1), 257–278. DOI: 10.1007/s00146-022-01499-8

Hort, M., Chen, Z., Zhang, J. M., Harman, M., & Sarro, F. (2022). Bias mitigation for machine learning classifiers: A comprehensive survey. *ACM Computing Surveys*, *55*(13s), 1–38. DOI: 10.1145/3583475

Howard, P. N., & Hussain, M. M. (2013). *Democracy's fourth wave?: Digital media and the Arab Spring*. Oxford University Press. DOI: 10.1093/acprof:oso/9780199936953.001.0001

Huang, K., Joshi, A., Dun, S., & Hamilton, N. (2024). AI regulations. In *Generative AI security: theories and practices* (pp. 61–98). Springer Nature Switzerland., DOI: 10.1007/978-3-031-54252-7_3

Hunt, A. K., Wang, J., Alizadeh, A., & Pucelj, M. (2024). Advancing a theoretical framework for exploring heuristics and biases within HR decision-making contexts. *Personnel Review*, *53*(7), 1823–1841. DOI: 10.1108/PR-03-2023-0192

Iaia, V. (2022). To be, or not to be … original under copyright law, that is (one of) the main questions concerning AI-produced works. *GRUR International: Journal of European and International IP Law*, *71*(9), 793–812. DOI: 10.1093/grurint/ikac087

Ienca, M., Fins, J., Jox, R., Jotterand, F., Voeneky, S., Andorno, R., Ball, T., Castelluccia, C., Chavarriaga, R., Chneiweiss, H., Ferretti, A., Friedrich, O., Hurst, S., Merkel, G., Molnár-Gábor, F., Rickli, J., Scheibner, J., Vayena, E., & Kellmeyer, R. (2021). Towards a Governance Framework for Brain Data. *Neuroethics*, *15*(2), 20. Advance online publication. DOI: 10.1007/s12152-022-09498-8

Ignas, K. (2019). *Algorithmic governance: Politics and law in the post-human era.* Palgrave Macmillan.

Israhadi, E. I. (2023). The Impact of Developments in Artificial Intelligence on Copyright and other Intellectual Property Laws. *Journal of Law and Sustainable Development*, *11*(11), 1–22. DOI: 10.55908/sdgs.v11i11.1965

Jackson, M. C. (2021). Artificial intelligence & algorithmic bias: The issues with technology reflecting history & humans. *Journal of Business & Technology Law*, *16*, 299–330.

Jain, R., & Jain, A. (2024, July). Generative AI in writing research papers: A new type of algorithmic bias and uncertainty in scholarly work. In *Intelligent Systems Conference* (pp. 656-669). Cham: Springer Nature Switzerland. DOI: 10.1007/978-3-031-66329-1_42

Janssen, M. (2025). Responsible governance of generative AI: Conceptualizing GenAI as complex adaptive systems. *Policy and Society*, *44*(1), 38–51. DOI: 10.1093/polsoc/puae040

Jerphagnon, L. (2025). Blaise Pascal French philosopher and scientist. https://www.britannica.com/biography/Blaise-Pascal

Jha, D., Durak, G., Sharma, V., Keles, E., Cicek, V., Zhang, Z., Srivastava, A., Rauniyar, A., Hagos, D. H., Tomar, N. K., Miller, F. H., Topcu, A., Yazidi, A., Håkegård, J. E., & Bagci, U. (2025). A Conceptual Framework for Applying Ethical Principles of AI to Medical Practice. *Bioengineering (Basel, Switzerland)*, *12*(2), 180. DOI: 10.3390/bioengineering12020180 PMID: 40001699

Jobin, A., Ienca, M., & Vayena, E. (2019). The global landscape of AI ethics guidelines. *Nature Machine Intelligence*, *1*(9), 389–399. DOI: 10.1038/s42256-019-0088-2

Jodha, D., & Bera, P. (2023). Copyright issues in the era of Ai-A critical analysis. *Res Militaris*, *13*(3), 1737–1748. https://resmilitaris.net/uploads/paper/4b260afa4323eb024822083157506b97.pdf

Joshi, H. (2025). AI Governance by Design for Agentic Systems: A Framework for Responsible Development and Deployment.

Judge, B., Nitzberg, M., & Russell, S. (2024). When code isn't law: Rethinking regulation for artificial intelligence. *Policy and Society*. Advance online publication. DOI: 10.1093/polsoc/puae020

Junaidi, J. (2023). "Artificial Intelligence in The Implementation of Criminal Case ettlement Law Using Restorative Justice Based on Progressive Law". *ICL-C Proceedings of the First International Cyber Law Conference*, ICL-C 2023, 198-204, DOI: 10.4108/eai.11-11-2023.2351325

Junaidi, J. (2025), "Transformation of Literary Culture in the Digital Era Through the Use of Artificial Intelligence as a Form of Copyright Protection", *in Role of AI in Producing, Evaluating, and Interpreting Literary Materials, IGI Global Scientific Publishing*, 177-204, DOI: 10.4018/979-8-3373-1057-2

Junaidi, J. (2020). Kajian Yuridis Pasal 362 Kitab Undang-Undang Hukum Pidana dan Pasal 30 Undang-Undang Nomor 11 tahun 2008 tentang Informasi dan Transaksi Elektronik Terhadap Pencurian Internet Wifi. *Jurnal Wajah Hukum*, *4*(2), 325–330. DOI: 10.33087/wjh.v4i2.190

Junaidi, J. (2025). Education Reform in the Application of Artificial Intelligence in Supporting Learning Media: AI and Education for Sustainable Development Goals (SDGs). In *AI, Policy, and the Future of Human-Centered Education* (pp. 57–84). IGI Global Scientific Publishing., DOI: 10.4018/979-8-3373-5781-2.ch003

Junaidi, J., & Fadzil, R. M.Rozlinda Mohamed Fadzil. (2024). Urgency of Legal Personal Data Protection in E-Commerce Transactions Involving Artificial Intelligence. *Walisongo Law Review (Walrev)*, *6*(2), 96–108. DOI: 10.21580/walrev.2024.6.2.25548

Junaidi, J., Pujiyono, P., & Fadzil, R. M. (2024). Legal Reform of Artificial Intelligence's Liability to Personal Data Perspectives of Progressive Legal Theory. *Journal of Law and Legal Reform*, *5*(2), 587–612. DOI: 10.15294/jllr.vol5i2.3437

Jungherr, A. (2023). Artificial Intelligence and Democracy: A Conceptual Framework. *Social Media + Society*, *9*(3), 20563051231186353. Advance online publication. DOI: 10.1177/20563051231186353

Kalenzi, C. (2022). Artificial intelligence and blockchain: How should emerging technologies be governed? *Frontiers in Research Metrics and Analytics*, *7*, 801549. DOI: 10.3389/frma.2022.801549 PMID: 35224423

Kamila, M. K., & Jasrotia, S. S. (2025). Ethical issues in the development of artificial intelligence: Recognizing the risks. *International Journal of Ethics and Systems*, *41*(1), 45–63. DOI: 10.1108/IJOES-05-2023-0107

Kantorowicz-Reznichenko, E., & Kantorowicz, J. (2025). Law & Economics at sixty: Mapping the field with bibliometric and machine learning tools. *Journal of Economic Surveys*, *39*(1), 201–245. DOI: 10.1111/joes.12613

Karagkouni, V., Kouroupis, S., Parlalis, I., Revolidis, M., Stylianidou, A., & Sykiotis, [First initial]. (2023). *The impact of the Covid-19 pandemic on human rights*. Logos Verlag Berlin GmbH.

Karell, D., Sachs, J., & Barrett, R. (2025). Synthetic duality: A framework for analyzing generative artificial intelligence's representation of social reality. *Poetics*, *108*, 101966. DOI: 10.1016/j.poetic.2024.101966

Karropoulos, E., & Psomiadi, A. (2026). Feeding the Future: AI-Driven Food Distribution for Aging Populations. In Stolakis, K., & Efthymiou, I. (Eds.), *AI for Geriatric Care in an Aging Society: Ethical, Clinical, and Policy Challenges* (pp. 345–370). IGI Global Scientific Publishing., DOI: 10.4018/979-8-3373-5462-0.ch012

Kashefi, P., Kashefi, Y., & Ghafouri Mirsaraei, A. (2024). Shaping the future of AI: Balancing innovation and ethics in global regulation. *Uniform Law Review*, *29*(3), 524–548. DOI: 10.1093/ulr/unae040

Katyal, S. K. (2023). Private accountability in the age of artificial intelligence. *UCLA Law Review. University of California, Los Angeles. School of Law*, *70*, 1–60.

Kazeeva, I. (2024). Sui generis intellectual property protection for works generated by AI systems. In *I*. Springer Nature Singapore., DOI: 10.1007/978-981-99-8897-6_5

Kazim, E., Dennehy, D., & Koshiyama, A. (2021). Systematizing audit in algorithmic recruitment. *Journal of Intelligence Studies in Business, 11*(1), 26–41. PMID: 34564294

Ke, L., Tong, S., Cheng, P., & Peng, K. (2025). Exploring the frontiers of llms in psychological applications: A comprehensive review. *Artificial Intelligence Review, 58*(10), 305. DOI: 10.1007/s10462-025-11297-5

Keller, P. (2023), A first look at the copyright relevant parts in the final AI Act compromise. Kluwer Copyright Blog. https://copyrightblog.kluweriplaw.com/2023/12/11/a-first-look-at-the-copyrightrelevant-parts-in-the-final-ai-act-compromise/

Kerr, A. D., & Scharp, K. (2022). The end of vagueness: Technological epistemicism, surveillance capitalism, and explainable artificial intelligence. *Minds and Machines, 32*(3), 585–611. DOI: 10.1007/s11023-022-09609-7

Kgoale, T. J., & Odeku, K. O. (2023). An analysis of the international and European Union legal instruments for holding artificial intelligence accountable. *Juridical Tribune, 13*(3).

Kim, J. J., Um, R. S., Lee, J. W., & Ajilore, O. (2025). Generative AI can fabricate advanced scientific visualizations: Ethical implications and strategic mitigation framework. *AI and Ethics, 5*(5), 4481–4493. DOI: 10.1007/s43681-024-00439-0

Kleinberg, J., Mullainathan, S., & Raghavan, M. (2017). Inherent trade-offs in the fair determination of risk scores. *Proceedings of the 8th Innovations in Theoretical Computer Science Conference (ITCS)*, 43:1–43:23.

Koniakou, V. (2022). From the "rush to ethics" to the "race for governance" in Artificial Intelligence. *Information Systems Frontiers, 25*(1), 71–102. DOI: 10.1007/s10796-022-10300-6

Kop, M. (2020). Regulating Transformative Technology in the Quantum Age: Intellectual property, standardization & sustainable innovation. *SSRN, 2*, 145–173. DOI: 10.2139/ssrn.3653544

Kouroupis, K. (2022). The AI Act in light of the EU Digital Agenda: A critical approach. *Journal of Data Protection & Privacy, 5*(3), 216–229. DOI: 10.69554/FQUZ5239

Kouroupis, K. (2023). *Digital transformation and the EU's Digital Agenda*. Logos Verlag.

Kouroupis, K., & Efthymiou, I. (Eds.). (2026). *Legal Challenges of AI Across Interdisciplinary Sectors*. IGI Global Scientific Publishing., DOI: 10.4018/979-8-3373-5452-1

Kreps, S., & Kriner, D. (2023). How AI threatens democracy. *Journal of Democracy, 34*(4), 122–131. https://www.journalofdemocracy.org/articles/how-ai-threatens-democracy/. DOI: 10.1353/jod.2023.a907693

Kretschmer, M.. (2021). Creativity, artificial intelligence, and the future of copyright. *Journal of the Copyright Society, 68*(3), 353–392.

Krishnan, R. (2025). "FROM POLICY TO PRACTICE: RESEARCH DIRECTIONS FOR TRUSTWORTHY AND RESPONSIBLE AI "BY DESIGN"."

Kriz, D. (2023). Artificial intelligence and the future of creative production. *Technology Innovation Management Review, 13*(2), 22–35.

Kroll, J. A., Huey, J., Barocas, S., Felten, E. W., Reidenberg, J. R., Robinson, D. G., & Yu, H. (2017). Accountable algorithms. *University of Pennsylvania Law Review, 165*, 633–705.

Kumar, D., & Suthar, N. (2025). Balancing Innovation and Regulation: Ethical Frameworks for AI Policy and Governance. In *Navigating the Intersection of AI Policy, Technology, and Governance* (pp. 171-200). IGI Global Scientific Publishing.

Kumari, N., Sharma, R., & Priyanka, . (2020). Intellectual property right (IPR). *Asian Journal of Pharmacy and Technology, 10*(4), 250–254. DOI: 10.5958/2231-5713.2020.00042.2

Kumar, V., Barik, S., Aggarwal, S., Kumar, D., & Raj, V. (2023). The use of artificial intelligence for persons with disability: A bright and promising future ahead. *Disability and Rehabilitation. Assistive Technology*. Advance online publication. DOI: 10.1080/17483107.2023.2288241 PMID: 38079540

Kusumaatmadja, M. (1970). *Fungsi dan Perkembangan Hukum dalam Pembangunan Nasional*. Fakultas Hukum, Universitas Padjadjaran.

Kusumaatmadja, M. (1970). *Fungsi dan perkembangan hukum dalam pembangunan nasional. Lembaga Penelitian Hukum dan Kriminologi*. Fakultas Hukum, Universitas Padjadjaran.

Kusumaatmadja, M. (1976). *Hukum, masyarakat dan pembinaan hukum nasional*. Binacipta.

Kusumaatmadja, M. (1986). *Pembinaan Hukum Dalam Rangka Pembangunan Hukum Nasional.* Lembaga Penelitian Hukum Dan Kriminologi Fakultas Hukum Universitas Padjajaran Cetakan Kedua.

Kus, Z. (2025). Digital citizenship and the future of AI engagement, ethics, and privacy. In *Digital citizenship and the future of AI engagement, ethics, and privacy*. IGI Global.

Lageson, S. E. (2020). *Digital punishment: Privacy, stigma, and the harms of data-driven criminal justice*. Oxford University Press. DOI: 10.1093/oso/9780190872007.001.0001

Lam, K., O'Neill, E., Candon, P., & Crow, D. (2024). A framework for assurance audits of algorithmic systems. *Proceedings of the 2024 ACM Conference on Fairness, Accountability, and Transparency*, 1564–1575. DOI: 10.1145/3630106.3658957

Larsson, S. (2019). The socio-legal relevance of artificial intelligence. *Droit Social, 103*(3), 573–593. DOI: 10.3917/drs1.103.0573

Larsson, S. (2020). On the Governance of Artificial Intelligence through Ethics Guidelines. *Asian Journal of Law and Society*, *7*(3), 437–451. DOI: 10.1017/als.2020.19

Latour, B. (2010). *The making of law: An ethnography of the Conseil d'État*. Polity Press.

Law, R., Ye, H., & Lei, S. S. I. (2025). Ethical artificial intelligence (AI): Principles and practices. *International Journal of Contemporary Hospitality Management, 37*(1), 279–295. DOI: 10.1108/IJCHM-04-2024-0482

Lazer, D. M. J., Baum, M. A., Benkler, Y., Berinsky, A. J., Greenhill, K. M., Menczer, F., Metzger, M. J., Nyhan, B., Pennycook, G., Rothschild, D., Schudson, M., Sloman, S. A., Sunstein, C. R., Thorson, E. A., Watts, D. J., & Zittrain, J. L. (2018). The science of fake news. *Science, 359*(6380), 1094–1096. DOI: 10.1126/science.aao2998 PMID: 29590025

Lechterman, T. M. (2024). *The concept of accountability in AI ethics and governance. The Oxford Handbook of AI Governance*. Oxford University Press.

Lecoeur, B., Mohsin, H., & Donaldson, A. F. (2023). Program reconditioning: Avoiding undefined behaviour when finding and reducing compiler bugs. *Proceedings of the ACM on Programming Languages, 7*(PLDI), 1801-1825. DOI: 10.1145/3591294

LeCun, Y., Bengio, Y., & Hinton, G. (2015). Deep learning. *Nature, 521*(7553), 436–444. DOI: 10.1038/nature14539 PMID: 26017442

Lee, K., Cooper, A. F., & Grimmelmann, J. (2024). "Talkin''Bout AI Generation: Copyright and the Generative-AI Supply Chain (The Short Version)". *In Proceedings of the Symposium on Computer Science and Law*, 48-63. DOI: 10.1145/3614407.3643696

Leiser, M. (2025). *Dark Patterns, Deceptive Design, and the Law: AI's Hidden Influence on Our Digital Experience*. Bloomsbury Publishing. DOI: 10.5040/9781509987122

Lemley, M. A., & Casey, B. (2023). Remedies for AI harms. *Texas Law Review*, *101*(6), 1191–1266.

Lepri, B., Oliver, N., Letouzé, E., Pentland, A., & Vinck, P. (2018). Fair, transparent, and accountable algorithmic decision-making processes. *Philosophical Transactions. Series A, Mathematical, Physical, and Engineering Sciences*, *376*(2128), 20170353. DOI: 10.1098/rsta.2017.0353 PMID: 30082300

Lessig, L. (1999). *Code and other laws of cyberspace*. Basic Books.

Lewandowsky, S., Ecker, U. K. H., Seifert, C. M., Schwarz, N., & Cook, J. (2020). Misinformation and its correction: Continued influence and successful debiasing. *Psychological Science in the Public Interest*, *13*(3), 106–131. DOI: 10.1177/1529100612451018 PMID: 26173286

Liang, C. J., Le, T. H., Ham, Y., Mantha, B. R., Cheng, M. H., & Lin, J. J. (2024). Ethics of artificial intelligence and robotics in the architecture, engineering, and construction industry. *Automation in Construction*, *162*, 105369. DOI: 10.1016/j.autcon.2024.105369

Liang, P., Bommasani, R., & Lee, T.. (2023). Holistic evaluation of language models. *Transactions of the Association for Computational Linguistics*, *11*, 531–551.

Li, F., Li, Y., & Li, F. (2023). Evaluating and mitigating bias in machine learning models: A practical review. npj. *Digital Medicine*, *6*, 170. DOI: 10.1038/s41746-023-00921-9

Lima, G., & Cha, M. (2020). Responsible AI and its stakeholders. *arXiv preprint arXiv:2004.11434*.

Lindsey, T.. (2022). *Hak Kekayaan Intelektual, Suatu Pengantar Edisi Kedua*. Alumni.

Lipton, Z. C. (2018). The mythos of model interpretability. *Communications of the ACM*, *61*(10), 36–43. DOI: 10.1145/3233231

Li, Q., Li, L., & Li, Y. (2024). Developing ChatGPT for biology and medicine: A complete review of biomedical question answering. *Biophysics Reports*, *10*(3), 152. DOI: 10.52601/bpr.2024.240004 PMID: 39027317

Liu, Y. C., Kuo, C. H., Wang, G. H., ... (2023). "Dialogues of Liu, Kuo & Wang creation: Collaborative content generation by the human author and ChatGPT and its impact on the evolving intellectual property landscape". In E. Y. Li, et al. (Eds23). *Proceedings of The International Conference on Electronic Business*, 23, 166-185. ICEB'23, Chiayi, Taiwan, October 19-23, 2023.

Liu, Y., Zheng, W., & Su, Y. (2024). Enhancing Ethical Governance of Artificial Intelligence Through Dynamic Feedback Mechanism. **, 105-121. DOI: 10.1007/978-3-031-57867-0_8

Liu, W., & Kim, H. G. (2025). The visual communication using generative artificial intelligence in the context of new media. *Scientific Reports*, *15*(1), 11577. DOI: 10.1038/s41598-025-96869-9 PMID: 40185848

Liu, X., Glocker, B., McCradden, M. M., Ghassemi, M., Denniston, A. K., & Oakden-Rayner, L. (2022). The medical algorithmic audit. *The Lancet. Digital Health*, *4*(5), e384–e397. DOI: 10.1016/S2589-7500(22)00003-6 PMID: 35396183

Livieri, G., Anastasopoulou, C., Briola, K., Nikolopoulou, P., & Efthymiou, I. P. (2024). Ethical issues arising from the use of AI in drug discovery. *Journal of Politics and Ethics in New Technologies and AI*, *3*(1), e37093–e37093. DOI: 10.12681/jpentai.37093

Longan, M. (2023). A System Out of Balance: A Critical Analysis of Philosophical Justifications for Copyright Law Through the Lenz of Users' Rights. *University of Michigan Journal of Law Reform. University of Michigan. Law School*, *56*(3), 779–826. DOI: 10.36646/mjlr.56.3.system

Lucchi, N. (2023). ChatGPT: A case study on copyright challenges for generative AI systems. *European Journal of Risk Regulation*, *15*(3), 1–23. DOI: 10.1017/err.2023.59

Luhur, W. H. (2024). Penerapan Teknologi Kecerdasan Buatan dalam Proses Penciptaan Karya [IKJ]. *Seminar Nasional Institut Kesenian Jakarta*, *2*, 325–335. DOI: 10.52969/seminarikj.v2i.77

Lund, B., Orhan, Z., Mannuru, N. R., Bevara, R. V. K., Porter, B., Vinaih, M. K., & Bhaskara, P. (2025). Standards, frameworks, and legislation for artificial intelligence (AI) transparency. *AI and Ethics*, *5*(4), 1–17. DOI: 10.1007/s43681-025-00661-4

Lv, Z. (2023). Generative artificial intelligence in the metaverse era. *Cognitive Robotics*, *3*, 208–217. DOI: 10.1016/j.cogr.2023.06.001

Maidanyk, L. (2021). Artificial Intelligence and Sui Generis Right: A Perspective for Copyright of Ukraine. *Justice in Eastern Europe*, *3*(11), 144–154. DOI: 10.33327/AJEE-18-4.3-n000076

Maity, S., & Saikia, M. J. (2025). Large Language Models in Healthcare and Medical Applications: A Review. *Bioengineering (Basel, Switzerland)*, *12*(6), 631. DOI: 10.3390/bioengineering12060631 PMID: 40564447

Mallikarjuna, B., & Chittemsetty, P. (2024). Generative artificial intelligence: Fundamentals and evolution. In Generative, A. I. (Ed.), *Current trends and applications* (pp. 3–17). Springer Nature Singapore., DOI: 10.1007/978-981-97-8460-8_1

Mantelero, A. (2022). Beyond data: Human rights, ethical and social impact assessment in AI systems. *Computer Law & Security Review*, •••, 45.

Manullang, S.O., et al., (2023), "Limits of the Concept of Fair Use in Law Number 28 of 2014 Concerning Copyright," *Jurnal Hukum dan HAM Wara Sains,* 2(3), 181-187, DOI: 10.58812/jhhws.v2i03.251

Mao, Q., & Xu, M. (2025). Navigating the Ethical Landscape of AI Innovation: Challenges and Opportunities. *International Journal of Digital Law and Governance*, (0).

Margoni, T. (2018). *The harmonisation of authorship in EU copyright law*. CREATe Working Paper. https://zenodo.org/record/1256996

Marif, M., & Nurhaedah, N. (2024). Teori Hukum Pembangunan dalam Eksistensi Mediasi Penal di Tingkat Penuntutan dalam Mewujudkan Penegakan Hukum Yang Berkeadilan. *Indonesian Journal of Legality of Law*, *6*(2), 217–224. DOI: 10.35965/ijlf.v6i2.4529

Marsal, A. P., & Hardyansah, R. (2025). Legal Implications of Digital Platform Models: Regulatory Approaches to Consumer Protection, Data, and Competition for Start-up Sustainability. *Bulletin of Science, Technology & Society*, *4*(2), 17–26.

Marwala, T. (2023). Artificial intelligence in politics. In *Artificial intelligence, game theory and mechanism design in politics* (pp. 1–20). Palgrave Macmillan., DOI: 10.1007/978-981-99-5103-1_4

Matthias, A. (2004). The responsibility gap: Ascribing responsibility for the actions of learning automata. *Ethics and Information Technology*, *6*(3), 175–183. DOI: 10.1007/s10676-004-3422-1

Mazzoleni, G. (2008). Populism and the Media. In Albertazzi, D., & McDonnell, D. (Eds.), *Twenty-First Century Populism*. Palgrave Macmillan., DOI: 10.1057/9780230592100_4

McCarthy, J., Minsky, M., Rochester, N., & Shannon, C. (1955). *A proposal for the Dartmouth summer research project on artificial intelligence*. Dartmouth College.

Md Ashraful, A., Gazi, M., & Sanjida Akter, S. (2025). Enhancing Regulatory Frameworks to Prevent Financial Crises. https://core.ac.uk/download/651277907.pdf

Megdad, M. M., Abueleiwa, M. H., Al Qatrawi, M., El-Tantaw, J., Harara, F. E., Abu-Nasser, B. S., & Abu-Naser, S. S. (2024). Ethics in AI: Balancing innovation and responsibility.

Meleouni, C., & Efthymiou, I. P. (2024). The Use of Artificial Intelligence (AI) in National Security: Defining International Standards and Guidelines. *Journal of Politics and Ethics in New Technologies and AI, 3*(1), e37847–e37847.

Meyer, S. (2024). Legal challenges of artificial intelligence and how to manage them. In Barton, T., & Müller, C. (Eds.), *Artificial intelligence in application* (pp. 9–30). Springer Fachmedien., DOI: 10.1007/978-3-658-43843-2_2

Mitchell, G. (2014). Alternative behavioral law and economics. *The Oxford handbook of behavioral economics and the law*, 167.

Mitchell, M., Wu, S., & Zaldivar, A.. (2019). Model cards for model reporting. In *Proceedings of the Conference on Fairness, Accountability, and Transparency* (pp. 220–229). ACM. DOI: 10.1145/3287560.3287596

Mitchell, T. M. (1997). *Machine learning*. McGraw-Hill.

Mittelstadt, B. (2019). Principles alone cannot guarantee ethical AI. *Nature Machine Intelligence, 1*(11), 501–507. DOI: 10.1038/s42256-019-0114-4

Mittelstadt, B. D., Allo, P., Taddeo, M., Wachter, S., & Floridi, L. (2016). The ethics of algorithms: Mapping the debate. *Big Data & Society, 3*(2), 1–21. DOI: 10.1177/2053951716679679

Mock, C. (2025). Tech CEOs, Trump and society's new AI age. *Australian Journalism Review, 47*(1), 7–15. DOI: 10.1386/ajr_00176_7

Mohamed, S., Png, M. T., & Isaac, W. (2020). Decolonial AI: Decolonial Theory as Sociotechnical Foresight in Artificial Intelligence. *Philosophy & Technology, 33*(4), 659–684. DOI: 10.1007/s13347-020-00405-8

Moon, M. J. (2023). Searching for inclusive artificial intelligence for social good: Participatory governance and policy recommendations for making AI more inclusive and benign for society. *Public Administration Review, 83*(6), 1496–1505. DOI: 10.1111/puar.13648

Morganti, P., & Valdes, R. C. (2025). Toward Fair Futures? Digital Transformation and Income Disparities in the Era of Industry 5.0. Springer Books. In Teran-Bustamante, A., & Salazar, M. V. (Eds.), *Digital Transformation and Innovation in Organizations* (pp. 45–76). Springer., DOI: 10.1007/978-3-031-90794-4_3

Morley, J., Elhalal, A., Garcia, F., Kinsey, L., Mökander, J., & Floridi, L. (2021). Ethics as a service: A pragmatic operationalisation of AI ethics. *Minds and Machines*, *31*(2), 239–256. DOI: 10.1007/s11023-021-09563-w PMID: 34720418

Moseley, A., & Thomann, E. (2021). A behavioural model of heuristics and biases in frontline policy implementation. *Policy and Politics*, *49*(1), 49–67. DOI: 10.1332/030557320X15967973532891

Muhaimin, (2020). *Metode Penelitian Hukum,* Mataram: Mataram University Press.

Muir, A., & Oppenheim, C. (2002). National information policy developments worldwide I: Electronic government. *Journal of Information Science*, *28*(3), 173–186. DOI: 10.1177/016555150202800301

Müller, V. C. (2020). *Ethics of artificial intelligence and robotics*

Munn, L., Magee, L., & Arora, V. (2024). Truth machines: Synthesizing veracity in AI language models. *AI & Society*, *39*(6), 2759–2773. DOI: 10.1007/s00146-023-01756-4

Murfianti, F. (2020). Hak Cipta dan Karya Seni di Era Digital. *Acintya Jurnal Penelitian Seni Budaya*, *12*(1), 44–58. DOI: 10.33153/acy.v12i1.3147

Myers, A. (2023, February 27). AI's powers of political persuasion. *Stanford HAI*. https://hai.stanford.edu/news/ais-powers-political-persuasion

Nafila Andriana Putri, N. A. (2023). Copyright Protection for Internet Memes: The Doctrine of Fair Use in Indonesia. *JIPRO: Journal of Intellectual Property*, *6*(2). Advance online publication. DOI: 10.20885/jipro.vol6.iss2.art3

Naqvi, Z. (2020). Artificial intelligence, copyright, and copyright infringement. Marq. Intell. Prop. L. Rev., 24, 15. https://scholarship.law.marquette.edu/iplr/vol24/iss1/4/

National Institute of Standards and Technology. (2022). *Artificial intelligence definition and characteristics*. https://www.nist.gov

National Institute of Standards and Technology. (2023). *AI risk management framework (AI RMF 1.0)*. https://www.nist.gov

Netanel, N. W. (2008). *Introduction, Copyright's Paradox.* Oxford University Press, UCLA School of Law Research Paper No. 08-06, Available at *SSRN*: https://ssrn.com/abstract=1099457

Newman, J. C. (2020). Decision Points in AI Governance: Three Case Studies Explore Efforts to Operationalize AI Principles. UC Berkeley, Center for Long-Term Cybersecurity. https://cltc.berkeley.edu/wp-content/uploads/2020/05/Decision_Points_AI_Governance.pdf 19

News, N. B. C. (2025), Matt Lavietes, Meta's new hate speech guidelines permit users to say LGBTQ people are mentally ill, https://www.nbcnews.com/tech/social-media/meta-new-hate-speech-rules-allow-users-call-lgbtq-people-mentally-ill-rcna186700

Nilsson, N. J. (2009). *The quest for artificial intelligence.* Cambridge University Press. DOI: 10.1017/CBO9780511819346

NIST. (2022). Towards a standard for identifying and managing bias in artificial intelligence (NIST Special Publication 1270). National Institute of Standards and Technology. DOI: 10.6028/NIST.SP.1270

Nordström, M. (2022). AI under great uncertainty: Implications and decision strategies for public policy. *AI & Society*, *37*(4), 1703–1714. DOI: 10.1007/s00146-021-01263-4 PMID: 34511737

Norris, P., & Inglehart, R. (2019). *Cultural backlash: Trump, Brexit, and authoritarian populism.* Cambridge University Press. DOI: 10.1017/9781108595841

Novelli, C., & Sandri, G. (2024). Digital Democracy in the Age of Artificial Intelligence. SSRN Electronic Journal. DOI: 10.2139/ssrn.4901264

Novelli, C., Hacker, P., Morley, J., Trondal, J., & Floridi, L. (2025). A robust governance for the AI Act: AI office, AI board, scientific panel, and national authorities. *European Journal of Risk Regulation*, *16*(2), 566–590. DOI: 10.1017/err.2024.57

Novelli, C., Taddeo, M., & Floridi, L. (2024). Accountability in artificial intelligence: What it is and how it works. *AI & Society*, *39*(4), 1871–1882. DOI: 10.1007/s00146-023-01635-y

Nyhan, B., & Reifler, J. (2010). When corrections fail: The persistence of political misperceptions. *Political Behavior*, *32*(2), 303–330. DOI: 10.1007/s11109-010-9112-2

O'Neil, C. (2016). *Weapons of math destruction: How big data increases inequality and threatens democracy*. Crown.

OECD. (2003). *Promise and Problems of e-Democracy. Challenges of Citizen Engagement*. OECD.

OECD. (2019). *OECD principles on artificial intelligence*. https://www.oecd.org/ai

OECD. (2023). *AI language models: Technological, socio-economic and policy considerations*. OECD Digital Economy Papers. https://www.oecd-ilibrary.org/science-and-technology/ai-language-models_13d38f92-en

OECD. (2023). *The state of implementation of the OECD AI Principles four years on*. OECD Publishing. https://www.oecd-ilibrary.org/science-and-technology/the-state-of-implementation-of-the-oecd-ai-principles_13d38f92-en

Ognyanova, K., Lazer, D., Robertson, R. E., & Wilson, C. (2020). Misinformation in action: Fake news exposure is linked to lower trust in media, higher trust in government when your side is in power. *Harvard Kennedy School Misinformation Review*, *1*(4). Advance online publication. DOI: 10.37016/mr-2020-024

Ooi, K. L. (2025). Modern Behavioural Finance Theories. In *Demystifying Behavioral Finance: Foundational Theories to Contemporary Applications and Future Directions* (pp. 59–69). Springer Nature Singapore.

Organisation for Economic Co-operation and Development (OECD). (2019). *Recommendation of the Council on Artificial Intelligence*. OECD Legal Instruments. https://legalinstruments.oecd.org/en/instruments/OECD-LEGAL-0449

Organisation for Economic Co-operation and Development. (2025). *Intellectual property issues in artificial intelligence trained on scraped data*. https://www.oecd.org

Pagallo, U., Bassi, E., & Durante, M. (2023). The normative challenges of AI in outer space: Law, ethics, and the realignment of terrestrial standards. *Philosophy & Technology*, *36*(2), 23. DOI: 10.1007/s13347-023-00626-7

Parker, L., Hayes, J., Loper, A. J., & Karakas, A. L. (2024). Navigating the unknown: anticipating concerns and gaps in generative AI research. In *General Aspects of Applying Generative AI in Higher Education: Opportunities and Challenges* (pp. 3-20). Cham: Springer Nature Switzerland. DOI: 10.1007/978-3-031-65691-0_1

Parrilli, D. M. (2025). *Informational Privacy for Service Design: An Ethical Framework for Designing Privacy-Oriented Services*. Springer Nature. DOI: 10.1007/978-3-031-76926-9

Pasechnyk, O. (2022). International aspects of intellectual property rights protection. *Baltic Journal of Economic Studies*, *8*(5), 146–157. DOI: 10.30525/2256-0742/2022-8-5-146-157

Pasquale, F. (2015). *The black box society: The secret algorithms that control money and information*. Harvard University Press. DOI: 10.4159/harvard.9780674736061

Pasricha, S. (2022). AI ethics in smart healthcare. *IEEE Consumer Electronics Magazine*, *12*(4), 12–20. DOI: 10.1109/MCE.2022.3220001

Patel, K. (2024). Ethical reflections on data-centric AI: balancing benefits and risks. Available at *SSRN* 4993089. DOI: 10.2139/ssrn.4993089

Perez-Morote, R., Rosa, C. P., & Cortes, E. A. (2022). Exploring the relation between the digital divide and government's effort to develop e-participation. In *Research anthology on citizen engagement and activism for social change* (pp. 997–1018). IGI Global., DOI: 10.4018/978-1-6684-3706-3.ch054

Pham, N., Kuo, R., & Huang, Y. (2025). Fairness for machine learning software in education: Mapping and review. *Computers & Education*, *220*, 105120. DOI: 10.1016/j.compedu.2024.105120

Pires Ting, C., Filippovna, V., & Araujo, M. M. (2024). AI and the Visualization of Paradise: Cultural Paradigms, Aesthetic Evolution, and Cognitive Exploration. *Logica Universalis*, *18*(4), 505–519. DOI: 10.1007/s11787-024-00365-z

Pistilli, G., Munoz Ferrandis, C., Jernite, Y., & Mitchell, M. (2023). *Stronger Together: on the Articulation of Ethical Charters*. Legal Tools, and Technical Documentation in ML.

Pistilli, G., Muñoz Ferrandis, C., Jernite, Y., & Mitchell, M. (2023, June). Stronger together: on the articulation of ethical charters, legal tools, and technical documentation in ML. In *Proceedings of the 2023 ACM Conference on Fairness, Accountability, and Transparency* (pp. 343-354). DOI: 10.1145/3593013.3594002

Politico. (2025). EU lawmakers blast Meta's elimination of fact-checkers in U.S. https://www.politico.eu

Poszler, F., Portmann, E., & Lütge, C. (2025). Formalizing ethical principles within AI systems: Experts' opinions on why (not) and how to do it. *AI and Ethics*, *5*(2), 937–965. DOI: 10.1007/s43681-024-00425-6

Presno Linera, M., & Meuwese, A. (2025). AI regulation and constitutional design in the EU. *European Constitutional Law Review*, *21*(1), 1–26.

Prunkl, C. (2024). Human autonomy at risk? An analysis of the challenges from AI. *Minds and Machines*, *34*(3), 26. DOI: 10.1007/s11023-024-09665-1

Psomiadi, A. (2026). Social Dynamics and AI-Driven Care for Older Adults. In *AI for Geriatric Care in an Aging Society: Ethical, Clinical, and Policy Challenges* (pp. 123-154). IGI Global Scientific Publishing.

Psomiadi, A., Diareme, K. C., & Chatzivasileiou, S. (2020). HAPSc Policy Briefs Series.

Puertas-Bravo, L., & Piedra, N. (2024). Regulation of artificial intelligence: Challenges and perspectives in the Andean Community. In Bianchi, I., & Dávila, G. A. (Eds.), *Knowledge management and artificial intelligence for growth* (pp. 221–244). Springer Nature Switzerland., DOI: 10.1007/978-3-031-65552-4_11

Puspasari, A. T., & Sardjono, A. (2023). Pembatasan Hak Cipta Terkait Remix Lagu Berdasarkan Doktrin Fair Use dan Undang- Undang Nomor 28 Tahun 2014 Tentang Hak Cipta. *Technology and Economics Law*, *2*(2), 265–281. DOI: 10.21143/TELJ.vol2.no2.1040

Putro, H. P.. (2023). *Development Of Artificial Intelligence Applications (Studi Kasus & Implementasi AI Menggunakan Berbagai Bahasa Pemrograman.* Jambi: PT. Sonpedia Publishing Indonesia. Qian, Y., Siau, K.L. & Nah, F.F., "Societal impacts of artificial intelligence: Ethical, legal, and governance issues". *Societal Impacts*, *3*(1000040), 1–5. DOI: 10.1016/j.socimp.2024.100040

Rachlinski, J. J. (2011). The psychological foundations of behavioral law and economics. *U. Ill. L. Rev.*, 1675.

Rachovitsa, A., & Johann, N. (2022). The human rights implications of the use of AI in the digital welfare state: Lessons learned from the Dutch SyRI case. *Human Rights Law Review*, *22*(2), ngac010. Advance online publication. DOI: 10.1093/hrlr/ngac010

Radanliev, P. (2025). AI ethics: Integrating transparency, fairness, and privacy in AI development. *Applied Artificial Intelligence*, *39*(1), 2463722. Advance online publication. DOI: 10.1080/08839514.2025.2463722

Raglin, A. (2024, June). Leveraging generative AI concepts with uncertainty of information concept. In *International Conference on Human-Computer Interaction* (pp. 230-238). Cham: Springer Nature Switzerland. DOI: 10.1007/978-3-031-60611-3_17

Raharjo, B. (2023). *Teori Etika Dalam Kecerdasan Buatan (AI)*. Penerbit Yayasan Prima Agus Teknik.

Rahman, P. (2025). *Exploring ethics and trust in persuasive technologies* (Doctoral dissertation, University of British Columbia).

Rahmayanti, S. D., Rismana, V., Salsabila, Y., & Syechnas, N. Z. Z. (2024). Pertarungan Hukum Hak Cipta di Era Digital: Studi Kasus dalam Konteks Indonesia. *Jurnal Dunia Ilmu Hukum, 2*(1), 16–21. DOI: 10.59435/jurdikum.v2i1.372

Rai, A. K. (2021). Intellectual property and artificial intelligence. *Chicago-Kent Law Review, 96*(1), 1–26.

Raji, I. D., & Buolamwini, J. (2019). Actionable auditing: Investigating the impact of publicly naming biased performance results. Proceedings of the 2019 AAAI/ACM Conference on AI, Ethics, and Society, 429–435. DOI: 10.1145/3306618.3314244

Raji, I. D., Smart, A., White, R. N., Mitchell, M., Gebru, T., Hutchinson, B., Smith-Loud, J., Theron, D., & Barnes, P. (2020). Closing the AI accountability gap: Defining an end-to-end framework for internal algorithmic auditing. *Proceedings of the 2020 Conference on Fairness, Accountability, and Transparency*, 33–44. DOI: 10.1145/3351095.3372873

Rama, B. G. A., Prasada, D. K., & Mahadewi, K. J. (2023). Urgensi Pengaturan *Artificial Intelligence* (AI) dalam Bidang Hukum Hak Cipta di Indonesia. *Jurnal Rechten, 12*(2), 209–224. DOI: 10.56013/rechtens.v12i2.2395

Ramadhani, M. (2025). The fair use doctrine in the digital age: Reassessing the balance between creators' rights and public access. *Journal of Law Science, 7*(4), 733–742. DOI: 10.35335/jls.v7i4.6749

Ramadika, I. K. G. A., & Wijaya, I. K. K. A.I Ketut Gede Adi RamadikaI Ketut Kasta Arya Wijaya. (2025). Legality of Copyright Protection on *Artificial Intelligence* Works. *Pemuliaan Keadilan, 2*(1), 105–114. DOI: 10.62383/pk.v2i1.460

Ramli, A. M. (2021). "Pentingnya Bangsa Ini Mulai Adaptif Dengan Industri 5.0", https://www.republika.co.id/berita/qynv2n371/pentingnya-bangsa-ini-mulai-adaptif -dengan-industri-50

Ramya, R., Priya, S., Thamizhikkavi, P., & Anand, M. (2024). The pillars of AI ethics: Transparency, accountability, and privacy. In *Responsible implementations of generative AI for multidisciplinary use*. IGI Global Publishing. DOI: 10.4018/979-8-3693-9173-0.ch004

Raso, F. A., Hilligoss, H., Krishnamurthy, V., Bavitz, C., & Kim, L. (2018). *Artificial intelligence & human rights: Opportunities & risks*. Berkman Klein Center Research Publication, 2018(6).

Rassolov, I., & Chubukova, S. (2022). Artificial Intelligence and Effective Governance: Legal Framework. *Kutafin Law Review*. DOI: 10.17803/2713-0525.2022.2.20.309-328

Rawls, J. (1971). *A theory of justice.* Harvard University Press. DOI: 10.4159/9780674042605

Raza, S., Qureshi, R., Zahid, A., Kamawal, S., Sadak, F., Fioresi, J., . . . Shoman, M. (2025). Who is responsible? the data, models, users or regulations? a comprehensive survey on responsible generative ai for a sustainable future. *arXiv preprint arXiv:2502.08650.* https://doi.org//arXiv.2502.08650DOI: 10.48550

Redi Pirmansyah, R., Junaidi, J., & Merta, M. M. (2021). OSC 2021-FHISIP Universitas Terbuka, Indonesia Theme 1: "*Law Transformation the Law of Personal Data Protection in The Pandemic Time Covid-19 On Electronic System*". https://www.cips-indonesia.org/post/siaran-pers-kebocoran-data-ingatkan-pentingnya-legislasi-ruu-perlindungan-data-pribadi

Reed-Berendt, R., Dove, E. S., & Pareek, M.UK-REACH Study Collaborative Group. (2022). The ethical implications of big data research in public health:"big data ethics by design" in the uk-reach study. *Ethics & Human Research, 44*(1), 2–17. DOI: 10.1002/eahr.500111 PMID: 34936235

Rejeki, H.P., Iriyanti & Dadang. (2023). "Pelanggaran Hak Ciptapada Penjualan Buku Hasil Bajakan Berdasarkan Undang-Undang Nomor 28 tahun 2014 tentang Hak Cipta", *Jurnal Surya Kencana Dua: Dinamika Masalah Hukum dan Keadilan*, 10(1), 52-68, DOI: 10.32493/SKD.v10i1.y2023.32254

Resseguier, A., & Rodrigues, R. (2021). Ethics as attention to context: Recommendations for the ethics of artificial intelligence. *Open Research Europe, 1*, 27. DOI: 10.12688/openreseurope.13260.2 PMID: 37645106

Ricketson, S. (2003). *SCCR/9/7: WIPO Study on Limitations and Exceptions of Copyright and Related Rights in The Digital Environment.*

Ricketson, S., & Ginsburg, J. (2022). *International copyright and neighbouring rights* (3rd ed.). Oxford University Press. DOI: 10.1093/oso/9780198801986.001.0001

Risnandar, A. & Zaenudin, K.M.I.S., "Development and Evolution of Indonesian Law from the Perspective of Development Law Theory", *Rechtsvinding*, 3(2), 9-24. https://doi.org/DOI: 10.59525/rechtsvinding.v3i2.893

Roberts, H., Cowls, J., Hine, E., Mazzi, F., Tsamados, A., Taddeo, M., & Floridi, L. (2021). Achieving a 'Good AI Society': Comparing the Aims and Progress of the EU and the US. *Science and Engineering Ethics, 27*(6), 68. DOI: 10.1007/s11948-021-00340-7 PMID: 34767085

Roberts, H., Cowls, J., Morley, J., Taddeo, M., Wang, V., & Floridi, L. (2021). The Chinese approach to artificial intelligence governance. *AI & Society*, *36*(1), 59–77. DOI: 10.1007/s00146-020-00992-2

Robles, P., & Mallinson, D. J. (2023). Artificial intelligence technology, public trust, and effective governance. *The Review of Policy Research*. Advance online publication. DOI: 10.1111/ropr.12555

Rocco, S. (2022). *Implementing and managing Algorithmic Decision-Making in the public sector* (No. ex93w_v1). Center for Open Science.

Rochim, A. A. (2024). Kecerdasan Buatan: Resiko, Tantangan dan Penggunaan Bijak pada Dunia Pendidikan. *Antroposen: Journal of Social Studies and Humaniora*, *3*(1), 13–25. DOI: 10.33830/antroposen.v3i1.6780

Rodríguez, N., Ser, J., Coeckelbergh, M., De Prado, M., Herrera-Viedma, E., & Herrera, F. (2023). Connecting the Dots in Trustworthy Artificial Intelligence: From AI Principles, Ethics, and Key Requirements to Responsible AI Systems and Regulation. *Information Fusion*, *99*, 101896. DOI: 10.1016/j.inffus.2023.101896

Ronchi, A. M. (2019). e-Democracy. Cham: Springer International Publishing.

Routray, S. K., Jha, M. K., Sharmila, K. P., Javali, A., Pappa, M., & Singh, M. (2024, October). Generative Artificial Intelligence: Principles, Potentials and Challenges. In *2024 2nd International Conference on Self Sustainable Artificial Intelligence Systems (ICSSAS)* (pp. 211-216). IEEE. DOI: 10.1109/ICSSAS64001.2024.10760992

Royal Society. (2024). Towards algorithm auditing: Managing legal, ethical and technical challenges. *Royal Society Open Science*, *11*(1), 231209. DOI: 10.1098/rsos.231209

Rumiantsev, A., Morozov, V., & Panikar, G. (2024). Innovation at the crossroads: How Public-private partnerships drive technological development in Europe. *Actual Problems of International Relations*, *1*(159), 94–108. DOI: 10.17721/apmv.2024.159.1.94-108

Russell, S., & Norvig, P. (2020). *Artificial intelligence: A modern approach* (4th ed.). Pearson.

Ryan, M., & Stahl, B. C. (2021). Artificial intelligence ethics guidelines for developers and users: Clarifying their content and normative implications. *Journal of Information. Communication and Ethics in Society*, *19*(1), 61–86. DOI: 10.1108/JICES-12-2019-0138

Saber, A., Shaheen, N., & Giammalva, S. (2024). "6 AI Cases And What They Mean For Copyright Law," Law360 (New York, 2024), https://www.crowell.com/a/web/7QtNejMH1FSM1n5Ddt6cdU/6-ai-cases-and-what-they-mean-for-copyright-law.pdf

Sahebi, S., & Formosa, P. (2025). The AI-mediated communication dilemma: Epistemic trust, social media, and the challenge of generative artificial intelligence. *Synthese*, *205*(3), 1–24. DOI: 10.1007/s11229-025-04963-2

Samuelson, P. (2023). Generative AI and copyright. SSRN *Electronic Journal*. https://papers.ssrn.com/sol3/papers.cfm?abstract_id=4374012 DOI: 10.1126/science.adi0656

Sandvig, C., Hamilton, K., Karahalios, K., & Langbort, C. (2014). *Auditing algorithms: Research methods for detecting discrimination on internet platforms*. Data & Society Research Institute.

Saputra, A. M. A., et al. (2023). *Teknologi Informasi: Peranan TI dalam berbagai Bidang*. Jambi: PT. Sonpedia Publishing Indonesia.

Sarnoff, L., & Burton, C. (2024, January 15). ABC News live fact-checks Trump-Harris debate. *ABC News*. https://www.abcnews.com

Saura, J., Ribeiro-Soriano, D., & Palacios-Marqués, D. (2022). Assessing behavioural data science privacy issues in government artificial intelligence deployment. *Government Information Quarterly*, *39*(4), 101679. DOI: 10.1016/j.giq.2022.101679

Schiff, D. (2022). Education for AI, not AI for education: The role of education and ethics in national AI policy strategies. *International Journal of Artificial Intelligence in Education*, *32*(3), 527–563. DOI: 10.1007/s40593-021-00270-2

Schiff, D. S. (2023). Looking through a policy window with tinted glasses: Setting the agenda for US AI policy. *The Review of Policy Research*, *40*(5), 729–756. DOI: 10.1111/ropr.12535

Schiff, D., Borenstein, J., Biddle, J., & Laas, K. (2021). AI ethics in the public, private, and NGO sectors: A review of a global document collection. *IEEE Transactions on Technology and Society*, *2*(1), 31–42. DOI: 10.1109/TTS.2021.3052127

Schneider, J. (2024). Explainable Generative AI (GenXAI): A survey, conceptualization, and research agenda. *Artificial Intelligence Review*, *57*(11), 289. DOI: 10.1007/s10462-024-10916-x

Schneier, B., & Sanders, N. E. (2023, July 28). Six ways that AI could change politics. *MIT Technology Review*. https://www.technologyreview.com/2023/07/28/1076756/six-ways-that-ai-could-change-politics/

Schuett, J. (2019). A legal definition of artificial intelligence. *Stanford Technology Law Review : STLR : An Online High-Technology Law Journal from Stanford Law School, 22*(1), 1–50.

Sedkaoui, S., & Benaichouba, R. (2024). Generative AI as a transformative force for innovation: A review of opportunities, applications and challenges. *European Journal of Innovation Management*. Advance online publication. DOI: 10.1108/EJIM-02-2024-0129

See, B. R. (2020). Peranan Hukum sebagai Sarana Pembaharuan Masyarakat Menuju Masyarakat Indonesia yang Sadar Hukum. *Jurnal Hukum Caraka Justitia, 1*(1), 42–50. DOI: 10.30588/jhcj.v1i1.700

Selbst, A. D., Boyd, D., Friedler, S. A., Venkatasubramanian, S., & Vertesi, J. (2019). Fairness and abstraction in sociotechnical systems. Proceedings of the ACM on Human-Computer Interaction, 3(CSCW), 1–36. DOI: 10.1145/3287560.3287598

Selbst, A. D. (2019). Negligence and AI's human users. *Boston University Law Review. Boston University. School of Law, 100*, 1315–1344.

Selbst, A. D., & Barocas, S. (2018). The intuitive appeal of explainable machines. *Fordham Law Review, 87*, 1085–1139.

Sellwood, M. (2017). The road to autonomy. *The San Diego Law Review, 54*, 829.

Senftleben, M. (2021). Copyright, creativity, and innovation. *Amsterdam Law School Research Paper*. https://ssrn.com/abstract=3032406

Sengar, S. S., Hasan, A. B., Kumar, S., & Carroll, F. (2025). Generative artificial intelligence: A systematic review and applications. *Multimedia Tools and Applications, 84*(21), 23661–23700. DOI: 10.1007/s11042-024-20016-1

Shafik, W. (2024). Generative AI for social good and sustainable development. In *Generative AI: current trends and applications* (pp. 185–217). Springer Nature Singapore., DOI: 10.1007/978-981-97-8460-8_10

Shah, Z., Shahzad, M. H., Saleem, S., Taj, I., Amin, S., Almagharbeh, W. T., & Durvesh, S. (2025). ETHICAL CONSIDERATIONS IN THE USE OF AI FOR ACADEMIC RESEARCH AND SCIENTIFIC DISCOVERY: A NARRATIVE REVIEW. *Insights-Journal of Life and Social Sciences, 3*(2), 183–189. DOI: 10.71000/jfesgv69

Sharma, M. (2023). Intellectual property rights and artificial intelligence. *Indian Journal of Law and Legal Research, 5*(2), 1–15.

Sharma, M. (2024). *The impact of artificial intelligence (AI) on the future of democracy and civic participation*. DOI: 10.2139/ssrn.4703439

Shefan, A. (2023). Copyright and Its Exceptions and Limitations in Ukraine. In Richter, H. (Ed.), *Competition and Intellectual Property Law in Ukraine. MPI Studies on Intellectual Property and Competition Law* (Vol. 31, pp. 507–536). Springer., DOI: 10.1007/978-3-662-66101-7_20

Sheikh, H., Prins, C., & Schrijvers, E. (2023). Artificial intelligence: Definition and background. In Mission, A. I. (Ed.), *Research for policy* (pp. 11–34). Springer., DOI: 10.1007/978-3-031-21448-6_2

Shidarta. (2006). *Karakteristik Penalaran Hukum Dalam Konteks Ke-Indonesiaan*. Jakarta: CV Utomo.

Singh, S., & Singh, A. (2023). "Intellectual Property Rights and Artificial Intelligence: Contemporary Convergence and Probable Challenges", *2023 10th IEEE Uttar Pradesh Section International Conference on Electrical, Electronics and Computer Engineering (UPCON)*, Gautam Buddha Nagar, India, 1279-1286, DOI: 10.1109/UPCON59197.2023.10434754

Slote, M. (2023). Agent-based virtue ethics. In *Philosophical Essays East and West: Agent-Based Virtue Ethics and other topics at the intersection of Chinese thought and Western analytic philosophy* (pp. 83–95). Springer Nature Switzerland. DOI: 10.1007/978-3-031-39955-8_6

Smuha, N. A. (2025). *The Cambridge Handbook of the Law*. Ethics and Policy of Artificial Intelligence.

Solaiman, I., Talat, Z., Agnew, W., Ahmad, L., Baker, D., Blodgett, S. L., Subramonian, A. (2023). Evaluating the social impact of generative ai systems in systems and society. *arXiv preprint arXiv:2306.05949*. https://doi.org/10.48550/arXiv.2306.05949

Solum, L. B. (1992). Legal personhood for artificial intelligences. *North Carolina Law Review*, *70*(4), 1231–1287.

Song, S. (2023). *China's First Case on Copyrightability of AI-Generated Picture*. https://www.kwm.com/cn/en/insights/latest-thinking/china-s-first-case-on-copyrightability-of-ai-generated-picture.html

Soraya, J., & Althafzufar, M. A. (2024). Intellectual Property Rights Protection for Actors in the Creative Economy Based on Intellectual Property Rights Law Number 28 of 2014 Concerning Copyright. *Realism: Law Review*, *2*(1), 39–53. DOI: 10.71250/rlr.v2i1.30

Sotiropoulos, L. (2025). Shaping the Future of Shipping: Legal and Policy Impacts of EU Emissions Trading and Fuel EU Maritime Regulations. In *AI, Sustainability and Solidarity in Advancing European Values in a Globalized Age* (pp. 95–106). Springer Nature Switzerland. DOI: 10.1007/978-3-032-06508-7_8

Sotiropoulos, L. D. (2023). Addressing Smart Contracts in the Insurance Sector: Institutional Framework and Practical Aspects. *ENTHA*, *20*, 22.

Soyer, B., & Tettenborn, A. (2022). Artificial intelligence and civil liability—Do we need a new regime? *International Journal of Law and Information Technology*, *30*(4), 385–397. DOI: 10.1093/ijlit/eaad001

Special Secretariat for Foresight. (2024). *A blueprint for Greece's AI transformation*. Government of Greece. https://foresight.gov.gr/en/studies/A-Blueprint-for-Greece-s-AI-Transformation/

Spindler, G. (2019), "Copyright Law and Artificial Intelligence," *IIC - International Review of Intellectual Property and Competition Law* 50, 1049-1051, https://doi.org/https://dx.doi.org/10.18196/iclr.v5i2.17927

Stamboliev, E., & Christiaens, T. (2025). How empty is trustworthy AI? A discourse analysis of the ethics guidelines of trustworthy AI. *Critical Policy Studies*, *19*(1), 39–56. DOI: 10.1080/19460171.2024.2315431

Stanford University DigiChina Project. (2023). Translation: Interim measures for the management of generative artificial intelligence services (China). DigiChina Research Brief. https://digichina.stanford.edu

Strubell, E., Ganesh, A., & McCallum, A. (2019). Energy and Policy Considerations for Deep Learning in NLP. *Proceedings of the 57th Annual Meeting of the Association for Computational Linguistics*, 3645–3650. DOI: 10.18653/v1/P19-1355

Summerfield, C., & Parpart, P. (2022). Normative principles for decision-making in natural environments. *Annual Review of Psychology*, *73*(1), 53–77. DOI: 10.1146/annurev-psych-020821-104057 PMID: 34555286

Sunstein, C. R. (2017). *Republic: Divided democracy in the age of social media*. Princeton University Press. DOI: 10.1515/9781400884711

Suresh, H., & Guttag, J. V. (2021). A framework for understanding sources of harm throughout the machine learning life cycle. *Proceedings of the 2021 ACM Conference on Fairness, Accountability, and Transparency*, 806–807. DOI: 10.1145/3465416.3483305

Susskind, R. (2019). *Online courts and the future of justice*. Oxford University Press. DOI: 10.1093/oso/9780198838364.001.0001

Suzor, N. P., West, S. M., Quodling, A., & York, J. (2019). What do we mean when we talk about transparency? Toward meaningful transparency in commercial content moderation. *International Journal of Communication*, *13*, 1526–1547.

Swan, A., & Aguilar, Y. (2025). RESPONSIBLE AI BY DESIGN, NOT FOR COMPLIANCE.

SWARGIARY. K. (2024). *Foundations and Frontiers in Ethics (A Comprehensive Study)*. EdTech Research Association, US.

Taddeo, M., & Floridi, L. (2018). How AI can be a force for good. *Science*, *361*(6404), 751–752. DOI: 10.1126/science.aat5991 PMID: 30139858

Taeihagh, A. (2025). Governance of generative AI. *Policy and Society*, *44*(1), 1–22. DOI: 10.1093/polsoc/puaf001

Taj, I., & Zaman, N. (2022). Towards industrial revolution 5.0 and explainable artificial intelligence: Challenges and opportunities. *International Journal of Computing and Digital Systems*, *12*(1), 295–320. DOI: 10.12785/ijcds/120128

Tambouris, Efthimios & Gorilas, S. & Boukis, G. & A, Archetypon. (2001). Investigation of Electronic Government.

Tampubolon, M. (2025). Decoding Legal Ambiguity: the Interplay between Law and Legal Semiotics in Modern Jurisprudence. *International Journal for the Semiotics of Law-Revue internationale de Sémiotique juridique*, 1-29.

Taylor, L. (2017). What is data justice? The case for connecting digital rights and freedoms globally. *Big Data & Society*, *4*(2), 1–14. DOI: 10.1177/2053951717736335

Teng, Z., Xia, H., & He, Y. (2025). Algorithmic Fairness and Digital Financial Stress: Evidence from AI-Driven E-Commerce Platforms in OECD Economies. *Journal of Theoretical and Applied Electronic Commerce Research*, *20*(3), 213. DOI: 10.3390/jtaer20030213

Thaler v. Comptroller-General of Patents, Designs and Trade Marks, [2023] UKSC 49. https://www.supremecourt.uk

Thaler v. Perlmutter, 999 F.3d ___ (D.C. Cir. 2025).

The Guardian. (2025), Meta is ushering in a ‘world without facts’, says Nobel peace prize winner, https://www.theguardian.com/world/2025/jan/08/facebook-end-factchecking-nobel-peace-prize-winner-maria-ressa

The Guardian. (2025). Meta's changes to policing will lead to clash with EU and UK, say experts. https://www.theguardian.com/technology/2025/jan/08/metas-changes-to-social-media-policing-will-lead-to-clash-with-eu-and-uk-say-experts

Times of India. (2024). *Bombay HC grants interim relief to Asha Bhosle against AI misuse.* https://timesofindia.indiatimes.com

Toh, A. (2020, February 6). Dutch ruling a victory for rights of the poor: Court orders government to stop use of welfare surveillance program. *Human Rights Watch.* https://www.hrw.org/news/2020/02/06/dutch-ruling-victory-rights-poor

Tomczak, J. M. (2024). Why deep generative modeling? In *Deep generative modeling* (pp. 1–13). Springer International Publishing., DOI: 10.1007/978-3-031-64087-2_1

Tufekci, Z. (2017). *Twitter and tear gas: The power and fragility of networked protest.* Yale University Press.

U.S. Copyright Office Guidelines. (2021) Compendium of U.S. Copyright Office practices, 3rd edn. https://www.copyright.gov/comp3/docs/compendium.pdf

U.S. Copyright Office. (2023). *Copyright registration guidance: Works containing AI-generated material.* https://www.copyright.gov/ai

Ubaydullayeva, A. (2023). Intellectual property in the era of artificial intelligence: Challenges and solutions. *Uzbek Journal of Law and Digital Policy*, *1*(3). Advance online publication. DOI: 10.59022/ujldp.93

Uddagiri, C., & Isunuri, B. V. (2024). Ethical and privacy challenges of generative AI. In *Generative AI: Current Trends and Applications* (pp. 219-244). Singapore: Springer Nature Singapore. DOI: 10.1007/978-981-97-8460-8_11

Uddin, M. Z. (2025). Trustworthy AI and explainability. In *Trustworthy Multimodal Intelligent Systems for Independent Living* (pp. 111–137). Springer Nature Switzerland., DOI: 10.1007/978-3-031-97359-8_4

Uddin, M., Arfeen, S. U., Alanazi, F., Hussain, S., Mazhar, T., & Arafatur Rahman, M. (2025). A Critical Analysis of Generative AI: Challenges, Opportunities, and Future Research Directions. *Archives of Computational Methods in Engineering*, •••, 1–31. DOI: 10.1007/s11831-025-10355-z

Ufert, F. (2020). AI Regulation Through the Lens of Fundamental Rights: How Well Does the GDPR Address the Challenges Posed by AI? *European Papers*, *5*(2), 1087–1097.

Ullah, A., Chan, T. C., & Xie, Z. Y. (2024). Improving CNC turning machine precision through vibration analysis for clamping error detection employing principal component analysis.

Umapathy, T. (2024). *Behavioural Economics: Understanding Human Decision-Making*. Academic Guru Publishing House.

UNESCO. (2021). *Recommendation on the ethics of artificial intelligence.*

UNESCO. (2021). *Recommendation on the ethics of artificial intelligence*. UNESCO Publishing. https://unesdoc.unesco.org/ark:/48223/pf0000377993

UNESCO. (2021). *Recommendation on the ethics of artificial intelligence*. United Nations Educational, Scientific and Cultural Organization. https://unesdoc.unesco.org

United Kingdom Supreme Court. (2023). Thaler v. Comptroller-General of Patents, Designs and Trade Marks [2023] UKSC 49. https://www.supremecourt.uk/cases/docs/uksc-2021-0201-judgment.pdf

United Nations. (1966). *International Covenant on Civil and Political Rights*. United Nations Treaty Series, 999, 171. https://www.ohchr.org

United States Copyright Office. (2021). *Copyright Law of the United States: and Related Laws Contained in Title 17 of the United States Code*, (P. 19-20).

United States Patent and Trademark Office. (2024). *Inventorship guidance for AI-assisted inventions*. https://www.uspto.gov

Vakili, S. (2025). A Study on the Intellectual Property Challenges of AI-Based Algorithms in International Law. *Legal Studies in Digital Age*, *4*(2), 1–8. DOI: 10.61838/kman.lsda.134

Vallor, S., & Bekey, G. A. (2017). *Artificial intelligence and the ethics of self-learning robots*

Vartiainen, H., Kahila, J., Tedre, M., López-Pernas, S., & Pope, N. (2025). Enhancing children's understanding of algorithmic biases in and with text-to-image generative AI. *New Media & Society*, *27*(9), 5342–5368. DOI: 10.1177/14614448241252820

Vaswani, A., Shazeer, N., & Parmar, N.. (2017). Attention is all you need. *Advances in Neural Information Processing Systems*, •••, 30.

Veale, M., & Binns, R. (2017). Fairer machine learning in the real world: Mitigating discrimination without collecting sensitive data. *Big Data & Society*, *4*(2). Advance online publication. DOI: 10.1177/2053951717743530

Veale, M., Van Kleek, M., & Binns, R. (2018). Fairness and accountability design needs for algorithmic support in high-stakes public sector decision-making. *Proceedings of the 2018 CHI Conference on Human Factors in Computing Systems*, 1–14. DOI: 10.1145/3173574.3174014

Veale, M., & Zuiderveen Borgesius, F. (2021). Demystifying the draft EU Artificial Intelligence Act. *Computer Law Review International*, *22*(4), 97–112. DOI: 10.9785/cri-2021-220402

Vecchione, B., Levy, K., & Barocas, S. (2021). Algorithmic auditing and social justice: Lessons from the field. *Proceedings of the 2021 ACM Conference on Fairness, Accountability, and Transparency*, 662–663. https://doi.org/DOI: 10.1145/3442188.3451338

Verhulst, S. G. (2021). Reimagining data responsibility: 10 new approaches toward a culture of trust in re-using data to address critical public needs. *Data & Policy*, *3*, e6. DOI: 10.1017/dap.2021.4

Voegtlin, C., Scherer, A. G., Stahl, G. K., & Hawn, O. (2022). Grand societal challenges and responsible innovation. *Journal of Management Studies*, *59*(1), 1–28. DOI: 10.1111/joms.12785

Vosoughi, S., Roy, D., & Aral, S. (2018). The spread of true and false news online. *Science*, *359*(6380), 1146–1151. DOI: 10.1126/science.aap9559 PMID: 29590045

Vousinas, G. L., Simitsi, I., Livieri, G., Gkouva, G. C., & Efthymiou, I. P. (2022). Mapping the road of the ethical dilemmas behind Artificial Intelligence. *Journal of Politics and Ethics in New Technologies and AI*, *1*(1), e31238. DOI: 10.12681/jpentai.31238

Wachter, S., & Mittelstadt, B. (2021). A right to reasonable inferences: Re-thinking data protection law in the age of big data and AI. *Columbia Business Law Review*, *2021*(2), 494–620.

Wachter, S., Mittelstadt, B., & Floridi, L. (2017). Why a right to explanation of automated decision-making does not exist in the GDPR. *International Data Privacy Law*, *7*(2), 76–99. DOI: 10.1093/idpl/ipx005

Wachter, S., Mittelstadt, B., & Russell, C. (2017). Counterfactual explanations without opening the black box: Automated decisions and the GDPR. *SSRN*, *31*(2), 841–887. DOI: 10.2139/ssrn.3063289

Waisbord, S. R. (2018). Truth is what happens to news: On journalism, fake news, and post-truth. *Journalism Studies*, *19*(13), 1866–1878. DOI: 10.1080/1461670X.2018.1492881

Wallsten, T. S. (Ed.). (2024). *Cognitive processes in choice and decision behavior*. Taylor & Francis. DOI: 10.4324/9781003469544

Walter, Y. (2024). Managing the race to the moon: Global policy and governance in Artificial Intelligence regulation—A contemporary overview and an analysis of socioeconomic consequences. *Discover Artificial Intelligence*, *4*(1), 14. DOI: 10.1007/s44163-024-00109-4

Walz, A., & Firth-Butterfield, K. (2019). Implementing ethics into artificial intelligence: A contribution, from a legal perspective, to the development of an AI governance regime. *Duke L. & Tech. Rev.*, *18*, 176.

Wang, H., Wang, Y., & Wang, Y. (2021). An interactive approach to bias mitigation in machine learning. *Proceedings of the 2021 CHI Conference on Human Factors in Computing Systems*, 1–14. DOI: 10.1109/ICCICC53683.2021.9811333

Wang, J., Huo, Y., Mahe, J., Ge, Z., Liu, Z., Wang, W., & Zhang, L. (2024). Developing an Ethical Regulatory Framework for Artificial Intelligence: Integrating Systematic Review, Thematic Analysis, and Multidisciplinary Theories. *IEEE Access : Practical Innovations, Open Solutions*, *12*, 179383–179395. DOI: 10.1109/ACCESS.2024.3501332

Weichert, J., Zhu, Q., Kim, D., & Eldardiry, H. (2025). Perceptions of AI Ethics Policies Among Scientists and Engineers in Policy-Related Roles: An Exploratory Investigation. *Digital Society : Ethics, Socio-Legal and Governance of Digital Technology*, *4*(1), 1–25. DOI: 10.1007/s44206-025-00176-9

Weidener, L., & Fischer, M. (2024). Proposing a principle-based approach for teaching AI ethics in medical education. *JMIR Medical Education*, *10*(1), e55368. DOI: 10.2196/55368 PMID: 38285931

Weidinger, L., Mellor, J., Rauh, M., ... (2022). Ethical and social risks of harm from language models. *arXiv preprint*. https://arxiv.org/abs/2112.04359

Weller, A. (2019). Transparency: Motivations and challenges. In *Explainable AI: Interpreting, Explaining and Visualizing Deep Learning* (pp. 23–40). Springer.

Wills, K. (2021). AI around the world: Intellectual property law considerations and beyond. *Journal of the Patent and Trademark Office Society*, *102*, 186.

Wolff, J. (2025). Policy approaches to defining and enforcing responsibilities for online platforms. In *Defeating disinformation*. Cambridge University Press. DOI: 10.1017/9781009438636.011

World Intellectual Property Organization. (2016). *Understanding Copyright and Related Rights.* (P. 5).

World Intellectual Property Organization. (2023). *Generative AI: Navigating intellectual property*. https://www.wipo.int

World Intellectual Property Organization. (2024). *Generative AI and intellectual property*. https://www.wipo.int

Xue, L., & Pang, Z. (2022). Ethical governance of artificial intelligence: An integrated analytical framework. *Journal of Digital Economy*. DOI: 10.1016/j.jdec.2022.08.003

Yanamala, A. K. Y., Suryadevara, S., & Kalli, V. D. R. (2024). Balancing innovation and privacy: The intersection of data protection and artificial intelligence. *International Journal of Machine Learning Research in Cybersecurity and Artificial Intelligence*, *15*(1), 1–43.

Yeung, K. (2018). Algorithmic regulation: A critical interrogation. *Regulation & Governance*, *12*(4), 505–523. DOI: 10.1111/rego.12158

Yigitcanlar, T., David, A., Li, W., Fookes, C., Bibri, S. E., & Ye, X. (2024). Unlocking artificial intelligence adoption in local governments: Best practice lessons from real-world implementations. *Smart Cities*, *7*(4), 1576–1625. DOI: 10.3390/smartcities7040064

Yousefi, Y. (2024, October). Approaching Fairness in Digital Governance: A Multi-Dimensional, Multi-Layered Framework. In *Proceedings of the 17th International Conference on Theory and Practice of Electronic Governance* (pp. 41-47). DOI: 10.1145/3680127.3680197

Yudha, F. M.. (2024). Kajian Deskriptif Analitis tentang Teori Hukum Pembangunan Prof. Dr. Mochtar Kusumaatmadja, S.H., LLM. *Ensiklopedia of Journal*, *7*(1), 362–368. DOI: 10.33559/eoj.v7i1.2627

Yu, P. K. (2022). Artificial intelligence, big data, and intellectual property. *Vanderbilt Journal of Entertainment and Technology Law*, *24*(4), 1–54.

ZadZiabari. A., & Tabatabaei, A. (2025). Ethics and Regulations in Generative AI. In *Application of Generative AI in Healthcare Systems* (pp. 197-211). Cham: Springer Nature Switzerland. DOI: 10.1007/978-3-031-82963-5_8

Zhang, J., & Zhang, Z. (2023). Ethics and governance of trustworthy medical artificial intelligence. *BMC Medical Informatics and Decision Making*, *23*(1), 7. Advance online publication. DOI: 10.1186/s12911-023-02103-9 PMID: 36639799

Zhang, W. (2024). AI fairness in practice: Paradigm, challenges, and prospects. *AI Magazine*, *45*(3), 386–395. DOI: 10.1002/aaai.12189

Zhong, L. (2023). A survey of prevent and detect access control vulnerabilities. arXiv preprint arXiv:2304.10600. Σιδηρόπουλος, Σ. (2024). *Παγκόσμια υγεία και κοινωνία των πολιτών: Χρηματοδοτικά εργαλεία και πολιτικές*. Εκδόσεις ΤΖΙΟΛΑ.

Zidouemba, M. T. (2025). Governance and artificial intelligence: The use of artificial intelligence in democracy and its impacts on the rights to participation. *Discover Artificial Intelligence*, *5*(1), 12. DOI: 10.1007/s44163-025-00229-5

Zuboff, S. (2019). *The age of surveillance capitalism: The fight for a human future at the new frontier of power*. PublicAffairs.

About the Contributors

Iris-Panagiota Efthymiou is a lecturer at the Institute for Study Abroad (IFSA) at the University of Greenwich. She has over 20 years of experience in entrepreneurship, international affairs, and executive leadership. She has previously taught at the University of East London and WSKZ in Poland and held senior academic and institutional roles, including as a college director. She combines academic teaching with strategic consultancy, specialising in public affairs, research development, and institutional management. She is the author of 19 books and more than 60 peer-reviewed articles, chapters, and monographs published by academic publishers such as Springer and IGI Global. She serves as Chief Editor of the Journal of Politics and Ethics in New Technologies and AI, and sits on several editorial and academic boards. She has delivered keynote speeches at high-level forums, including the United Nations Headquarters in Geneva, and has worked with policymakers, diplomats, entrepreneurs, academics, and journalists on cross-sector initiatives. Iris holds a PhD in Behavioural Economics (summa cum laude), a Master's in Health Economics and Management, and a Bachelor's in Economics. She is a Fellow of the Higher Education Academy (FHEA), with formal recognition of her teaching excellence in UK higher education. She is a Board Member of the Hellenic Association of Political Scientists (HAPSc) and serves on the academic board of the Laboratory of Health Economics and Management (LabHEM) at the University of Piraeus. She is also an Advisory Council Member of the Harvard Business Review, a member of the UK All-Party Parliamentary Group on Artificial Intelligence, and a partisan expert for CLAIRE AI in France. In 2020, she was awarded the title Exceptional Woman of Excellence by the Women Economic Forum. Her academic and professional work reflects a consistent commitment to equal opportunities, transparency, and the acceptance of human imperfection.

Konstantinos Kouroupis is an Associate Professor of European and IT Law and Vice-Head of the Department of Law of Frederick University, in Cyprus. His area of expertise is online protection of privacy, security and data rights, artificial intelligence, digital law, human rights and European Law. He has participated as a speaker and keynote lecturer on data rights, privacy and EU issues in numerous international conferences. Some of his studies have been uploaded to the digital library of EUROJUST, in The Hague and are accessible from a large number of prosecutors, judicial officers and administrative staff of the European agency. In 2018 he delivered a speech on "National and European legal framework on Cybersecurity and data protection" as an expert and on behalf of the Republic of Cyprus, during an international exercise under the supervision of the European Defence Agency, held at Zenon Coordination Centre, in Larnaka, Cyprus. He has participated in European projects and actions funded by the EU concerning the consolidation of a European area of justice and human rights. He has served as a Management Committee Delegate, representing Cyprus, in the framework of the European Cost action entitled "Global Digital Human Rights Network", under the auspices of the European Cooperation in Science and Technology (COST). He is a member of the editorial board of esteemed legal journals. He is also member of the pool of experts of the European Data Protection Board in the field of Data Protection and New Technologies, upon evaluation. Since August 2023, he has been appointed as a member of the Cyprus National Bioethics Committee upon a decision of the Council of Ministers of the Republic of Cyprus for 4 years.

***

Azamat Ali is an Associate Professor in the School of Liberal and Creative Arts (Journalism and Mass Communication) at Lovely Professional University, Punjab, India. He earned his Ph.D. from the Central University of Gujarat, Gandhinagar, and a M.Sc. in Science Communication from Devi Ahilya Vishwavidyalaya, Indore. His research interests span emerging technologies, development communication, public policy, and health and science communication. Dr. Ali research is directed toward bridging the divide between scientific innovation and societal application, with a particular focus on harnessing the transformative potential of emerging technologies to drive innovation, address developmental challenges, and enhance societal wellbeing.

Pankaj Bhambri is an Associate Professor in the Department of Information Technology at Guru Nanak Dev Engineering College, Ludhiana, with over twenty-one years of teaching and research experience. He holds a Ph.D. in Computer Science and Engineering from I.K.G. Punjab Technical University, Jalandhar, and is currently pursuing a Post Doctorate Fellowship at Lincoln University College, Malaysia. Recognized as a Top 2% Scientist by Stanford University and Elsevier in 2025, his research focuses on areas including machine learning, AI-driven digital twins, blockchain, and sustainable technologies. He has an extensive publication record in Scopus-indexed books and journals and has guided several Ph.D. candidates to completion.

Junaidi is a legal practitioner, writer and lecturer at the Faculty of Law at Sjakhyakirti University in Palembang. He was born in Palembang in 1978. He completed primary school in 1992, junior high school in 1995 and senior high school in 1998, all in Palembang. He completed a Bachelor of Laws at the Faculty of Law, Bhayangkara University, Jakarta (2013) and a Master of Laws in Business Law at Krisnadwipayana University, Jakarta (2016). He completed the Legal Auditor Education programme at the Jimly School of Law and Government (JSLG) in 2014 (batch 13). He also serves as a resource person in various national seminars. He is actively involved in writing scientific works that have been published in various national journals, national and international conference proceedings, and book chapters in collaboration with various universities and research institutions. He is also a reviewer for several national law journals. The author is also an active book editor in various fields, particularly law.

Kumar Priyanshu Ranjan is pursuing a BA LL.B. (Honours) course in the School of Law, CHRIST (Deemed to be University), India. His academic work focuses on the intersection of constitutional law, data regulation and law theory, with special reference to the regulation of artificial intelligence, data protection, and emergent regulations. His work has been experimental on the subject of legal and ethical issues related to foundational models and systemic AI, carefully reviewing the questions of accountability, transparency, and regulation based on risk. As his work shows, he is a strong advocate of comparative and international approaches to AI regulation, and there is a need to incorporate ethical values into the legal systems that are enforceable. His vision is to contribute meaningfully to the worlds of technology law and digital governance in the area of scholarship and policy development.

Rachna Rana is an Associate Professor in the Department of Computer Science and Engineering at Ludhiana Group of Colleges, Ludhiana with over fifteen years of teaching and research experience. Her research focuses on areas including machine learning, wireless sensor networks, and network security. She has an extensive publication record in Scopus-indexed books.

Nehal Saxena is currently pursuing a 3rd year B.A., LL.B. (Hons.) at CHRIST (Deemed to be University), Delhi NCR. Their academic journey has been marked by active participation in moot court competitions, research paper presentations, and national essay writing contests, each of which has contributed to refining my legal reasoning and research aptitude. In addition, They have volunteered at several academic and co-curricular events and participated in diverse seminars and conferences, enabling them to engage with contemporary legal and policy issues. Their research interests lie primarily in the areas of law, policy, and social justice.

Alanamol Shinto is a law student at CHRIST (Deemed to be University), Delhi NCR, pursuing a BBA LLB (Honors) with hands-on experience in various legal fields through internships, research, drafting, and case analysis.

Index

A

B

C

D

E

F

www.ingramcontent.com/pod-product-compliance
Lightning Source LLC
LaVergne TN
LVHW081535300926
849161LV00006B/1438